Black Pepper

Cardamom (Small)

Cardamom (Large)

Ginger

Turmeric

Chilli

Vanilla

Nutmeg

Cinnamon

Cassia

Tejpat

Clove

Cambodge

Kokam

Curry Leaf

Tamarind

Pimento

Star-anise

Caper

Sweet Bay

Juniper

Coriander

Cumin

Fennel

Fenugreek

Anise

Celery

Dill

Mustard

Poppy

Caraway

Ajowan or Bishop's Weed

Pomegranate

Marjoram

Savory

Tarragon (Estragon)

Thyme

Origanum

Sage

Basil

Pepper Mint

Parsley

Rosemary

Lovage

Garlic

Asafoetida

Greater Galanga

Saffron

Sweet Flag

Long Pepper

Horse Radish

Hyssop

Spices

Horticulture Science Series — 5

Other Volumes of the Series

HORTICULTURE SCIENCE SERIES

Chief Editor : Prof. K.V. Peter

Former Vice Chancellor and Presently Professor of Horticulture

Kerala Agricultural University, Kerala

1. Aromatic Plants
   Baby P. Skaria
2. Medicinal Plants
   Kurian Alice and Asha Sankar
3. Fruit Crops
   Radha T. and Lyla Mathew
4. Vegetable Crops
   T.R. Gopalakrishnan
6. Propagation of Horticultural Crops
   Rajan S. and Baby Lizzy Markose
7. Post Harvest Technology of Horticultural Crops
   Sudheer K.P. and V. Indira
8. Management of Horticultural Crops
   Kumar Pradeep, T.R. Gopalakrishnan and N. Kumar
9. Flowering Trees
   P.K. Valsala Kumari
10. Flowers for Trade
    M.N. Sheela
11. Ornamental Flowers
    Rajeevan
12. Plantation Crops
    Kurian Alice and K.V. Peter
13. Tuber Crops
    P. Palaniswamy
14. Coastal Horticulture
    Kanthaswamy
15. Basics of Horticulture
    K.V. Peter

Horticulture Science Series Vol. 5

Chief Editor :
Prof. K.V . Peter, Former VC and *Presently* Professor of Horticulture
Kerala Agricultural University, Thrissur - 680 656, Kerala

Spices

Authors :
E.V. Nybe and N. Mini Raj
Deptt. of Plantation Crops and Spices
College of Horticulture
Kerala Agricultural University
Thrissur - 680 656, Kerala
&
Prof. K.V. Peter
Professor of Horticulture
Kerala Agricultural University
Thrissur - 680 656, Kerala

New India Publishing Agency
Pitam Pura, New Delhi- 110 088

ISBN 978-81-89422-44-8

Published by:
Sumit Pal Jain *for*

New India Publishing Agency

101, Vikas Surya Plaza, CU Block, L.S.C. Mkt.,
Pitam Pura, New Delhi- 110 088, (INDIA)
Phone: 011-27341717, Fax: 011-27341616
E-mail: newindiapublishingagency@gmail.com
Web: www.bookfactoryindia.com

National Academy of Agricultural Sciences

Foreword to the Series

Prof. M.S. Swaminathan
President

A second green revolution aimed to meet the food requirements of about 1.8 billion people in India by 2010 is on the anvil. It is, however, imperative that the attainment of the food and nutritional security of this vast population should be combined with the preservation of clean environment. Horticulture encompassing fruits, vegetables, ornamentals, plantation crops, spices, tuber crops, medicinal and aromatic plants, besides harvest and post harvest technologies has a special role in providing food and nutritional security, rural poverty alleviation, employment generation as well as improving the quality of life in our country sides.

India, incidentally, is the world's leading producer of fruits and is followed by Brazil; in vegetables it occupies the second place after China. India is also the land of spices producing an array of such crops. Indeed, the varied eco-climatic conditions prevailing in different parts of this country make it possible to grow a disparate range of fruits, vegetables and ornamental plants at any time of the year.

NASC Complex
DPS Marg, P.O. Pusa
New Delhi - 110 012

Tel. 91-11-25846055
91-11-25846051
91-11-25846052
91-11-25846053
Fax 91-11-25846054
E-mail naas@vsnl.com
Website
//education.vsnl.com/naas
P.O. Box No. 11325

With the opening up of the world markets under the WTO regime, however, international competition for Indian horticultural products has reached unprecedented levels, necessitating strategies for quality control, branding and product development. In particular, production of organic vegetables, fruits, cut flowers, pepper, ginger, tea, coffee, rubber and cardamom has of late become demand-driven.

Among the inputs for higher productivity in horticultural crops, seeds and quality planting materials account for 20 to 30% of the potential productivity. Transgenic crops are a reality now; yet the technology for production of transgenics is a far cry in many situations. Furthermore, water is a limiting factor in several environments. Hence, water saving crops and those with high water use efficiency should be preferred over other crops and management practices. Likewise, the use of biodegradable plastics, biological control of pests and diseases, and new product development are receiving increasing attention, especially in the arena of horticultural crop management. Marketing, village knowledge centres, packaging industry, zero-energy cool chambers and a variety of economically viable technologies, however, remain to be streamlined.

The present series on **Horticulture Science**, edited by Professor K.V. Peter, Vice-Chancellor, Kerala Agricultural University, is a good compilation of the scientific knowledge and experience gained over the past 30 years in this important sector of Indian economy. Each volume in the series is authored by accomplished researchers such as Dr. T. Radha and Dr. Lyla Mathew (*Fruit Crops*), Dr. T.R. Gopalakrishnan (*Vegetable Crops*), Dr. P.K. Valsalakumari (*Flowering Trees*), Dr. P.K. Rajeevan (*Ornamental Flowers*), Dr. Alice Kurian and Dr. K.V. Peter (*Plantation Crops*), Dr. Alice Kurian and Dr. M. Asha Sankar (*Medicinal Plants*), Dr. Baby P. Skaria (*Aromatic Plants*), Dr. P. Palaniswamy (*Tuber Crops*), Dr. E.V. Nybe and Dr. N. Mini Raj and Dr. K.V. Peter (*Spices*), Dr. S. Rajan and Dr. Baby Lizzy Markose (*Propagation of Horticultural Crops*), Dr. M.N. Sheela (*Flower for Trade*), Dr. Kanthaswamy (*Coastal Horticulture*), Dr. K.P. Sudheer and Dr. V. Indira (*Post Harvest Technology of Horticultural Crops*), Dr. K. Pradeep, N. Kumar and Dr. T.R. Gopalakrishnan (*Management of Horticultural Crops*) and Professor K.V. Peter (*Basics of Horticulture*). The New India Publishing Agency, New Delhi deserves special credit for publishing this useful series, especially in view of the fact that the Government of India is currently implementing the National Horticultural Mission.

Prof. M.S. Swaminathan, FNA, FRS
President, National Academy of Agricultural Sciences
and Chairman, National Commission on Farmers
New Delhi 110 012

INDUSTRIAL CHEMICALS LIMITED
Synthite Valley, Kolenchery-682 311, Kerala, India SYNTHITE

Foreword

Spices have been virtually indispensable in the culinary art of flavoring foods since antiquity. Many of them possess properties as colourant, odourant, preservative and/ or nutraceutical. Apart from their application as flavourings, spices are now widely used in health, personal care and hygiene areas and thus, play a significant role in the economy of the producing country. The international trade in spices is estimated at about 906,700 tonnes valued at $ 2125 million. India commands a formidable position in world spice trade with a share of 37 % in volume and 23 % in value.

Acclaimed as the 'Home of Spices', India is the biggest producer and consumer of spices in the world. Besides whole spices and powders, oils and oleoresins are the primary categories of spices exported from the country. These spices are of exceptional quality and have a wide variety to suit every palate. India's crop management programmes and post harvest processing technologies are in line with contemporary global practices. Export of spices from India during 2004-05 touched an all time high of 335500 tonnes valued at $ 500 million. Major items of export were chilli, mint, turmeric and black pepper based products. There was also an increase in the export of seed spices (both in terms of quantity and value) in the same period. With stiff global competition from other producing countries, there is an increasing need to focus on cost reduction and quality upgradation.

Research and development activities on spices in India are undertaken by government institutions, regulatory and trade bodies, and the industry. Prominent among them are: Indian Institute of Spices Research (Calicut), National Research Centre on Seed Spices (Ajmer), a number of State Agricultural Universities, Central Food Technological Research Institute (Mysore), Regional Research Laboratories (Jammu, Trivandrum) and the Indian industry. These bodies are actively involved in spice research for improvement in production, productivity, quality, as well as for new application development and marketing. These efforts would certainly help the industry and with continued focus on specific goals, business growth could be faster and more sustainable.

Value addition and creation of opportunities in the industry exist via., better supply chain management and with the constant upgradation of processing and formulation technologies. Developing novel functional ingredients for specialty applications based on natural spices would be of increasing relevance in future years as numerous spice ingredients are being sought after globally, due to their acceptance as nutraceuticals. Though the health food industry in India is at its infancy, its immense potential would ensure that the momentum would increase in the coming years. The overall growth of the Indian spice industry would be linked to the ability to tap the existing traditional markets viz., for seed spices , spice oils and oleoresins as well as their expanding and emerging markets in non traditional application areas such as natural colourants and health promoters. Consequently, in addition to providing adequate R & D inputs to support the existing business, the industry in collaboration with the government agencies and institutions should allocate additional resource to explore potential non-traditional opportunities for these plant derived natural products. International market for organic spices is on the increase and popularisation of biofertilizers and biocontrol of pests and diseases are called for in this context.

The book on "Spices" provides a comprehensive account on economic importance, botany, chemical composition, crop improvement, agrotechnology, post harvest technology and end uses of 52 spices. This volume forms a part in the series on Horticulture Science edited by Dr. K. V. Peter, Vice- Chancellor, Kerala Agricultural University, Vellanikkara.

"Spices" is organized into six separate sections by grouping them under classes such as major spices, tree spices, seed spices, herbal spices and other spices. Dr. E. V. Nybe, Head and Dr. N. Mini Raj, Associate Professor, Department of Plantation Crops and Spices, College of Horticulture, Kerala Agricultural University with their vast experience in tropical spices have brought their scientific and practical experience in collating and presenting the information quite concisely.

This narration of diverse spices in a solitary volume would be quite informative and useful to a wide spectrum of readers. It gives me immense pleasure to present this book to all those who are passionate about spices – farmers, technologists, industrialists, extension workers, policy makers, researchers, teachers and students.

I congratulate the authors and the series editor for their earnest efforts in bringing out this publication.

C. V. Jacob
Chairman
Synthite Industrial Chemicals Ltd.
Synthite Valley, Kolenchery
Ernakulam, Kerala

Preface

India, considered as the "Home of Spices" is one of the major spice producing and exporting countries of the world contributing about 20 - 25 % of the world trade in spices. Besides, huge quantities of spices are also consumed within the Country for seasoning foods and for other purposes. Out of the 109 spices recognized by the International Organization for Standardization (ISO) world over, 52 are grown in India and promoted by Spices Board. In view of the potential demand for quality spices globally, research and development programmes have been intensified in the Country. There is increasing interest in industry, academia and health sciences for spices. Despite the tremendous scope of spices, we do not have a comprehensive publication which includes information on all the 52 spices.

This book on 'Spices' brings together the currently scattered knowledge on all aspects of spices – botany, crop improvement, chemistry, production technology, pests and diseases, harvesting, processing, products and uses. We have tried to collect and collate as much as information as possible about the aforesaid spices. The book is organized into six chapters as introduction, major spices, tree spices, seed spices, herbal spices and other spices. Major, tree and seed spices have been dealt in detail whereas herbal and other spices in which information available is scanty are discussed in brief.

This book is not claimed to be very exhaustive by any means. The authors are fully conscious of their limitations due to paucity of space. For more detailed information, the interested reader can refer the bibliography given at the end.

The book is targetted to the students and researchers in the area of Horticulture, Agriculture, industry people, processors, exporters and end users of spices. Prof. K.V. Peter, the general editor of the series has kindly commissioned us to write this book. He has been a source of help and inspiration and supported us by critical suggestions and corrections. We are also very much thankful to the publishers for entrusting the responsibility of writing this book with us. We are extremely grateful to Mr. H.R. Shankar, Mrs. A. Sincy and Ms. Leesha, Research Assistants for their services rendered in the exhaustive literature search, manuscript preparation, computer work *etc.* In the preparation of this book we have made use of the published information from diverse sources and by many authors giving full citation. We acknowledge with gratitude all those workers. For the chapter on tree spices, Dr. T. Thangaraj and Dr. R. Richard Kennedy from Tamil Nadu Agricultural University provided valuable information. We thank them all profusely. We place on record our thanks and gratitude to the Kerala Agricultural University for permitting us to undertake the writing of this book. We place on record our sincere thanks to M/s JMJ Computers for the neat and timely typing of the manuscript. Finally, we thank all our well wishers and all those who helped in the preparation of this book.

Authors

Contents

Chapter 1

Introduction

The term 'Spices and Condiments' applies to such natural plant or vegetable products or mixtures thereof, used in whole or ground form, mainly for imparting flavour, aroma and pungency to food. They are low volume high value crops which play a pivotal role in our national economy and provide a strong footing in the international market. According to the latest report of the International Organization for Standardization (ISO), there are about 109 spices grown in different parts of the world. India grows about 52 of them. The list of those spices with botanical and English names which come under the purview of Spices, Board, Ministry of Commerce, Government of India is given in Table 1.

Of these, a few are grown commercially in India and remaining grown on a small scale in gardens but have good international market potential or are in great demand and are even imported (clove, nutmeg, cinnamon, long pepper, sage, thyme, marjoram, asafetida *etc.*). Five major spices (black pepper, cardamom, chillies, ginger and turmeric) together account for about 75 – 89 % of the total annual Indian foreign exchange earnings from spices (Pruthy, 2001).

Status of spices production

India, the land of spices is the largest producer, consumer and exporter of spices in the world. According to Spices Board, 52 spices are grown in the country. However, commercial cultivation is limited to about a dozen crops which have importance in the domestic market and international trade. More than 90 % of the spices produced in the country is used for domestic consumption and the rest only is being exported as raw and value added forms. The trend in world spice production furnished in Table 2 shows that volume recorded an increase of 4.7, 4.8 and 4.9 million tonnes in 1998, 1999 and 2000, respectively. The largest share was from India (203 million tonnes) and China was the second largest producer (0.58 million tonnes in 2000) which also showed an increasing trend. Total volume of spice from Indonesia was 0.26 million tonnes in the year 2000.

Table 1. Nomenclature of spices and condiments

S. No.	Common Name	Botanical Name	Family	Part Used
1	Cardamom (Small)	*Elettaria cardamomum*	Zingiberaceae	Seed
	Cardamom (Large)	*Amomum subulatum*	Zingiberaceae	Seed
2	Black pepper	*Piper nigrum*	Piperaceae	Seed
3	Chilli			
	Bird eye	*Capsicum frutescens*	Solanaceae	Fruit
	Capsicum	*Capsicum annuum*	Solanaceae	Fruit
	Chilli	*Capsicum annuum*	Solanaceae	Fruit
	Paprika	*Capsicum annuum*	Solanaceae	Fruit
4	Ginger	*Zingiber officinale*	Zingiberaceae	Rhizome
5	Turmeric	*Curcuma longa*	Zingiberaceae	Rhizome
6	Coriander	*Coriandrum sativum*	Apiaceae	Leaf & Seed
7	Cumin	*Cuminum cyminum*	Apiaceae	Fruit
8	Fennel	*Foeniculum vulgare*	Apiaceae	Fruit
9	Fenugreek	*Trigonella foenum-graecum*	Fabaceae	Seed
10	Celery	*Apium graveolens*	Apiaceae	Fruit
11	Anise	*Pimpinella anisum*	Apiaceae	Fruit
12	Ajowan	*Trachyspermum ammi*	Apiaceae	Fruit
13	Caraway	*Carum carvi*	Apiaceae	Fruit
14	Dill	*Anethum graveolens*	Apiaceae	Fruit & Seed
15	Cinnamon	*Cinnamomum verum*	Lauraceae	Bark
16	Cassia	*Cinnamomum aromaticum*	Lauraceae	Bark
17	Garlic	*Allium sativum*	Liliaceae	Bulb
18	Curryleaf	*Murraya Koenigii*	Rutaceae	Leaf
19	Kokum	*Garcina indica*	Clusiaceae	Peel of fruit
20	Pepper mint	*Mentha piperita*	Lamiaceae	Leaf
21	Mustard	*Brassica juncea*	Brassicaceae	Seed
22	Parsley	*Petroselinum crispum*	Apiaceae	Seed
23	Pomegranate	*Punica granatum*	Punicaceae	Seed
24	Saffron	*Crocus sativus*	Iridaceae	Stigma of flower
25	Vanilla	*Vanilla planifolia*	Orchidaceae	Pod
26	Tejpat	*Cinnamomum tamala*	Lauraceae	Bark & Leaf
27	Long pepper	*Piper longum*	Piperaceae	Fruit
28	Star anise	*Illicium verum*	Illiciaceae	Fruit
29	Sweet flag	*Acorus calamus*	Araceae	Rhizome
30	Greater galanga	*Alpinia galanga*	Zingiberaceae	Rhizome
31	Horse radish	*Armoracia rusticana*	Brassicaceae	Rhizome
32	Caper	*Capparis spinosa*	Capparidaceae	Fruit/root

33	Clove	*Syzygium aromaticum*	Myrtaceae	Unopened flower bud
34	Asafoetida	*Ferula asafoetida*	Apiaceae	Oleogum resin from rhizome and thickened root
35	Cambodge	*Garcinia gummi-gutta*	Clusiaceae	Pericarp lobes
36	Hyssop	*Hyssopus officinalis*	Lamiaceae	Leaf
37	Juniper	*Juniperus communis*	Cupressaceae	Berry
38	Sweet bay	*Laurus nobilis*	Lauraceae	Leaf
39	Lovage	*Levisticum officinale*	Apiaceae	Leaf
40	Marjoram	*Marjorana hortensis*	Lamiaceae	Leaf & flower top
41	Nutmeg	*Myristica fragrans*	Myristicaceae	Seed
42	Mace	*Myristica fragrans*	Myristicaceae	Aril of fruit
43	Basil	*Ocimum basilicum*	Lamiaceae	Leaf
44	Poppy	*Papaver somniferum*	Papaveraceae	Seed
45	All-spice	*Pimenta dioica*	Myrtaceae	Fruit & Seed
46	Rosemary	*Rosemarinus officinalis*	Lamiaceae	Leaf
47	Sage	*Salvia officinalis*	Lamiaceae	Leaf
48	Savory	*Satureja hortensis*	Lamiaceae	Stem, leaf and flowering top
49	Thyme	*Thymus vulgaris*	Lamiaceae	Friut & leaf
50	Oreganum	*Origanum vulgare*	Lamiaceae	Leaf and flowering top
51	Tarragon	*Artemisia dracunculus*	Asteraceae	Leaf
52	Tamarind	*Tamarindus indica*	Fabaceae	Fruit

Table 2. World spice production

Country	Production (mt)		
	1998	1999	2000
India	2,163,600	2,243,700	2,255,800
China	523,828	556,288	584,871
Indonesia	258,167	251,802	260,902
Pakistan	179,451	160,755	208,328
Nigeria	140,500	141,500	146,500
Bangladesh	139,000	138,000	144,000
Ethiopia	113,150	114,150	110,960
Vietnam	97,008	113,500	123,752
Mexico	92,364	91,853	91,755
Turkey	73,799	77,050	74,100
Hungary	68,619	37,105	44,000
Egypt	67,000	67,000	67,000
Thailand	62,500	72,600	77,200
World	4,724,967	4,802,892	4,939,990

Source: FAOSTAT Database

According to world spice production statistics 2003-04 by FAO, the contribution of India was 86% followed by China which contributed only 4%, Bangladesh (3%), Pakistan and Turkey (2% each) and Nepal (1%). With respect to black pepper, Vietnam has recently become the largest producer and exporter with production of 85000 t in 2003 followed by Indonesia (67000 t), India (65000 t), Brazil (35000 t), Malaysia (22000 t) and Sri Lanka (12750 t). Vietnam has dominated the export market (82000 t) with negligible quantity being used for domestic purpose whereas export from India was 17888 t only in 2003.

World production of cardamom is estimated at 30000 t. Presently, major producer is Guatemala with an annual production of 18000 – 20000 t. India is the second largest producer with an average production of 11000 – 12000 tonnes. Indian cardamom is considered superior in quality, but has always been out priced by Guatemala where the domestic consumption is negligible. India is the second largest consumer of small cardamom after Saudi Arabia. Global consumption of small cardamom is estimated as 15000 – 24000 t.

Global ginger production is looking bright, the area and production had increased since 1999. The area under ginger was 321732 ha in 2001 with a production of 835173 t. Nigeria has the largest area (174000 ha) under ginger followed by India 80000 ha. However, the largest production comes from India (275000 t) followed by China (240000 t) and Nigeria (90000 t).

India is the largest producer, consumer and exporter of turmeric in the world, 85 % production comes from Asian countries and 15 % from Europe and North America. World production of nutmeg is estimated to be 10000 – 12000 t with an annual world demand of 9000 t. Production of mace is estimated at 1500 – 2000 tonnes. Indonesia (75%) and Grenada (20%) dominate in production and export of nutmeg. Other producers include India, Malaysia, Papua New Guinea and Sri Lanka.

The share of spices in the total agriculture export of India is around 6 %. Indias' share of world spices trade is estimated 45-50 % by volume and 25-30 % by value. With respect to export, the quantity exported has registered a 45 % increase during the last four years (2.30 lakh tonnes in 2000-01 to 3.35 lakh tonnes in 2004-05). The average export earnings during the aforesaid period were Rs.1690 crores and Rs. 2200 crores which recorded 30 % increase. Chillies fetch the highest export earnings (Rs. 499 crores) followed by oils and oleoresin (Rs. 464 crores) and mint oil (Rs. 408 crores) during 2004-05 (Table 3).Area, production and productivity of major and tree spices are furnished in Tables 4 and 5. Among the spices cultivated in India, chilli ranks first in area (8.83 lakh t) with a productivity of 1266 kg/ha. The area has recorded an increase from 8.27 lakh ha in 1978-79 to 9.15 lakh ha in 1999-00 which again has slightly reduced to 8.83 lakh ha in 2001-02. India is the major producer and exporter of chilli and among the various spices produced in the Country, per capita consumption is the highest for chilli. Andhra Pradesh followed by Karnataka are the major chilli producing states in India (Table 6). Considering the increasing demand for diversified uses like natural colourant and medicinal properties, there is great scope for expanding chilli cultivation in the State.

Table 3. Export of spices from India

Item	Quantity (mt)		Value (Rs. lakhs)	
	2003-04	2004-05	2003-04	2004-05
Black pepper	16635	14150	14277	12140
Cardamom (S)	757	650	3692	2390
Cardamom (L)	924	950	1234	1134
Ginger	4696	13000	2275	5950
Turmeric	37044	43000	13112	15650
Chillies	86575	138000	36688	49901
Coriander	21018	33750	7201	8266
Cumin	7957	13750	5884	10190
Fennel	5007	7100	2211	2530
Fenugreek	6932	13750	1555	2661
Garlic	3691	2250	1423	560
Celery	4815	4100	1520	1301
Other seed spices	14031	11100	3397	2614
Other spices	7214	8500	4097	3960
Curry powder	8318	7750	6805	6610
Mint oil/products	10110	9300	39436	40777
Spice oils and oleoresins	5133	5600	37992	46375
Total	**254382**	**335488**	**191160**	**220000**

Source: Spices Board, Kochi

Table 4. Trends in area and production of spices in India

Spices	Area (000 ha)		Production (000 t)	
	1978-79	1999-2000	1978-79	1999-2000
Black pepper	85	192	22	58
Cardamom (s)	93	72	4	9
Ginger	41	77	76	263
Turmeric	90	161	190	653
Chillies	827	915	566	1018
Coriander	384	547	145	290
Cumin	181	264	79	109
Garlic	49	114	167	495
Other seed spices	49	85	60	99
Tree spices	-	28	-	11
Vanilla	-	1600	-	60
Others	-	45	-	18
Total	**1799**	**4100**	**1309**	**3083**

Source: Spices Board, Kochi

Table 5. Area and production of spices in India (2001-02)

S. No.	Spices	Area (ha)	Production (mt)
1	Pepper	218217	61452
2	Cardamom (small)	72663	11365
3	Cardamom (large)	30008	5851
4	Ginger	85920	323059
5	Turmeric	164457	558847
6	Chilli	882875	1185800
7	Vanilla	1600	60
8	Nutmeg	7850	1986
9	Cinnamon	338	Neg
10	Clove	1885	1040
11	Tamarind	61336	121294
12	Coriander	433381	318700
13	Cumin	526634	206410
14	Fennel	39924	38530
15	Fenugreek	115603	136641
16	Celery	746	299
17	Dill	12297	10003
18	Mustard	2228552	2319154
19	Poppy	7200	3557
20	Ajowan	13884	9411
21	Garlic	78816	310003
22	Saffron	2713	3
	Total	**4986899**	**5556252**

Source: Spices Board, Kochi

Black pepper, popularly known as the "king of spices" is one of the most important dollar earning crops next to chilli. Kerala is in the foremost position contributing about 95 % of area and production of the crop in the Country. The area has increased from 1.80 lakh ha in 1996-97 to 2.18 lakh ha in 2001-02, with a corresponding increase in production from 55590 t to 61300 t. However, the crop registers a low productivity of 280 kg/ha (Table 6) as compared to other pepper producing countries (3-4 t /ha). Vietnam has emerged as a strong competitor for us in pepper production and export. The production and productivity of pepper are highly influenced by oscillating prices which is more pronounced recently. At its peak, in November 1999, black pepper was sold at Rs. 270 /kg which was nose dived to the present price of Rs. 65 to 70 / kg.

Table 6. State-wise area, production and productivity of major spices in India (2001-02)

State	Pepper		Cardamom		Ginger		Turmeric		Chilli		Seed spices	
	A	P	A	P	A	P	A	P	A	P	A	P
AP	-	-	-	-	2.6	6.3	62	249	225	591	61	24
Gujarat	-	-	-	-	1.1	16.3	0.7	12	14	12	449	406
Karnataka	9.7	2.2	26	2.1	10.8	14.2	6.7	36	197	167	11	2
Kerala	204	58	41	8.4	10.8	40	3.6	7.9	0.7	0.7	-	-
MP	-	-	-	-	5	7	0.4	.04	40	22	104	38
Maharashtra	-	-	-	-	1.2	1.2	6.8	8.5	107	59	-	-
Orissa	-	-	-	-	13	25	27	66	70	59	135	45
Rajasthan	-	-	-	-	0.3	1.1	0.1	0.5	32	49	248	243
TN	4.1	1.0	5	0.9	0.5	10	24	118	73	42	31	7
Others	0.5	0.1	30	6.0	45.7	209	33	61	125	117	106	85
Total	**218.3**	**61.3**	**102.4**	**17.4**	**91.0**	**329.6**	**164.5**	**559.3**	**883.7**	**1118.7**	**1145**	**850**
Productivity	280		170		3622		3400		1266		900	

A- Area (000 ha) *P- Production (000 t)* *Productivity in kg / ha* *Source: Spices Board, Kochi*

Cardamom, the "queen of spices" enjoys a premium preference in the international market. The area under cardamom in India has increased from 72400 ha during 1999-00 to 102400 ha in 2001-02. However, the production has registered an increase of 87 % during the same period (ie. 9300 t to 17400 t). Kerala is the largest producer of cardamom in the Country. However, the area under cardamom in Kerala is almost static (41288 ha in 2001-02 and 41412 ha in 2002-03). The production has slightly increased during the above period (8400 t in 2001-02 and 8680 t in 2002-03). Record price was realized in January, 2002 which was as high as Rs. 680 / kg. However, the price is highly oscillating and has declined to Rs. 300-350 /kg. The productivity has increased from 145 kg /ha to 170 kg /ha during the last pentinnium, mainly as a result of improved varieties and hitech crop management practices.

Ginger and turmeric are the other major spices of importance to India. Being annuals, area and production are highly influenced by price fluctuations. The area under ginger has increased from 83940 ha in 1999-00 to 91000 ha in 2001-02. So also the production has registered an annual growth rate of 7% during the aforesaid period. Even though the area under ginger cultivation is more in Kerala (8998 ha in 2002-03) as compared to Meghalaya (7941 ha), the highest production is registered by Meghalaya (45655 t) as against 32412 t in Kerala. Devastating diseases like soft rot and bacterial wilt coupled with high cost of cultivation are the major factors limiting production of ginger. Area and production of turmeric in India has registered an increasing trend during the last five decades. The annual growth rate recorded were 3.7 % and 9.1 % with respect to area and production respectively. The present area is 1.65 lakh ha with a production of 5.6 lakh t. Andhra Pradesh is the State which tops the list with respect to area and production of turmeric followed by Orissa and Tamil Nadu. However, the area in Kerala has decreased from 4127 ha (2000-01) to 3140 ha (2002-03). So also, the production has registered a decrease from 9037 t to 6938 t during the same period. The crop is gaining importance in the international market because of its medicinal properties, colouring pigment curcumin and spicy flavours. Kerala produces the best quality turmeric which is exported under the trade name Alleppy Finger Turmeric (AFT).

Tree spices like nutmeg, clove, cinnamon and tamarind are also important spices grown in India, especially in South India. Monocrop of tree spice is seldom seen in the Country. However, these are integral part of every homestead in Kerala. Nutmeg cultivation is showing an increasing trend whereas clove and cinnamon are almost static. In India, nutmeg is grown in an area of 7850 ha with an annual production of 1986 tonnes and productivity of 255 kg/ha. Kerala is the leading state with 97 % of the area and 95 % of production in the Country. With respect to clove, total area under the crop in India is 1885 ha with a production of 1040 tonnes and productivity 473 kg/ha. Tamil Nadu has the highest area (777 ha) followed by Kerala (754 ha). Cinnamon is not usually cultivated on commercial scale in India, the total area being only 338 ha with 285 ha in Kerala. Tamarind is another major tree spice with an acreage of 61336 ha and production of 1.21 lakh tonnes with a productivity of 3078 kg/ha. Tamil Nadu leads in area (21353 ha) whereas production is more in Karnataka (75064 t).

Advancements, challenges and opportunities

Production technologies have been standardized in most of the spice crops and large growers are invariably following those agrotechniques. However, the adoption is not satisfactory in the case of small and marginal farmers. Many a time, the limiting factor is the non-remunerative price of the produce and non-availability of credit in time. If prices are assured, no doubt the farmers will be ready to produce the required quantity of a given spice in a given time.

Spice research in India was initiated in the late 1940's at a few centres under the Department of Agriculture in the erstwhile states of Travancore, Cochin, Madras and Hyderabad. During 1947, Pepper Research Scheme in Panniyur and Cardamom Research Station in Mudigere and Pampadumpara were established. Spices research has attained new dimension and momentum with the establishment of All India Co-ordinated Cashew and Spices Improvement Project in 1971. With the concerted research efforts at various central and state institutes, various production technologies in spices could be evolved. Crop improvement is one of the areas wherein we could achieve commendable results which helped a lot to improve productivity of our spices.

Though India is the homeland of many spices, productivity level of these crops is very low. As the international market becomes increasingly competitive, it is imperative to enhance the production and productivity and bring down cost of cultivation to make the Indian products globally competitive. Low productivity and high production cost makes our products costlier in the world market. Several reasons like existence of uneconomic and poor varieties, incidence of serious diseases like *Phytophthora* foot rot of pepper, viral diseases of cardamom, soft rot and bacterial wilt of ginger etc. are attributed to low productivity of spices.

India is the leading producer of black pepper, but our productivity is very low (280 kg/ha) compared to other pepper producing countries like Thailand (4079 kg/ha) or Malaysia (1955 kg/ha). In India, pepper is grown as a mixed crop on live standards with an average plant density of 560 vines/ ha whereas in Malaysia, it is being grown on dead standards with a density of 5000 vines/ ha. Labour cost which accounts for more than 60 per cent of total variable cost has increased by about 400 per cent in the predominantly spice growing state of Kerala between 1980 and 1995 while the increase was only 47 per cent in Malaysia during the same period.

Cardamom is another commodity in which the Country has lost its competitiveness in the international trade because of increased production cost. Guatemala has an advantage over India from cost of cultivation point of view also because of higher productivity and lower wages. Similarly, Indian ginger is costlier compared to Chinese and Nigerian produce. The Country can sustain the production and recapture the international market only by attaining reduction in unit cost of production by increasing productivity.

Promotional activities for organic spices production have to be strengthened to capture the growing organic world market. The cost of certification of organic produce is prohibitively high and beyond the capacity of an average Indian farmer. This has to be brought down to a reasonable and affordable level. Import of low grade, low priced produce from other countries to India and re-export under Indian label may destroy the quality image of Indian spices in the international market and may detrimentally affect the demand of the same in future.

The changed economic order in the context of globalization and liberalization of world trade in agriculture has opened up new vistas of growth. Spices sector is one of the key areas in which India has an inherent strength to dominate the global markets. However, the present situation raises many challenges for survival of spice industry in the Country.

Our spice farmers are in the habit of selling the produce in the primary form. With respect to export also, about 85 – 90 % go out as primary produce. No doubt, there exists good scope for value added products in the global markets. Therefore, the first step is to penetrate into super market chains with diversified high quality value added products for substantial gains. Another area which offers great potential is the complete utilization of various byproducts generated in the processing industry.

With the ban on artificial colours in food industry and increasing health awareness, natural colours and other products from spices attain much importance. Use of curcumin has become important in food industry, pharmaceuticals, preservatives and body care. The increasing demand for natural colours can also be exploited in the case of high colour Indian paprika.

Exploration of medicinal and neutraceutical properties of our traditional spices will pave a long way in the development of Indian spice industry. The antioxidant property of curcumin in addition to its wider use as a natural colourant is well established. The antibiotic property of hydroxy citric acid in garcinia is also well documented. Piperine from pepper is a potent inhibitor of drug metabolism and acts as a bio-availability enhancer. Chillies usually known as an irritant spice have got counter irritant properties also at very low concentration and are used in skin ointments.

Organic spices are gaining great momentum in the world market. The very good scope for production and export of organic spices in India, could be tapped fully for the treatment of Indian spices sector.

Chapter 2

Major Spices

Black Pepper
(*Piper nigrum* L., F. Piperaceae)

Black pepper, christened as "King of spices" and "Black gold" is the most important and the most widely used spice in the world occupying a position that is supreme and unique. It was the lure of Indian pepper that brought the western world to Indian sub continent.

The black pepper of commerce is the dried mature fruit of the tropical perennial climbing plant *Piper nigrum*. Pepper originated in the tropical evergreen forests of the Western Ghats. The Malabar coast of India was the centre of pepper trade from time immemorial. Black pepper was essentially a minor forest produce in the past, people collected it from forests where it abundened. Domestication of pepper appears to be a much later event. The recorded history shows that Arabs, followed by the Portuguese, the Dutch and the British traders took pepper to all over the world and spread its use and cultivation. Initially black pepper was taken from the Malabar coast to the Indonesian islands, then it spread to various Pacific islands, South East Asian countries and later to tropical Africa and America. Currently, pepper is grown in about 26 countries, major being India, Indonesia, Brazil, Malaysia and Srilanka. In India, pepper is grown mainly in Kerala, Karnataka, Tamil Nadu and to a little extent in Goa, Orissa, Assam and Andaman group of islands. Kerala is the original home of pepper. This state accounts for 95% of country's area and production.

Chemical composition

Quality of pepper is contributed by two components. Piperine that contributes the pungency and volatile oil that is responsible for the aroma and flavor. Piperine is the major alkaloid present in pepper. The essential oil of black pepper is a mixture of large number of volatile compounds. The main groups are monoterpene hydrocarbons (69.4-84%), sesquiterpene hydrocarbons (15-27%) and the rest were oxygenated constituents. (Lewis et al., 1969). Dominating monoterpene hydrocarbons were a-pinene (5.9%-12.8%), b-pinene (10.6%) and limonine (22-31.1%). b-caryophylline (10.3-22.4%) constituted the major sesquiterpine hydrocarbon.

Botany

Pepper belongs to family Piperaceae and genus *Piper*. Apart from black pepper (*P. nigrum*), the genus also includes economically important species like

P. longum - long pepper

P. betle - betel leaf

P. chaba - Java long pepper and

P. cubeba - Cubeb, Tailed pepper

Apart from these economically important species, other closely related indigenous species identified by Ravindran (2000) are *P. colubrinum, P. attenuatum, P. argyrophyllum, P. bababudani, P. hymenophyllum, P.galeatum, P. pseudonigrum, P. sugandhi, P. wightii, P. barberi, P. hapnium, P. mullesua* and *P. silentvalleyensis*. Among these *P. longum, P. hapnium, P. mullesua* and *P. silentvalleyensis* possess erect spikes and all other species bear pendant spikes

Piper nigrum is a perennial climber, climbing by means of roots which adhere to the support tree. Old stem becomes thick, produces numerous lateral branches.

Runner shoots arise from base of vine. Leaves thick, coriaceous, glabrous shape much variable, commonly ovate, elliptic or elliptic lanceolate, size varies from small to large, base round, acute or cordate, tip acuminate, upper surface dark green to light green, lower surface dull green; petioles short or long, grooved, leaf margins wavy or even. Wax glands present on under surface of leaves and on young shoots and petioles. Spike filiform; pendulous, young ones green or whitish green, or light purple, mature ones green, pale purple or pale yellow. Peduncle glabrous, bracts oblong, decurrent, sessile-with upper margin developing into a hollow cup in female spikes, rachis and bracts glabrous, stamens two, anther dithecous, carpel single, ovary spherical, style absent, stigma 3-5 lobed, papillate. Fruit a drupe, green when young, changes to red on ripening, seed mostly spherical and pungent. Wild forms usually dioecious, cultivated ones are bisexual.

Morphology

Black pepper exhibits dimorphic growth pattern. The main stem of orthotropic shoot climbs up the support. It has indefinite growth and undergoes branching, producing lateral fruiting branches or plagiotrops. Each node provides clinging roots with help of which plant anchors on to support tree. These roots when come in contact with a rooting medium develop into normal underground roots. The plagiotrops have sympodial growth habit. As shoot grows, the terminal bud gets modified into a spike and the growth is further continued by axillary bud. The lateral branches do not produce aerial roots. Orthotropic shoots occasionally produce axillary branches that exhibit monopodial growth. They do not produce branches and roots, are positively geotropic, grow downwards and are called hanging shoots. Pepper plants also produce adventitious runner shoots from base especially during rainy season. These shoots have prominent aerial root initials on nodes and upon coming in contact with soil strike root easily. These shoots trail on ground and climb up when they come in contact with a support. These runner shoots are the common planting materials used extensively for clonal propagation.

Biosystematic studies

Cluster analysis and principle component analysis were carried out by Ravindran *et al.* (1992) using ten species and seven accessions of *P. nigrum* and thirty characters for establishing similarity / dissimilarity among the taxa. They reported highly significant correlation among certain characters, and analysis led to six character clusters as follows:

1. Leaf length, leaf breadth, leaf size index
2. Fruit taste, presence of gall forming thrips
3. Leaf length / leaf breadth index, number of ribs on leaf, ecological distribution, growth habit
4. Spike length, peduncle length, spike orientation, fruit shape
5. Leaf length / spike length ratio and spike shape
6. Leaf shape, leaf base

Clustering of taxa carried out by centroid linkage method led to six clusters. They are:

Cluster A - *P. attenuatum, P. argyrophyllum*
B - *P. schmidtii, P. galeatum, P. trichostachyon*
C - *P. nigrum, P. wightii*
D - *P. hymenophyllum*
E - *P. silentvalleyensis, P. mullesua*
F - *P. longum*

A comparative study using flavanoid profile by Ravindran (1991) and Ravindran and Nirmalbabu (1994) indicated following chemical affinities.

P. galeatum - *P. trichostachyon* 87 %

P. attenuatum - *P. argyrophyllum* 79%

P. argyrophyllum - *P. hymenophyllum* 87 %

P. galeatum - *P. sugandhi* 82 %

P. longum - *P. mullesua* 69 %

P. longum - *P. silentvalleyensis* 35 %

P. mullesua - *P. silentvalleyensis* 57 %

These chemical relationships by and large support the taxonomical relationships arrived at by conventional tools.

Sebastian *et al.* (1996), on grouping 11 species of *piper* based on iso enzyme similarity, reported that *P. nigrum, P. pseudonigrum, P.bababudani* and *P. galeatum* formed one group, *P. argyrophyllum* and *P. attenuatum*, second group, P. chaba, *P. hapnium*, and *P. colubrinum* third group. *P. betle* and *P. longum* showed their distinctness from rest of species. Lowest similarity was observed between *P. colubrinum* on one side and *P. pseudonigrum* and *P. bababudani* on other side (similarity index 0.05).

Ambili (1997) carried out detailed morphological studies of eight species of *Piper* to find out relationship among species and a key for identification of these species was proposed.

In a study on molecular characterization of *Piper* species using RAPD techniques, Murugan (2002) obtained four clusters, comprising the first, two accessions of *P. longum* in the second, *P. colubrinum* and *P. attenuatum* forming the third cluster and *P. chaba, P. betle* and *P. arboreum* forming the fourth. Babu *at al* (2002) utilized RAPD and AFLP techniques for evaluation of relatedness among nine *Piper* species. The similarity coefficient in both the dendrograms ranged between 0.50 and 0.16 indicating high variability among the species. Exotic species *P. colubrinum* and *P. arboreum* were highly distinct. *P. argyrophyllum, P. attenuatum, P. bababudani* and *P. nigrum* were closely related. *P.betle, P. chaba* and *P. longum* stood independently.

Flushing and flowering

The pepper inflorescence is leaf opposite and is produced on lateral plagiotropic branches. These flowering lateral branches have sympodial mode of growth, the apical bud develops into inflorescence and growth is continued by activity of axillary bud. As axillary bud develops, inflorescence is pushed out so that it becomes leaf opposed.

Physiological and biochemical aspect of flowering, berry set and development in black pepper were studied in detail by Satheesan (2000) at Kerala Agricultural University using two varieties, Panniyur 1 and

Karimunda. Flushing of vines commenced within a week after continuous rains during monsoon in June and the variety Panniyur 1 was earlier to flush. The leaf area development completed in about 35 days in both the varieties, fresh and dry weight of leaf increased till 50 days after flushing and thereafter these parameters registered wide fluctuation upto berry maturity. Leaf production was more in Karimunda and upper canopy levels produced more leaves. The plastochron index was 23 days on an average in both varieties.

There was synchronous flushing and flowering and emergence of spike was delayed up to 6-9 days after flushing between varieties. Spike growth was completed within 35-43 days after emergence. Spike development followed a linear pattern. Spike attained a maximum length in 31.67 and 29.26 days in Panniyur 1 and Karimunda, respectively. Fresh weight of spikes increased up to 150 days after berry set. Number of spikes produced per lateral and per vine was significantly higher in Karimunda and spike production was more in the upper canopy levels. Percentage of aborted spikes ranged from 21 to 26.31.

Period of anthesis (from first to last flower in a spike) was 6 to 10 days (mean 8.12) in Panniyur 1 and 4 to 11 days (mean 9.21) in Karimunda. Anther dehiscence extended for a period of 4 to 9 (mean 7.72) and 3 to 8 days (mean 5.00) respectively in Panniyur 1 and Karimunda. The protogynic stage existed for 5 to 8 and 3 to 7 days (mean 6.52 and 5.27) in the two cultivars studied. In Panniyur 1, the spike had hermaphrodite (97.18%) and pistillate flowers (2.82%), while in Karimunda all the flowers were hermaphrodite (100%).

Anthesis started between 18.00 and 18.30 hours and attained a peak between 18.30 to 24.30 hours. Anther dehiscence, which commenced from 14.30 hours, was maximum between 14.30 and 15.30 hours. Stigmatic receptivity lasted for seven days after anthesis, with maximum for first three days. Rainwater was a major factor in pollination.

Pollen grains had a mean diameter of 9.86 µm. Pollen production/flower (7000 to 13000) varied from standard to standard. Sucrose at concentrations of 2, 4, 6, 8 and 10 %, boric acid and calcium nitrate at concentrations of 10, 20, 30 and 40 ppm each promoted pollen germination. A combination of three (4% sucrose, 20 ppm boric acid and 10 ppm calcium nitrate) gave maximum germination (95.77%). Pollen remained viable for 1 to 3 days in water under different treatments.

Spike shedding which occurred during almost all months from spike emergence to harvest was maximum in June and minimum in May. No shedding was observed in September. Eight stages of shedding could be observed, with maximum shed during first month of set. Percentage of shedding varied significantly among months and among standards.

Fresh weight and volume of berries increased up to 165 to 180 days after berry set and increase in dry weight continued for another fortnight. The extension growth of laterals ceased after a period of 13 to 14 weeks.

The leaf and spike production in laterals were more in Karimunda and leaf area was more in Panniyur 1. Leaf production, leaf area and spike production increased towards upper canopy levels.

The total fresh weight and dry weight of lateral during the season was more in Panniyur 1 and was higher in upper canopy levels. Biomass partitioning to the spikes at berry maturity period also showed superiority in Panniyur 1. Partitioning of fresh weight and dry weight were in the decreasing order of spike, leaf and stem in Panniyur 1, while this was in the order of leaf, spike and stem in Karimunda. The partitioning of fresh weight and dry weight to spikes was also more in the upper canopy levels.

Spike length, berry size, number of developed berries and total number of berries/spike, average weight and green and dry pepper yield were more in Panniyur 1. However, more number of spikes/lateral and/vine were observed in Karimunda.

Specific leaf weight and spike to leaf ratio were higher in Karimunda and leaf area was more in Panniyur 1. The variety Panniyur 1 recorded a higher average leaf photosynthetic rate compared to Karimunda. Maximum photosynthetic rate coincided with flushing and flowering and spike bearing laterals recorded higher photosynthetic rate.

Stomatal conductance, leaf surface temperature, transpiration, vapor pressure deficit and PAR (Photosynthetically Active Radiation) at leaf surface recorded by the leaves were more in Panniyur 1. The sub-stomatal CO_2 level and stomatal resistance of leaves were higher in Karimunda. Leaves of upper canopy recorded higher stomatal conductance, sub stomatal CO_2 level, transpiration rate, leaf surface temperature, PAR and VPD. Stomatal conductance, sub stomatal CO_2 and stomatal resistance declined gradually towards berry maturity. Transpiration rate, leaf surface temperature and VPD increased towards berry maturity.

Chlorophyll development was completed in 35 days after flushing and chlorophyll 'a' and total chlorophyll content of leaves were higher in Karimunda. Middle canopy recorded more chlorophyll content and spike bearing laterals had higher chlorophyll content of leaves.

Carbohydrate content was more in leaves of Panniyur 1 while phenolic content was higher in Karimunda. There was no significant difference in amino acid content between varieties. Carbohydrate accumulation was more in lower canopy while upper canopy levels recorded higher phenolic and amino acid contents. Carbohydrate and amino acid content of leaves fluctuated with season while phenolic content of leaves was almost stable except at berry maturity. Spike bearing laterals recorded more foliar carbohydrates, phenols and amino acids. Electrophoretic pattern of proteins in berries showed variation between varieties. Number of bands seen was six and eight respectively for Karimunda and Panniyur 1.

Pattern of partitioning of labelled photosynthates among component organs showed that in Karimunda, maximum amount of photosynthates was fixed by the fed leaf itself during flushing, flowering and berry set. In Panniyur 1, except at flushing stage, berries partitioned the maximum labelled photosynthates. During all stages, percentage of labelled photosynthates partitioned by spikes was higher in Panniyur 1 compared to Karimunda.

Fresh weight and dry weight of lateral, fresh weight and dry weight of spikes and number of developed berries in a spike showed significant correlation with dry pepper yield. Leaf photosynthetic rate, chlorophyll and carbohydrate content of leaves also were significantly associated with economic yield in black pepper.

Bud differentiation in pepper

The flower bud differentiation in black pepper was influenced considerably by rainfall and relative humidity, nature of association being positive. However, with sun shine hours, relationship was found negative. Prior to the peak bud differentiation period, carbohydrate content and C/N ratio build up were higher. Phosphorous and potassium did not show any significant correlation.

A comprehensive analysis of climatic, nutritional and hormonal factors at time of peak differentiation period (June-July) showed that temperature, rainfall, RH and inhibitors were crucial factors influencing differentiation process. Peak flower bud differentiation was from middle of June to end of July in Panniyur 1 and slightly advanced in Karimunda (beginning of June to mid July).

Histological studies carried out in two types of laterals and in new shoots failed to show flower bud differentiation in laterals. Differentiation of vegetative buds was observed in laterals throughout, though at varying degrees. Maximum differentiation of vegetative buds occurred during June-July (40.0 to 67.5%) and minimum, during January to March (5.0 to 15.0 %). Three stages were identified in the development of vegetative buds. Appearance of conical undifferentiated vegetative primordia surrounded by leaf sheaths marked the initiation. During second stage, elongation of primordia was observed. Differentiation of tissues into outer dermatogen like layer and inner periblem - like zone occurred at third stage indicating completion of vegetative bud differentiation.

Flower bud differentiation was observed in shoots arising from two types of laterals. During dry months of January, February and March, percentage of buds differentiating into flower buds was negligible (5.0 to 12.5%). A spurt in flower bud differentiation activity was observed immediately after receipt of pre monsoon showers. Maximum flower bud differentiation (40.0 to 95.0%) occurred in June-July.

Five stages were identified in development of flower buds. During first stage, two undifferentiated conical primordia surrounded by leaf sheath were observed, indicating commencement of flower bud differentiation process. Towards later half of first stage, one of the primordia broadened and the third stage indicated floral initiation and a shape resembling pepper spike could be clearly observed. During fourth stage, differentiation of floral parts was observed. Completion of differentiation process was indicated by appearance of stamens and the ovary during fifth stage.

Pollination

Studies conducted in Kerala Agricultural University ruled out role of insects in pollination. Wind pollination is also practically absent under Indian condition. Pepper is adapted for pollination by gravitational descending of pollen grains combined with action of rain water or dew drops (Geitonogamy). Pendant hanging spikes, spiral arrangement of flower, sequential ripening of stigmas and prolonged receptivity of stigmas are all favourable to geitonogamy. Protogynous condition of pepper affects fruit set. Temporal separation of female and male phases can vary very much in bisexual cultivars from 0-30 days or even more and it was found to be season and climate dependent. In cultivars, where male and female phases mature simultaneously, autogamy may result. Inter-spike pollination may also be of frequent occurrence. Self pollination is the predominant form of fertilization in cultivated bisexual pepper. Though protogyny occurs, it appears ineffective to prevent selfing as the pendent spike is abundantly assured of pollen from upper flowers. Occurrence of apomixis is also reported in pepper.

Pepper fruit takes 6-8 months for full maturity from flowering, depending upon variety, the average being seven months. Fruit ripening is very much uneven. Fruits at top of spike turn yellow and then red on full ripening.

Spike shedding

Shedding of spikes (especially young spikes) is common in pepper. It could be due to diseases like fungal pollu or *Colletotrichum* attack, insects or due to other abiotic factors. Geetha and Nair (1989) attributed spike shedding to inadequate pollination, low soil moisture and low N and K in shoots. It was maximum in June and minimum in May. In general, two phases of shedding are observed the first lasting for 2-3 months occurring during early spike development and the second lasting for 1.5-2 months preceding harvest.

Application of IAA (50 ppm), planofix (50 ppm), zinc (0.5 %) and 2,4-D (5 ppm) reduced spike shedding by 63.6, 52.2, 48.4 and 35.3 % respectively.

Growth and development studies

Menon and Nair (1987) reported that shoot growth in Panniyur-1 started in late May, the highest growth was in June while highest flowering was in July.

Jayasree *et al.* (1988) studied root activity pattern of pepper trailed on *Erythrina* and teak pole, employing ^{32}P soil injection technique. In both cases 90 % of the root activity was found within an area of 30 cm radius around the plant. Vertical distribution of feeder roots of plants trailed on *Erythrina* was more or less uniform upto a soil depth of 40 cm. On other hand, root activity of plants trailed on teak pole was more at 40 cm depth than in upper layers. *Erythrina* was a surface feeder compared to pepper and its feeder roots extended over 60 cm into rhizosphere of pepper plants laterally.

Mathai and Nair (1990) reported that low yielding cultivars had large canopy surface area and accumulated 80 % of dry matter in laterals, whereas high yielding cultivars accumulated only 50 % of dry matter in laterals. Panniyur-1 has more laterals, spikes and berries, has higher mean berry weight, higher rate of photosynthesis and translocation and also higher yield compared to other cultivars tried. The efficient dry matter partitioning capacity of high yielding cultivars was strongly influenced by their total biomass production.

Geetha and Aravindakshan (1992) reported that leaf area/plant in bush pepper (developed from plageotropic fruiting laterals) depends primarily on increase in leaf area, while in vine pepper, it is primarily dependent on the number of leaves produced. Number of leaves, total leaf area and total and aerial biomass production differ significantly between vine pepper and bush pepper and are dependent on nitrogen levels.

Crop improvement

The greatest contribution to pepper research in India is from the Pepper Research Station at Panniyur, a constituent institution of Kerala Agricultural University. Right from it's inception in 1949, the station is actively involved in pepper research.

The important objectives of pepper breeding are higher yields, resistance to diseases (mainly foot rot and virus diseases), higher quality, tolerance to abiotic stresses and evolving low input responsive varieties. Pepper, blessed with twin advantages of vegetative propagation and viable sexual reproduction offer much scope for crop improvement work, especially for selection, breeding and exploitation of hybrid vigour. Existing genetic variability can be exploited through clonal selection, as there are more than 100 pepper cultivars known to exist in India with substantial amount of inter and intra cultivar variability.

Germplasm resources

Diversity of pepper is great in evergreen forests of Western Ghats. Out of 17 species of *Piper* occurring here, 11 are endemic (Ravindran and Peter, 1995). Through the process of natural evolution, domestication and selection, a lot of cultivars originated in its centre of origin. Cultivar diversity is the richest in Kerala, followed by Karnataka. Most of the cultivars are bisexual forms unlike their wild counter parts. Most of these cultivars are known by their major growing tracts (Arakkulam Munda, Perambra Munda, etc.) or by typical feature of cultivar (Kalluvally, Kuthiravally, Neelamundi, Uthirankotta etc.). Cultivar diversity is so great that a breeder gets any desirable character from them for breeding. As part of the five year multidisciplinary research project on "Technology Mission on black pepper carried out by the Kerala Agricultural University during 1999-2004, a survey was conducted in farmers fields at five selected districts of Kerala (Ernakulam, Kottayam, Pathanamthitta, Kollam and Idukki) for identifying location specific varieties. It was observed that Karimunda was the most popular one in southern districts of Kerala. The other important varieties are Panniyur 1, Neelamundi, Kaniyakkadan, Narayakodi and Perumkodi. Coverage by the variety Karimunda is 98.1 % in Ernakulam, 100 % in lower elevation of Idukki, 90.5% in higher elevation of Idukki, 89.1 % in Kollam and 87.3 % in Kottayam. Under high elevation, the maximum coverage was recorded by variety Neelamundi (91.6%). Varieties Karimunda, Neelamundi, Kottanadan Kotta, Karuvali and Vadakkan munda were identified as low input responsive. Many of these cultivars disappear due to various reasons. Initial efforts to collect and evaluate these cultivars resulted in evolving improved selections mainly for yield. Since present global situation demands other qualities like berry quality, resistance to biotic and abiotic stresses etc. the thrust is to conserve existing cultivar wealth in black pepper.

Germplasm of *Piper nigrum* and related species are maintained mainly at two centres in Kerala Agricultural University viz. Pepper Research Station, Panniyur and College of Horticulture, Vellanikkara. Apart from these two centres Regional Agricultural Research Station, Ambalavayal, Wynad and Cardamom Research Station, Pampadumpara also maintain valuable collections. Over 15 species and 250 accessions are maintained in these centres. Several thousands of seedlings from controlled crosses and open pollinated seedlings are also maintained and being evaluated. In addition to Kerala Agricultural University, Indian Institute of Spices Research, Kozhikode also maintains 727 collections of pepper. Horticulture Research Station, Yercaud,

Pepper Research Station, Sirsi (Karnataka) and Horticulture Research Station, Chintapally (Andhra Pradesh) also maintain pepper germplasm as part of AICRP on Spices.

Protocol for *in vitro* conservation by slow growth method was standardised in *P. nigrum, barberi, colubrinum, betel* and *longum* (Nirmalbabu *et al.*, 1996). The conserved materials of all the species showed normal rate of multiplication, when transferred to multiplication medium after storage. The normal sized plantlets when transferred to soil established with over 80 % success. They developed into normal plants and were morphologically similar to mother plants. Production of synthetic seeds was reported by Sajina *et al.* (1996) using shoot buds.

Evaluation of germplasm

Many of the popular pepper cultivars exhibit considerable intracultivar variability. In Kerala, the most popular cultivar is Karimunda. At IISR, Kozhikode two varieties Sreekara and Subhakara were released based on yield and quality traits, from Karimunda cultivars (Table 7). They have the yield potential of 4200 kg and 4487 kg/ha of dry berries respectively on fifth year planting (Ratnambal, 1990). Later, two clonal selections Panchami (clonal selection from Aimpiriyan) and Pournami (clonal selection from Ottaplackal) were released during 2001 from IISR (Ravindran *et al.*, 1992). Panchami has medium long spikes, high spiking intensity, high percentage of hermaphrodite flowers and high fruit set. Spikes have twisted nature in cv. Aimpiriyan. It is a late maturing variety. Pournami was selected based on good yield potential and tolerance to root knot nematode. PLD-2 is a selection from cv. Kottanadan by Central Plantation Crops Research Institute, Research Centre, Palode. Another selection, IISR Thevam (from cv. Thevanmundi) tolerant to foot rot disease has been identified at IISR (Sasikumar *et al.*, 2004).

Selection in OP progenies

Pepper being heterozygous and propagated mainly through cuttings, there exist segregation in open pollinated and selfed progenies. Selection in OP progenies was carried out at Pepper Research Station, Panniyur. Two varieties Panniyur-2 and Panniyur-5 were developed through selection from OP progenies of cvs. Balankotta and Perumkodi respectively (Table. Panniyur-2 has an average yield of 2570 kg/ha (dry) and Panniyur-5, 1098 kg/ha (dry). Panniyur-2 is a high piperine variety (6.6%). Panniyur-7 was released later in 2000 from open pollinated progeny of cv. Kalluvally. It is a regular bearing hardy variety tolerant to adverse climatic conditions. At IISR, Kozhikode a promising selection (IISR-Sakthi) was identified from OP progenies of cv. Perambra mundi. It has tolerance to *P. capscici* (Johny and Ravindran, 2005a).

Table 7. High yielding varieties of black pepper and their characteristics

Variety / Year of release	Pedigree/ Parentage	Institution/ University	Av. yield kg/ha (dry)	Salient features
Panniyur 1 1971	Inter-cultivar hybrid of Uthirankotta x Cheriyakaniyakadan	Pepper Research Station (KAU), Panniyur, Karimbam 670 143, Kerala	1242 (Potl yield 8800)	Vigorous growing vine. Donot tolerate shade. Long spikes, close setting of berries, bold berries, oleoresin 11.8%, piperine 5.3%, essential oil 3.5%, dry recovery 35.3%
Panniyur 2 1991	Clonal selection from open pollinated progeny of Balankotta	-do-	2570 (Potl. yield 3313)	Shade tolerant, suitable for intercropping, medium maturity group, medium quality, oleoresin 10.9%, high piperine (6.6%), essential oil 3.4%, dry recovery 35.7%
Panniyur 3 1991	Inter-cultivar hybrid of Uthirankotta x Cheriyankaniyakadan	-do-	1953 (Potl. yield 3269)	Suitable for all pepper growing regions, performs well under open situation. Late maturity group. Long spikes and bold berries, piperine 5.2%, oleoresin 12.7%, essential oil 3.1%, dry recovery 27.8%
Panniyur 4 1991	Clonal selection from Kuthiravally type II	-do-	1277 (Potl. yield 2443)	Stable yielder, performs well under adverse condition also, tolerant to shade, later maturity, 4.4% piperine, 9.2% oleoresin, 2.1% essential oil and 34.7% dry recovery
Panniyur 5 1996	Clonal selection from open pollinated progeny of Perumkodi	-do-	1110 (Potl. yield 2248)	Suitable for both mono and mixed cropping in coconut / arecanut gardens, shade tolerant, medium maturity, tolerant to nursery disease. Long spikes, piperine 5.3%, oleoresin 12.33%, essential oil 3.8% and dry recovery 35.7%
Panniyur 6 2000	Clonal selection from Karimunda type III	-do-	2127 (Potl. yield 3359	A vigorous vine. Tolerant to drought and adverse climatic conditions, stable and regular bearer, medium maturity group. Suitable for open condition as well as partial shade, spike 6-8 cm, more number of spikes/unit area, close setting and attractive bold berries, piperine 4.9%, oleoresin 8.27% and essential oil 1.33% and 33.0% dry recovery
Panniyur 7 2000	Open pollinated progeny of Kalluvally	-do-	1410 (Potl. yield 2770)	Vigorous vine and a regular bearer, long spike, a hardy type vine, tolerates adverse climatic condition, suitable for open and shaded conditions, very long spike (16-24 cm) high piperine content (5.6%), oleoresin 10.6%, essential oil 1.5% and 34.0% dry recovery

Contd...

Sreekara 1990	Clonal selection from Karimunda	Indian Institute of Spices Research, ICAR, Calicut - 673 012 Kerala	2677 (Potl. yield 4200)	Suitable for all pepper growing regions including high elevations as well as for intercropping, medium maturity, gives high quality pepper with 5.1% piperine, 13.0% oleoresin, 7.0% essential oil and 35.0% dry recovery	Kerala, South Karnataka and Tamil nadu
Subhakara 1990	Clonal selection from Karimunda	-do-	2352 (Potl. yield 4487)	Wide adaptability to all pepper growing tracts. Suitable for intercropping as well as for high elevations, high quality, medium maturity, piperine 3.4%, oleoresin 12.4%, 6.0% essential oil and 35.0% dry recovery.	Kerala, South Karnataka and Tamil nadu
Panchami 2001	Clonal selection from Aimpiriyan	-do-	2828(Potl. yield 6528)	High yielding, spikes twisting, late maturing, suitable for high elevations, excellent fruit set, piperine 4.7%, oleoresin 12.5%, 3.4% essential oil and 34.0%dry recovery.	Kerala and Karnataka
Pournami 2001	Clonal selection from Ottaplackal	-do-	2333(Potl. yield 5326)	Tolerant to root knot nematode and shade. Medium maturity, suitable for intercropping with arecanut anf banana. Piperine 4.1%, oleoresin 13.8%, essential oil 3.4% and dry recovery 31.0%	Kerala and Karnataka
IISR Thevam	Clonal selection from germplasm of Thevanmundi	-do-	2437(5.00 kg/vine fresh)	Stable yielding, grow vigorously, field tolerant to *phytopthora*, medium maturity, suitable for high altitude areas of South India up to 3000 ft MSL in coffee and tea estates. Piperine 1.6%, oleoresin 8.15%, essential oil 3.1% and dry recovery 32.5%	Kerala and high altitude areas of South India up to 3000 ft MSL
IISR Malabar Excel	A hybrid of Cholamundi x Panniyur - 1	-do-	1453 (3.00 kg/vine fresh)	Recommended for rain fed condition including coffee and tea plantaton. Oleoresin 11.7%, piperine 2.4%, essential oil 2.8%, dry recovery 32.3%. Suitable for high elevations and plains	Kerala and high altitude of South India
IISR Girimunda	A hybrid of Cholamundi x Panniyur -1	-do-	2880 (6.00 kg/vine/ fresh)	Recommended for rain fed conditions including tea and coffee esates. A medium maturity group. Piperine 2.2%, oleoresin 9.65%, essential oil 3.4% and dry recovery 32.0%.	Kerala and other high altitude areas of S. India
IISR-Sakthi (P-24)	Open pollinated seedling progeny of cv. Perambra mundi a clonal cultivar	-do-	(5.2kg/vine, fresh) (Potl. yield 5755)	Tolerant to *P. capsici*, piperine 3.3%, oleoresin 10.2%, essential oil 3.7% and driage 43.0%.	Kerala and Karnataka plains and high ranges
PLD - 2 1996	Clonal selection from Kottanandan	NRC for Oil Palm, Regional Station, ICAR, Palode, Pacha (Post) 695562, Kerala	2475 (Potl. yield 4732)	Late maturity high quality cultivar, contains piperine 3.0%, oleoresin 15.45%, essential oil 4.8%. Suitable for plains and higher elevations.	Kerala

Potl. yield – Potential yield

Source: Johny and Ravindran (2005)

Hybridization

Great variability that exists among cultivars is utilized in hybridization. Hybridisation work in pepper started at Pepper Research Station, Panniyur as early as 1959 and the first ever hybrid pepper variety in the world, Panniyur-1 was released from this station in 1971. It is a selection from F_1, of a cross between cv. Uthirankotta x Cheriyakaniakkadan. This variety was received with good response from all over and outside India and continue to be the most popular hybrid variety even now. The average yield of this variety is 1242 kg/ha of dry berries. Performance of this variety is the best under open condition. Another hybrid variety Panniyur-3 was also developed from F_1 population involving same parents during 1991. At IISR, Kozhikode two promising hybrids were identified for high altitude regions of South India. They are IISR Malabar Excel, a hybrid of Cholamundi x Panniyur 1 with an average yield of 3 kg/ vine of fresh berries and IISR Girimunda, a hybrid between Narayakodi x Neelamundi with an average yield of 6.00 kg of fresh berries / vine.

A large number of inter cultivar hybrids are under different stages of testing for yield, quality and reistance to diseases, mainly *Phytophthora* foot rot at various pepper research centres. At College of Horticulture resistance breeding against *Phytophthora* included released varieties viz. Panniyur 1, Panniyur 2 and Panniyur 5 as susceptible parents and Narayakodi, Kalluvally, Uthirankotta and Balankotta as tolerant parents (Nybe, 2001). Progenies are under various stages of evaluation.

Interspecific hybridization was reported by Sasikumar *et al.* (1999) who got successful crosses between *P. nigrum* x *P. attenuatum* and *P. nigrum* x *P. barberi*. Polyploidy and mutation breeding were also attempted in black pepper with no desirable results.

Breeding for quality

In pepper, quality is mainly decided by content of piperine, essential oil and oleoresin. Piperine is responsible for pungency. Piperine content varies from 2-7.4 % among the cultivars. Essential oil ranges from 0.4-7 %. Now, the property of piperine as a bioavailability enhancing agent is being exploited by pharmaceutical industry. Similarly, diversification of value added products from pepper also demands high quality varieties. The existing cultivar diversity can be utilized for locating high quality varieties which can be subjected to selection as well as intercultivar hybridization. Gopalan and Ravindran (1987) categorized pepper cultivars on the basis of piperine, oleoresin and essential oil contents. Among cultivars, Kottanadan had the highest oleoresin content. Similarly cv. Thommenkodi, Kuthiravally and Karimunda also offer scope for breeding for quality. Among high yielding varieties released for cultivation, Panniyur varieties have high piperine content (5-6.6%); Panniyur-2 having the maximum (6.6%). Subakara and Sreekara from IISR have high essential oil content (6.0 and 7.0% respectively). PLD-2 possesses high oleoresin (15.5%).

Resistance breeding

With an objective to develop a black pepper variety resistant/tolerant to *Phytophthora* foot rot disease, breeding trials were undertaken at the College of Horticulture, Vellanikkara. Around 75,000 open pollinated seeds of 55 germplasm accessions were sown and seedlings evaluated. In addition to this, seeds from 14 controlled crosses were also evaluated. Inter specific crosses involving Panniyur 1, Panniyur 2 and Panniyur 5 as female parents and *P. colubrinum* as resistant male donor were also attempted. In inter specific crosses, even though seeds were set and seedlings were obtained using *P. nigrum* as female parent, seedlings were found apomictic and not hybrids. Inter specific crosses involving *P. colubrinum* as female parent had not resulted in seed set.

Seedlings of controlled inter-variant crosses and open pollinated seedlings were raised in large number every year. This enabled screening of maximum recombinants possible. Seedlings were initially field screened in sick plots without giving fungicidal sprays. Seedlings which survived field screening were artificially inoculated with culture disc of *Phytophthora capsici*. For artificial screening, recently matured leaves of seedlings were collected. Pure isolates of *P. capsici* were cultured in carrot agar medium. Inoculated leaves were incubated and observed for symptoms after 24, 48 and 72 h after inoculation. So far 66 seedlings were obtained which survived field screening and artificial inoculation. Most of the seedlings did not show any symptoms after 72 h of inoculation.

Apparently resistant seedlings were field planted on *Garuga pinnata* standard. They are being maintained as per Package of Practices of Kerala Agricultural University. Disease tolerance/resistance of field planted seedlings is yet to be confirmed.

Investigations were undertaken in five released varieties of black pepper viz. Panniyur 1 to 5 and three cultivars viz. Kalluvally, Cheriakaniyakkadan and Karimunda for *in vitro* screening against *Phytopthora*. Establishment of callus cultures of above varieties/cultivars, *in vitro* induction of mutation through γ irradiation, *in vitro* screening of callus cultures using toxic metabolite(s) of *Phytophthora capsici* and *in vitro* testing of regenerants for disease reaction were undertaken. One month old calli of five released varieties and three cultures were subjected to γ irradiation of 10 Gy from cobalt 60 source. Kalluvally exhibited good callus growth in the two doses tried. Shoot morphogenesis was recorded in different varieties and cultures. Panniyur 5 and Karimunda recorded higher percentage of shoot morphogenesis with more number of sprouts (8.5-9) followed by Panniyur 4 and Panniyur 2. From field evaluation trial of already regenerated calli/clones of black pepper, five calli/clones of cultivar Cheriakaniyakkadan (CC58, CC 56, CC 57, CC 43 and CC 60) were highly tolerant to *P.capsici*, superior in yield, quality and tolerant to *Phytophthora* foot rot.

Heritability studies

Pepper Research Station, Panniyur conducted detailed studies on genotypic and phenotypic variability, heritability and genetic advance in hybrids and open pollinated progenies. Spike yield followed by spike number showed the highest phenotypic coefficient of variation (Ibrahim *et al.*, 1988) and these characters were amenable for improvement. Heritability values ranged from 28 to 81 %, the highest for fruit weight followed by spike length. Spike yield and spike number had low heritability indicating that these characters are highly influenced by environmental fluctuations. Spike yield recorded the highest value for genetic advance. They also found that expression of spike character was markedly influenced by season. Fruits/ spike was more prone to seasonal variation (CV = 22.73%) than spike length (CV = 16.3%). The cv. Arakkulam Munda showed high stability of expression over seasons for spike length. Ravindran *et al.* (1992) found that shoot tip colour in pepper is controlled by two pairs of genes having complementary gene action. Leaf shape in pepper is controlled by multiple alleles (IISR, 1997).

Somaclonal variation

High amount of somaclonal variation is reported from callus cultures of black pepper. The cultivar Kalluvally exhibited higher levels of somaclonal variation, compared to other cultivars like Cheriyakaniyakkadan, Karimunda and Balankotta (Shylaja *et al.*, 1996). Variability among calli clones were assessed based on morphological, yield and quality attributes and reaction to *Phytophthora* foot rot disease. Five elite clones with desirable traits were identified and based on overall attributes, the calliclone CC60 was selected as superior one with high yield and tolerance to *Phytophthora* foot rot. Extensive screening of callus cultures of five released varieties (Panniyur 1, 2, 3, 4 and 5) and three cultivars viz. Kalluvally, Cheriyakaniyakkadan and Karimunda for *Phytophthora* foot rot tolerance is in progress at Kerala Agricultural University, Vellanikkara.

Protoplast culture

Shaji *et al.* (1996) reported high frequency isolation of viable protoplasts from *in vitro* derived leaves of both *P. nigrum* and *P. colubrinum*; plants could be regenerated only from *P. colubrinum* protoplasts. In *P. nigrum* and *P. colubrinum* protoplast fusion was not successful (Paul, 2000).

PR proteins

SDS-PAGE analysis of PR proteins extracted at pH 2.8 in two varieties Panniyur-1 and Kalluvally of *P. nigrum* and *P. colubrinum* showed distinct banding pattern. Two additional bands were expressed in *P. nigrum* varieties upon infection with *Phytophthora*. The relatively tolerant variety Kalluvally recorded early expression of these proteins in resistant species *P. colubrinum* indicating their significance in defence mechanism (Parab, 2000). Among pathogenesis related enzymes studied, b-1, 3-glucanase and phenylamine ammonialyase had a positive role in defence mechanism of pepper in relation to *Phytophthora* foot rot disease.

Genetic transformation

Investigations on genetic transformation in black pepper using *Agrobacterium tumefaciens* were carried out at Kerala Agricultural University and at IISR, Kozhikode. Trials have not yet succeeded to final level. Preliminary reports are available on *Agrobacterium* mediated gene transfer system in *P. nigrum* (Sasikumar and Veluthambi, 1996a, 1996b).

Molecular markers

Molecular markers like RAPD, AFLP and ISSR polymorphism were used to characterize important cultivars, varieties and related species of black pepper to develop fingerprints and to study the inter relationships (Pradeepkumar *et al.*, 2001). A mapping population was developed for preparation pf genetic map of black pepper (Nirmalbabu, 2003).

Crop management

Black pepper is cultivated in twenty six countries, all located in tropical belt of globe - mainly India, Indonesia, Malaysia, Srilanka, Thailand and Vietnam.

Climate and soil

In tropics, pepper cultivation is concentrated in humid region. Characteristic climatic requirements for pepper are high rainfall, uniform temperature and high relative humidity. It comes up well from plains to undulating hill slops of 750 to 1500 m. The crop adapts well in temperature range of 15-35°C, rainfall of 1250 to 2500 mm and an RH of 60-95 %. Effective rainfall received during different periods of year is more vital than total rainfall received during any particular period of the year. It is grown widely in a variety of soils. Well-drained, red, lateritic and alluvial soils rich in humus are ideal for cultivation. The major pepper growing soils of India can be broadly classified into red loam (Alfisol), forest loam (Mollisol) and laterite (oxisol).

Propagation

Black pepper develops four types of aerial shoots namely primary stem with long internodes, runner shoots which originate from base of vine, hanging shoots and fruit bearing lateral branches. Cuttings are raised mainly from runner shoots. Cuttings from lateral branches are used to develop bush pepper.

The hanging shoots are not used as they give weak lanky growing plants. In countries other than India, orthotropic climbing shoot is used as the planting material. They give vigorous plants, develop fruiting laterals from base itself and start yielding early. Propagation studies conducted at Kerala Agricultural University have proved that middle 1/3 portion of runner shoot was the best planting material, followed by top 1/3 portion of runner shoots (KAU, 2003). In high altitude regions, March is the best time for rooting of runners followed by last fortnight of July. Two nodded cuttings are sufficient for propagation. Last fortnight of June and second fortnight of April are the best for rooting of laterals. Runner shoots from high yielding and healthy plants are kept coiled on wooden pegs at base of vine to prevent shoots from coming in contact with soil and striking roots. They are separated from mother plant during Jan. - Feb., dipped in fungicides like copper oxychloride or Bordeaux mixture for one minute, after trimming leaves, cut into pieces having 2 to 3 nodes and planted in polythene bags filled with potting mixture. Two noded cuttings dipped in IBA (1000 ppm) for 45 seconds resulted in the highest rooting (KAU, 2003). Adequate shade, frequent irrigation and plant protection measures are necessary.

Rapid multiplication technique

This technique originally developed in Srilanka was standardised under Indian conditions. A trench of 60 cm deep and 40 cm wide and of convenient length is made and is filled with rooting medium. Split halves of bamboo with the septa intact or split halves of PVC pipes having 1.25-1.5 m length and 8-10 cm diameter provided with plastic septa are fixed at 45° angle on a strong support, touching one another. Rooted cuttings, one each per bamboo split are planted. Lower septa of bamboo / PVC are filled with rooting medium and growing vine is tied to bamboo in such a way as to keep roots pressed into rooting medium. Vines are irrigated daily. As vines grow up, fill up bamboo with rooting medium. When vine reaches top of bamboo (after 3-4 months), terminal bud is nipped off and stem is crushed at about three nodes above base, to activate axillary buds. After about 10 days, each vine is cut at crushed point and removed from rooting medium and each node is separated. Nodal cutting with bunch of roots intact is planted in polybags filled with potting mixture, kept under shade and irrigated daily. Buds start developing in about three weeks. Stumps regenerate giving out one or two shoots each and they can be trained as in earlier case. Thus continuous regeneration and harvesting of rooted cuttings are possible. About four harvests can be taken in a year giving a multiplication rate of about 1:40 in a year. Advantage of this method are rapid multiplication rate, better root system, better field establishment and vigorous growth. For rapid growth, nutrient solution of urea (1 kg), super phosphate (0.75 kg), muriate of potash (0.5 kg) and magnesium sulphate (0.25 kg) in 250 l of water is recommended @ 0.25 litre per vine at intervals of two weeks. Similarly *Trichoderma* @ 1 g and VAM @ 100 cc/kg of soil is recommended for adding into potting mixture (Sreenivasan and Rema, 2002).

Bush pepper

By planting lateral branches (plagiotrops) of pepper, vine can be grown as a bush. One year old lateral branches with 3-5 nodes are to be planted in nursery during March-April for pre-rooting after treating cut ends with solution of IBA (1000 ppm) for 45 seconds. Rooting percentage of laterals is less than 30. Sujatha *et al.* (2004) reported that rooting of laterals could be improved (60-70%) by using 2-3 node lateral cuttings with a bit of orthotropic portion intact. Only well-rooted and established plants are to be used for transplanting. Rooted cuttings are to be planted at 3-5/pit or /pot. Fertilizers can be applied @ 1.0, 0.5 and 2.0 g / pot of N, P_2O_5 and K_2O respectively at bimonthly interval. Alternatively application of 15 g groundnut cake (or) 33 g neemcake can also meet N requirement of crop. Bushy nature of plant will have to be ensured by proper pruning of hanging shoots. Potted plants are to be kept preferably under partial shade. It is necessary that re-potting is carried out after every two years.

Micropropagation

A viable protocol was developed for *in vitro* mass multiplication of black pepper (Nazeem *et al.*, 1993, Nirmalbabu *et al.*, 1993, Joseph *et al.*, 1996, Lissamma *et al.*, 1996). Techniques for direct and indirect organogenesis were standardised. For rapid multiplication, the most suitable explant was nodal segments. Total time required for *in vitro* multiplication from nodal segments was 24 months including 15 serial subculturing at multiplication stage. Number of healthy plants that could be recovered was 4000/explant used and approximate cost was worked out to be Rs 8/ plant (Babu *et al.*, 2004). Micropropagated plants were hardened and successfully established in the field with > 80 % survival and later field establishment. TC derived plants were also evaluated in farmers fields all over Kerala. No significant difference was noticed among TC regenerants and also when compared with conventional propagules. Bearing habit was regular and yields stable. Quality parameters *viz.* oil and oleoresin contents were high in TC derived plants (2.23 and 14.66% respectively) compared to conventional plants (2.0 and 11.5% respectively) (Babu *et al.*, 2004).

Preparation of land and planting standards

When pepper is raised on sloppy land it is advisable to avoid slopes facing south to protect vines from sun scorching during summer. In slopes, soil conservation measures like terracing should be done.

Standards for pepper

Pepper needs a support for trailing. Standards can be live or dead. In India, it is invariably trailed on live standards whereas in Malaysia and Indonesia, dead standards are used.

In India, pepper is trailed on a variety of trees. For large scale planting, *Erythrina indica, E. lithosperma, Garuga pinnata, Grevillia robusta* and *Ailanthus malabaricum* are used. When grown as an intercrop, it is trailed on coconut and arecanut and in homegarden agroforestry system pepper is trailed on mango, jack and glyricidia.

Studies proved that dead standards (teak poles) gave higher yield compared to live ones (Nybe, 2001), but investment on dead standards is high and hence it is not commonly adopted. Availability is also low. In a study including seven tree standards involving pepper cultivar Karimunda, *Ailanthus malabaricum* was the best live support for obtaining maximum yield and it was on par with *Erythrina* (Nybe, 2001). Sadanandan *et al.* (1992) reported that pepper plants trailed on *Erythrina indica* yield better than those trailed on other standards. But this is highly susceptible to stem and root borer and often they do not last long. At higher elevations *Grevelia robusta* is the ideal shade tree.

Planting and after care

In India, planting is undertaken with onset of South West monsoon. Stem cuttings (about two meter long) of standards are planted at required spacing. After they were established (preferably after one year) pepper vines are planted. Pits of 50 x 50 x 50 cm size are taken, filled with top soil and organics and 2-3 rooted cuttings of pepper are planted on north or north east side of support. Plant density varies very much among different cultivation methods and depending upon standards used. In India, 2.5 - 3 x 2.5 - 3 m is commonly followed accommodating about 1100 to 1600 plants/hectare. In case of unrooted cuttings, about 4-5 cuttings/pit are to be planted, at least one node of cutting should go below soil for proper growth. As pepper plant grows, it is tied to support as and when required. Young plants are protected from hot sun during summer by artificial shade. Mulching was also found advantageous during summer. Sawdust, polythene sheet, arecanut husk and dried leaves are good mulches (Nybe, 2001). Coconut husk is not found suitable. Adequate mulch should be given towards end of north-east monsoon. Regulation of shade by lopping branches of standards is necessary not only for providing optimum light to vines but also for enabling standards to grow straight. Excessive shading during flowering and fruiting period result in poor flowering and yield. Prunings from standards can also be used for mulching.

Intercultivation in pepper plantations is a point of controversy. Clean cultivation was recommended in earlier years. But due to spread of dreaded foot rot disease, maintenance of grass cover and practice of minimum tillage are advocated (Sarma *et al.*, 1991; Thomas and George, 1992). Disease spread is faster in gardens where clean cultivation is practiced.

Water management

In India irrigation during summer is beneficial. A field experiment conducted to study effect of irrigation (on Karimunda and Panniyur-1) at the Pepper Research Station, Panniyur, Kerala showed that irrigation at IW/ CPE ratio of 0.25 significantly increased productivity of pepper by over 90 %, effect was more significant in Karimunda than in Panniyur-1 (Satheesan *et al.*, 1997). A study at IISR showed that irrigating pepper vines at seven to ten litres of water/day/vine during summer recorded higher yield of green pepper (4.07 kg/vine/year) compared to control (1.33 kg/vine/year) (IISR, 1997). Thanuja (2003) found that summer irrigation of pepper vines @ 10 litres/ vine/week by basin method from December to March improved yield by 48 %.

Pot culture experiments on varietal reaction to soil moisture was conducted at Kerala Agricultural University (Thankamany, 2000). Both bush and vine types of Panniyur-1 were very sensitive to water stress. Panniyur 5 survived water stress. Increased activity of peroxidase, catalase and superoxide dismutase were noticed in response to water stress in varieties Panniyur-5 and Padarpan. Epicuticular wax content was the highest in Panniyur-5 and the lowest in Panniyur-1.Amino acid content increased in response to water stress in varieties Poonjarmunda, Panniyur-1, Panniyur-5 and Padarpan.

Pruning

Pruning is practiced in all pepper growing countries except India. Pruning is done mainly in young vines, when they are about six months old, terminal shoot is pruned back to 3-4 nodes from ground. Terminal shoots are then selected and trailed on support.

Studies conducted on six year old bearing vines showed that pruning of hanging shoots had desirable effect on growth (Kurien and Nair, 1988). In certain pockets of India, Indonesia and Srilanka, there is practice of coiling one year old vine around standard under moist soil. Usually this is done in next rainy season. Before burying the stem under soil, all leaves are removed which help in early sprouting of axillary buds, many shoots come up and cover standard quickly.

Nutrition

Nutritional requirements of black pepper are standardised for various agro ecological conditions.

Foliar diagnosis

Rajagopal and Sadanandan (1985) standardised the youngest mature leaf and next matured leaf from fruiting laterals as the index leaves for foliar diagnosis in black pepper. May was found as the best time for sampling. Sushama *et al.* (1989) reported first mature leaf of fruiting laterals just before flushing as the most suitable for foliar analysis.

Nybe *et al.* (1989) reported that foliar content of N, K, Ca and Mg increased in the rainy season following fertilizer applications. The nutrient elements P, K, Ca, Mg and S are found to exert direct and indirect effects on yield, based on simple correlation and path coefficient analysis. Among the nutrients, P and K have greater influence on yield. The leaf nutrient status of elements for a few cultivars are given in the following table.

Sadanandan and Hamza (1998) worked out DRIS (Diagnosis and Recommendation Integrated System) norms for black pepper. They reported that 1.65-2.79 % N, 0.11-0.26 % P, 1.18-2.84 % K, 1.42-3.33 % Ca, 0.40-0.69 % Mg, 0.09-0.29 % S, 126-1145 ppm Fe, 109-721 ppm Mn, 21-67 ppm Zn and 16-120 ppm Cu in pepper leaves are required for getting optimum yield.

Nutrient removal studies showed significant relationships between yield and nutrients removed by spikes by way of harvest (Nybe, 2001). Regression analysis of nutrient-yield relationships showed that quantities of nutrients removed from the system through one kilogram of harvested produce were in the descending order of N, K, Ca, Mg, P, S, Fe, Mn and Zn at rate of 6.35 g, 6.33 g, 1.11 g, 0.47 g, 0.44 g, 0.29 g, 42.89 mg, 34.45 mg and 4.28 mg respectively. These quantities may be considered as additional requirement by vine for one kilogram increment in yield. It was observed that irrespective of yield potential, quantity of each nutrient required for production of 1 kg of green pepper is fairly constant and vines with higher production potential require relatively and proportionately higher dose of nutrients. This shows that there is need for amending fertilizer recommendation of black pepper based on yield potential of vine. The primary limiting factors for yield in pepper were high native content of Fe and Mn which will have to be ameliorated. Secondly, at low yield level, yield is limited by a relative deficiency of Zn caused by high P calling for a temporary skipping of P; supplemented with N and K which can take yield upto 5 kg $plant^{-1}$. Excess S through incidental application of prophylactic sprays of Bordeaux mixture and shortage of Ca and Mg appear to be limiting factors at higher yield class. Amelioration of acidity by supplying Ca + Mg carbonate, substitution of S containing fungicides and application of N and K shall constitute the recommendation, beyond which P may have to be applied.

Organic nutrition

Biomass production in black pepper was significantly increased following incorporation of organic materials into soil compared to control vines. When different levels of organic sources were compared, total biomass production of vine decreased at the highest level (3%) of application of coffee and black pepper leaves. This was attributed to allelopathic effect of decaying material. There was steady increase in biomass production with increasing level of garuga leaf application.

Significant increases were noticed in N, P and K concentrations of leaf and stem, Mg concentration of leaf, S concentration of leaf and stem, Fe content of stem and foliar Mn content of vine following organic matter treatments as compared to control vines. The vine removed significantly higher quantities of N, P, K, Mg, S, Fe and Mn following soil application of leaf materials as compared to control vines. Average nutrient removal by a six month old black pepper vine to produce 11.19 g of shoot dry matter was 200.19 mg N, 18.92 mg P, 432.53 mg K, 155.89 mg Ca, 19.44 Mg, 12.33 mg S, 1517 µg Fe and 3546 µg Mn. Contribution of leaf to total nutrient removal was more compared to that of stem. Significant increase in major and micronutrient availability in soil was noticed following incorporation of leaf materials. On an average, rate of organic carbon loss in laboratory incubation was the slowest in soil with coffee or silver oak leaves as organic source loss was the fastest when garuga was used. Half life values for organic carbon loss varied from 91 to 193 days. The half live values obtained for coffee, erythrina and black pepper leaves in field decomposition study were 40, 44 and 53 days respectively.

To study influence of organic manures and biofertilizers, in comparison with inorganic fertilizers, in improving growth and yield of black pepper, trials were undertaken in farmers' fields in different pepper growing regions of Kerala and at KAU campus, Vellanikkara. Results revealed that chemical fertilizer use in black pepper can be minimized considerably by substituting with organic manures and biofertilizers. The promising manurial schedules, which vary with locations, include substitution of 50% of inorganic fertilizer dose with FYM on N equivalent basis and remaining dose met either through inorganic P + *Azospirillum*, inorganic N and P, inorganic N + *Phosphobacterium* + VAM or composite culture of three biofertlizers (Nybe, 2004).

Another study was designed to formulate agronomic technologies, which require only minimum investment but can ensure comparatively high yield and profit in black pepper. The trials were taken up in gardens of progressive farmers in major pepper growing districts of Southern Kerala viz. Kollam, Pathanamthitta, Kottayam, Idukki and Ernakulam. Results revealed that growing cowpea in interspaces and basins, applying composite culture of *Azospirillum, Phospobacterium* and VAM, application of zinc sulphate and skipping of phosphorus are a few of the promising technologies that can lower input use in black pepper in different locations (Nybe, 2004).

Influence of organics (FYM, Neem cake), biofertilizers (*Azospirillum,* Phosphate solubilising bacteria and Arbuscular Mycorrhizal Fungi (AMF) on yield and quality of black pepper varieties Panniyur 1 and Panniyur 2 were studied. Fresh and dry yields of berries/vine as well as drying percentage of berries failed to produce any significant difference among treatments with respect to quality attributes viz essential oil, oleoresin and piperine contents, an integrated nutrient management with bio fertilizers, organic and inorganic fertilizers was found significantly superior. In general, treatments involving complete organic and biofertilizer + organic + inorganic combinations exhibited high values of all soil nutrients except S, Ca and Zn. Biofertilizers applied plots registered higher soil microbial population of fungi, bacteria and actinomycetes (Stephen, 2002; Stephen *et al.*, 2004).

Critical levels of nutrients

Highly significant positive correlations were shown between P, K, Ca and Mg with yield. Critical level of S was 0.15 %. Tentative critical levels suggested for other elements are as follows (Nybe *et al.*, 1989):

N - 2.1 to 2.4%	P - 0.19 to 0.20 %
K - 1.8 to 1.9%	Ca - 2.6 to 2.9%
Mg - 0.5 to 0.6%	Fe - 156 to 169 ppm
Mn - 77 to 88 ppm	Zn - 30 to 32 ppm

The two most important nutrient elements which are highly essential for production of pepper could be identified as K and P in view of their high direct and indirect effects on yield.

Nutrient uptake

Considerable variability existed among black pepper varieties with respect to growth and nutrient uptake. Of eleven varieties evaluated, biomass production was the lowest in Kuthiyanikkodi and the highest in Aimpiriyan. Super phosphate was the best P source for biomass production. Utilization of P sources was the lowest in Kuthiyanikkodi and the highest in Kuthiravally. When the total P absorbed was considered, Kuthiravally derived 40.37 % of its P requirement from the fertilizer P sources. Super phosphate and amophos were preferred by black pepper.

Varieties differed significantly with respect to concentration of P, Ca, Mg, Mn and Zn. Karimunda recorded more P concentration in the leaf and Ca in the stem. Kuching had more foliar Ca concentration. Magnesium concentration in leaf was higher in Kottakodi and that in the stem in Kottanadan. Leaf and stem concentrations of Mn were more in Aimpiriyan and Kottanadan, respectively, whereas those of Zn were more in Panniyur 1 and Neelamudi, respectively. P sources also influenced concentrations of Ca, Mg, Mn and Zn.

Table 8. Leaf nutrient status of some pepper varieties

Variety	N	P	K	Ca	Mg	Fe	Mn	Zn	Cu	Al	Mo
	(g 100 g^{-1})					(mg kg^{-1})					
Arakkulammunda	2.31	0.15	2.77	1.56	0.19	78	161	30	631	68	1.24
Balankotta	2.42	0.13	3.15	2.37	0.24	125	227	19	248	87	4.54
Karimunda	2.64	0.16	2.92	2.30	0.24	129	542	27	800	85	3.40
Kottanadan	2.12	0.13	3.55	1.40	0.15	94	293	26	425	60	2.34
Kuching	2.40	0.14	3.13	1.48	0.20	91	298	26	390	59	3.24
Kuthiravally	2.32	0.14	3.00	1.93	0.27	127	327	27	501	68	4.00
Kalluvally	2.18	0.14	3.14	1.61	0.20	139	446	28	608	86	3.10
Neelamundi	1.75	0.15	2.79	1.51	0.22	111	310	25	317	92	3.26
Panniyur-1	3.12	0.13	2.70	2.13	0.25	94	285	19	405	82	3.40
Thommankodi	3.02	0.14	2.94	2.02	0.17	92	199	29	627	78	3.28
Valiakaniakadan	1.93	0.12	2.61	2.33	0.23	92	311	29	335	71	4.72
Thevanmudi	2.54	0.12	3.52	1.69	0.20	84	273	14	356	53	1.77
CD (p=0.05)	0.24	0.02	0.33	0.23	0.04	14.6	89	5.2	126	13	0.70

Source: KAU (2003)

Significant differences were observed in uptake of P, S, Fe, Mn and Zn among varieties. Influence of P sources was also evident in uptake of Ca, Mg, S, Mn and Zn in black pepper. Superphosphate influenced uptake of Ca, S and Zn, but it was on par with nitrophosphate for Mg uptake and amophos for Mn uptake.

Leaf was found as a better accumulator of N, K, Ca, S, Fe and Mn in black pepper whereas P and Mg were more accumulated in stem. Concentration of Zn did not differ in both the parts. Order of nutrients removed by black pepper was K > N > Ca > Mg > P > S > Fe > Mn > Zn. Recovery of soil applied ^{32}P in leaves of pepper vines indicated that irrespective of type of standards used, fertilizer application in black pepper is to be restricted to 30 cm wide basin around the vine.

Nutrient deficiencies

Nutrient disorders and deficiencies are often encountered in the field. Nybe and Nair (1986; 1987a; 1987b and 1987c) conducted a series of studies to assess nutrient deficiency symptoms of major and micro elements.

Deficiency symptoms of macronutrients except Ca and S were first manifested on older leaves while that of micronutrients on younger leaves. Symptoms of N deficiency were expressed as uniform yellowing followed by necrosis whereas purple to bronze yellowing with ash coloured necrotic areas were symptoms of P deficiency. Potassium deficiency symptom was characterized by tip and marginal necrosis which later progressed to the two - thirds distal portion of lamina. Calcium deficiency symptoms appeared as tiny brown necrotic spots on chlorotic area near margins which later enlarged to form black necrotic areas. Visible symptoms of Mg deficiency were oval interveinal chlorotic area followed by black necrotic patches. Sulphur deficiency was manifested as uniform yellowing with brown necrotic spots. There was profound reduction in vegetative growth due to deficiency of macronutrients. Reduction in shoot growth and leaf area index was maximum in case of deficiency of N (56 and 63 % respectively) followed by S (48 and 17 % respectively). Reduction in root growth was quite high due to deficiency of Ca (61%), P (45%) and N (39%).

Interveinal chlorosis was the initial symptom of deficiency of all micronutrients. However, symptoms were specific to concerned nutrients. Iron chlorosis was characterized by presence of green bands along major veins whereas bronze yellow color of the intervienal area was the specific symptom of Mn deficiency. Bronze color of entire lamina with necrotic tips and margins were symptoms of Cu deficiency. Zinc deficiency was unique with little leaf and resetting. Due to B deficiency, leaves become large, thick and brittle with orange yellow mottles on upper surface and grey brown intervienal patches on lower surface. Unlike macronutrients, there was no marked reduction in vegetative growth due to deficiency of micronutrients except in iron and boron which recorded 35 and 22 % respectively, in total dry matter production. Boron deficient plants registered 18 % increase in leaf area index.

Growth of vine was completely arrested at comparatively early stage (4 ½ months after treatment) due to S deficiency followed by N (6th month), Fe (7th month), B (12th month), P (13th month) and Cu (14 ½ months after treatment). There was no cessation of growth in case of deficiencies of other nutrients.

Visual symptoms of deficiencies were concurred with a marked reduction in foliar levels of concerned elements. Antagonistic effects among K, Ca and Mg were also observed. In all other cases, deficiency of one element failed to influence foliar level of others. Deficiency symptoms could be recovered by application of deficient nutrient element which provided valid information for confirmation of deficiency symptoms.

Fertilizer requirements

Application of chemical fertilizers to pepper in general was very low in all southern districts of Kerala surveyed by the Kerala Agricultural University. Farmers usually resort to use of organic manures (89.1 % in Idukki). Use of bio-fertilizers is found on the increase in all the districts surveyed. Maximum adoption of bio-fertilizer application was in Pathanamthitta district (13.6 %) (Nybe, 2004).

Quantity of NPK required to get high yield varies largely among locations. Location specific recommendations for pepper advocated by Kerala Agricultural University are given below (KAU, 2003).

Generation recommendation : 50:50:50 ($N:P_2O_5: K_2O$ g/vine/year)

For Panniyur and similar areas : 50:50:200 "

For Kozhikode and similar areas : 140:55:275 "

Apply 1/3 dose for one year old plants and ½ dose for two year old plants. Fertilizer may be applied in two split doses, first in May-June with receipt of a few soaking rains and second in Aug.-September. Apply fertilizers in a circle of radius 30 cm around vine. Sadanandan (1994) by fitting response equation for N, P and K application on yield of pepper in var. Panniyur-1 and working out economics of yield arrived at a level of 140 kg N, 55 kg P_2O_5 and 270 kg K_2O/vine/year as optimum. The N and K fertilizers are to be applied in two equal split doses, first half in June and second half in September, while entire P is applied in one dose.

Inter cropping

Black pepper requires a live or dead support for trailing and is ideal inter/ mixed crop in coconut and coffee plantations. In coconut gardens, it suits well in third phase of growth of coconut palm i.e. 20-25 years onwards. By this time the palm attains a height of about 6 m, when 35-40 % of light is available below the palm canopy. Black pepper grows as tall as coconut and arecanut when trailed on them. But height of pepper vine is to be restricted to six meter by pruning for convenience in harvesting of both crops. In coffee, tea and cardamom estates at higher altitudes, black pepper is trailed on live standards viz. dadap (*Erythrina lithosperma*) and silver oak (*Grevelia robusta*) or on the existing shade trees like jack, glyricidia, erythrina, garuga, ailanthus and other jungle tress. Improved varieties emerging from breeding programmes are to be regularly evaluated for their performance as intercrop. Potty *et al.* (1979) evaluated performance of six varieties of black pepper in a multistoreyed cropping system involving coconut, pepper, cocoa and pineapple and suggested that Karimunda and Panniyur-1 perform better. CPCRI initiated field-screening trials of the latest pepper varieties / hybrids as mixed crops in coconut garden.

In an experiment conducted at Coconut Research Station, Bhatye, Ratnagiri, during 1979 to 1988, where pepper vines trained on glyricidia were grown as intercrop, yield of coconut was increased by 23.59 % (Nagawekar *et al.*, 2002). Average net profit/ha for four years was Rs.39,398 whereas monocrop of coconut gave a net profit of Rs. 30475.

Venkitaswamy (2002) from studies on mixed cropping of black pepper in coconut in Pollachi tract of Tamil Nadu, reported that varieties Panniyur 1, 2 and 5 established well as mixed crops.

Black pepper is a successful component of multistoried cropping system in which crops having canopies of varying heights are chosen. The most profitable multistoried system with coconut as main crop established at CPCRI, Kasaragod had black pepper, cocoa and pineapple as component crops.

Gopalakrishnan (2004) suggested that in west coast of India, Karimunda, Panniyur 1, Panniyur 2 and Panniyur 5 are suitable for mixed cropping in coconut.

However, black pepper is not recommended as a mixed crop with cocoa, papaw and cardamom in Srilanka as these crops are attacked by foot rot caused by same pathogen (Bandara *et al.*, 1985). Setepu and Kasim (1991) suggested avoiding intercropping pepper with cocoa and capsicum in Indonesia.

Pugalendhi *et al.* (1997) evaluated high yielding varieties of black pepper in coffee plantations of Shevroy hills in Tamil Nadu (Table 9).

Table 9. Performance of black pepper varieties trailed on silver oak (age-7 years) in coffee plantations

Variety	Yield per vine (kg green)	Setting (%)	Dry recovery (%)
Panniyur 1	2.2	96.0	35.3
Panniyur 2	3.5	74.2	35.7
Panniyur 3	4.4	89.2	27.8
Panniyur 4	2.3	85.2	34.7
Subhakara	4.2	68.0	35.5
Sreekara	4.0	63.4	35.0
Panchami	5.2	82.0	34.0
Pournami	4.2	68.0	31.0

The variety Panchami outyielded other varieties with a setting percentage of 82 and 34 drying %. Better cultural practices coupled with judicious fertilizer application can still improve productivity.

Diseases and pests

Black pepper is affected by several diseases caused by fungi, bacteria, virus and mycoplasma besides nutritional disorders, and a variety of pests. Crop losses due to diseases and pests are identified as major causes of low productivity of pepper in India. In India, Indonesia and Malaysia foot rot caused by *Phytophthora capsici* is the most devastating disease. Other major diseases include slow decline, anthracnose and viral diseases. Among insect pests, pollu beetle, top shoot borer, leaf gall thrips and scale insects infest the crop on a major scale. In a survey conducted in the southern districts of Kerala, it was found that adoption of control measures against pests and diseases are low in general (Nybe, 2004). Soil drenching against foot rot was adopted by

61.7% of pepper farmers in Idukky district. In other districts, adoption ranged from 16. 3 to 34. 6 %.

Field experiments were carried out to evolve an effective biocontrol method for management of *Phytophthora* foot rot of black pepper. Various treatment combinations were tried in existing pepper gardens and in newly planted gardens. Out of the 32 fungi isolated from various pepper fields, the fungi viz. *Trichoderma* spp., *Aspergillus* sp., *Pencillium* sp., *Fusarium* sp., *Pythium* sp., *Cunnighamella* sp., and *Rhizopus* sp., showed antagonistic property against foot rot pathogen *Phytophthora capsici*. The local antagonists which showed maximum inhibition against the foot rot pathogen identified were *Trichoderma koningii* (Kollam); *Trichoderma longibrachiatum* (Idukki) and *Trichoderma polysporum* (Thrissur & Ernakulam). The local antagonists were compared with standard cultures in pepper fields. In the existing pepper gardens, where disease incidence was moderate, treated plants which received *Trichoderma* local isolate and potassium phosphonate (0.3 %) showed minimum disease incidence. In severly affected grown up plants, drenching Fytolan and spraying Bordeaux mixture gave maximum control. In newly planted pepper garden, minimum wilt incidence was noticed in all the five districts in the plot which received *Trichoderma* local isolates and potassium phosphonate (0.3 %). Quality analysis of pepper viz. oleoresin and piperine content were estimated as per the standard procedure. The oleoresin content varied from 8.8 to 11.9 %, whereas piperine from 3.6 to 4.5 % which shows the quality remains good and unchanged due to various treatments.

A trial was carried out with objective of developing a suitable bio-control against *Phytophthora* disease in pepper nursery using antagonistic microflora. Antagonists *Trichoderma viride, T. harzianum* and *T. longibrachiatum* were used in 20 different treatments. The pepper variety Panniyur 1 was used for the study. Soil was solarised for a period of 30 days before planting.

At 45 DAP, the treatment T20 (soil solarization for 30 days + *T. longibrachiatum* + potassium phosphonate) recorded maximum percentage of sprouting (90.33 %). Incidence of disease in pepper nursery was recorded at four intervals - 14, 15, 16 and 17 weeks after planting. At 17 weeks after planting, minimum disease incidence was 17.67 % which was recorded by treatment T2 (disease control as per package of practices of KAU). The maximum plant height (19 cm) was observed in T7 (soil solarisation for 30 days + *T. viride*) and maximum leaf number (20.67) was in T8 (soil solarization for 30 days + *T. longibrachiatum*).

The microbial population in soil was recorded for 3 months at monthly intervals. The fungal count was maximum (15 x 10^2 cfu/g of soil) in T2. (Disease control as per package of practices of KAU). Maximum *Trichoderma* population was (5.33 x 10^2 cfu/g of soil) in T6 (soil solarization for 30 days + *T. harzianum*) and T15 (T. *harzianum* + potassium phosphonate). Maximum bacterial population (153 x 10^4 cfu/g of soil) was in T11 (*T. longibrachiatum* + Ridomil MZ). Actinomycetes count was maximum in T13. (soil solarization for 30 days + *T. viride* + Ridomil MZ).

Compatibility of three different *Trichoderma spp.* with commonly used fungicides, insecticides and fertilizers was evaluated under *in vitro* conditions. *T. harzianum* showed maximum compatibility with the fungicide captaf 0.2 %, insecticides phorate (1, 1.5 and 2 kg a.i./ha), and with the fertilizer MoP at 2 % concentration. *T. viride* was compatible with the fungicides Indofil M-45 (0.2 %) and Ridomil MZ (0.2 %), with insecticides phorate (1, 1.5 and 2 kg a.i./ha) and with the fertilizer MoP at 2.0 and 2.5 % concentrations. *T. longibrachiatum* was compatible with Akomin 40 (0.2 %). The insecticide and fertilizer compatibility was similar to that of *T. viride*.

Major diseases and pests affecting black pepper, their causal organisms and management practices are presented in the following table:

Disease / pest	Causal organism	Management practices
Diseases		
Phytophthora foot rot	*P. capsici*	Phytosanitation, providing drainage, removal of excess shade, minimum intercultivation, mulch in the form of grass or legume to avoid rain splashes, pre-monsoon application of 1 kg lime and 2 kg neem cake/ standard. Apply neem cake four weeks after lime application. Biological control with *Trichoderma*, VAM and *Pseudomonas fluorescens* in the nursery and mainfield. Soil solarization in nursery, chemical control using 0.2% copper oxy-chloride drenching of soil and foliar spray with 1% Bordeaux mixture just at beginning of South West and North East monsoons. or Soil drenching with 0.3% potassium phosphonate and a foliar spray of 0.3% potassium phosphonate at the break of monsoons.
Fungal pollu	*Colletotrichum gloeosporioides*	Shade regulation, chemical control using 1% Bordeaux mixture.
Stunted disease	Virus	Use disease free planting materials, collect and destroy affected plants, quarantine measure to check the spread of the disease.
Pests		
Pollu beetle	*Longitarsus nigripennis*	Chemical control using dimethoate, quinalphos or monocrotophos (0.05%), thrice at the time of spike emergence, berry formation and at berry maturity stage.
Top shoot borer	*Cydia hemidoxa*	Spraying monocrotophos or dimethoate.(0.05%)
Leaf gall thrips	*Liothrips karnyi*	Spray monocrotophos or dimethoate (0.05%)
Soft scale	*Lecanium* sp.	Spray quinalphos (0.05%)

Harvesting and processing

Pepper flowering and harvesting depend on climatic factors, the most important being rainfall. In India, pepper plants start flowering during May-June with onset of south west monsoon and harvesting is usually done in Dec.-January. At higher altitudes, the fruit growth and maturity get delayed. Generally harvesting is done when one or two berries in a few spikes turn orange or red. Care should be taken to harvest only mature spikes. Harvesting is done using single pole ladder kept leaned on to the support tree. Harvested spikes are collected in clean sacs or bags tied at back of worker.

Harvested spikes are spread on a clean floor and threshed manually by trampling with legs or by using mechanical thresher. Manual threshing is more common in India. The harvested spikes are piled up in a heap to initiate browning and then threshed. It makes threshing easy and also gives good colour to finished product.

Black pepper

Black pepper of commence is produced from whole, ripe but fully developed berries. After threshing, they are spread on suitable drying floor for sun drying. Berries are raked to ensure uniform colour and to avoid mould development. Drying takes about 3-5 days by which time moisture content will be brought down to 10-12 %. The dried berries are garbled, graded and packed in double lined gunny bags.

Blanching berries in boiling water for one minute prior to drying accelerates browning process as well as rate of drying. It also gives a uniform lustrous black colour to finished product and prevents mouldiness of berries. But prolonged blanching should be avoided, since it can deactivate enzymes responsible for browning process (Nybe, 2001). The black colour that pepper acquires on drying is due to oxidation of colourless phenolic compounds present in skin. Dry recovery varies from 29 to 38 % among cultivars. Solar driers and mechanical driers are now available for drying pepper.

White pepper

White pepper is prepared from ripe berries or by decorticating black pepper. Bright red berries after harvest are detached from stalk and packed in gunny bags. Bags are allowed to soak in slow running water for about one week during which bacterial rotting occurs and pericarp gets loosened. Then the berries are trampled under feet to remove any adhering pericarp, washed in water and then sun dried to reduce moisture content to 10-12 % and to achieve a cream or white colour. White pepper is garbled, sorted and packed in gunny bags. Approximately 25 kg white pepper is obtained from 100 kg ripe berries.

Central Food Technology Research Institute, Mysore, India improved the method in which fully mature but unripe berries are harvested and boiled in water for 10-15 minutes to soften the pericarp. After cooling, skin is rubbed off either manually or mechanically, washed and sun dried to obtain white pepper. Since no retting operation is involved, product will be free from any unpleasant odour. But white pepper produced by this method gives pepper powder of light green colour due to gelatinisation of starch in contrast to pure white powder obtained by traditional method.

Decorticated black pepper

This is a form of white pepper produced by mechanical decortication of outer skin of black pepper. Appearance of decorticated kernel is inferior to traditionally prepared white pepper, but is satisfactory when ground. Also the milling operation requires considerable skill to avoid excessive volatile oil loss.

Pepper oil

Black pepper is crushed to coarse powder and steam distilled to obtain 2.5 to 3.5 % colourless to pale green essential oil which becomes viscous on ageing. It is used in perfumery and in flavouring. Oil can also be distilled from white pepper but high price of white pepper and low oil yield do not favour its commercial production.

Pepper oleoresin

Extraction of black pepper with organic solvents like acetone, ethanol or dichloro ethane provides 10-13 % oleoresin possessing the odour, flavour and pungent principles of spice. Piperine content of oleoresin is 35 to 50 %. One kilogram of oleoresin when dispersed in an inert base can replace 15 to 20 kg of spice for flavouring purpose.

Table 10. Quality attributes of Panniyur varieties of pepper

Properties	**Varieties**						
	P_1	**P_2**	**P_3**	**P_4**	**P_5**	**P_6**	**P_7**
Drying %	35.3	35.7	27.8	34.7	35.7	32.9	33.6
Piperine %	5.30	6.50	4.80	4.40	5.30	4.94	5.57
Oleoresin %	11.78	12.20	10.40	11.30	12.33	8.27	10.61
Essential oil %	3.31	3.40	3.12	3.12	3.80	1.33	1.50

Source: KAU (2003)

Garbling and grading

Before packing, dried pepper is cleaned to get rid of extraneous matters like dirt, grit, stones, stalks, leaves etc. and berries are graded according to size or density. In India, generally garbling and grading are done at exporters premises. Garbling machines remove dust, chaff and grade pepper according to densities. Manual cleaning is also done by hand picking the contaminants such as plant debris and other extraneous matter. Good garbled pepper should have a bulk density of 500-600 gram/litre. Light berries should be less than 10 % and pin heads (unfertilized) less than 4 %. Malabar Garbled (MG), Malabar Ungarbled (MUG), Tellichery Garbled Black Pepper Special Extra Bold (TGSEB), Tellicherry Garbled Extra Bold (TGEB), Garbled Light Pepper (GL) and Pin Heads (PH) are the important grades of black pepper.

Uses

Black pepper is the widely used spice and occupies proud place in the cusines of both West and East. It contributes towards flavour, taste, anti-fungal, anti-microbial and anti-oxidant properties and hence pepper is a multifunctional spice. It is an essential ingredient of most curry powder and spice extracts.

Cardamom (Small)
(*Elletaria cardamomum* Maton., F. Zingiberaceae)

Small cardamom, popularly known as 'Queen of spices' is a tall perennial herbaceous plant in the family Zingiberaceae. Cardamom of commerce is the dried capsule of this shade loving plant indigenous to the evergreen forests of Western Ghats of South India. It is grown extensively in hilly regions of South India at elevations of 800-1300 m as an understorey crop in forest lands. Cardamom is also grown in Sri Lanka, Papua New Guinea, Tanzania and Guatemala. In recent times, Guatemala emerged as a large competitor to India in international cardamom trade. Cardamom is known to be in use in India from ancient times. Early European colonizers were more interested in procuring black pepper and ginger and cultivation of these crops picked up considerably during 16th-17th century (Ravindran and Madhusoodhan, 2002).

But, cardamom was a minor forest produce and this system of cardamom collection from naturally growing plants continued till 1803 at least, but in later years, demand increased and large scale organized cultivation started in India and Ceylon (Present Srilanka) by the beginning of 19th century.

In the erstwhile states of Travancore and Cochin cardamom was a monopoly of respective governments. The King of Travancore made it compulsory that all the produce should be sold to his official who forwarded it to main depot in Alleppy, the then most important port of Travancore. At Alleppy, cardamom was sold by auction. Main buyers were Muslim merchants and the best quality (designated as Alleppy Green) was reserved for export.

In Kerala, the major cardamom growing state in India, cardamom is mainly grown in the erstwhile Travancore area in the present Idukky district, known as Cardamom Hill Reserve (CHR) which was specially reserved, denotified and removed from the control of the forest department by orders of Travancore Government. An area of 86,506 ha was earmarked for cardamom cultivation and rules (Travancore) were framed for assignment of Government lands for cultivation of cardamom. Almost through 19th century, Travancore-Kochi Governments held an unruled monopoly over trade in cardamom. After the turn of century, this monopoly was given up. Planters were allowed to sell their produce at will. There was rapid expansion of area under cardamom. Rampant deforestation, extremely destructive agricultural practices with constantly changing crop combinations for maximizing profit, climate changes and total disregard even for a minimum of ecological safe guards in all developmental activities have together precipitated an extraordinarly severe environmental crisis in the CHR (Miniraj and Murugan, 2000). Efforts to sustain cardamom cultivation in the CHR are underway.

India has been the world's largest cardamom producer until 1979-80, when Guatemala came to the scene as a major rival. At present, Guatemala is the world leader in production. Cost of production in Guatemala is reported to be only half of that in India and India finds it difficult to compete with Guatemala in the international market. Cardamom growing regions of South India are Idukki and Wayanad in Kerala, Coorg, Chikmagalore and Hassan districts in Karnataka and Pulney and Kodai hills of Tamil Nadu.

Botany

The genus *Elettaria* consists of about seven species distributed in India, Sri Lanka, Malaysia and Indonesia. Only *E. cardamomum* occurs in India, and this is the only economically important species. Closely related *E. ensal*, native to Sri Lanka, is a much larger and slender plant known as Sri Lankan cardamom.

Cardamom is a tall herbaceous perennial with branched subterranean rhizomes from which arise several erect leafy shoots. Leafy shoots (up to 5 m height, 10-20 number) are composed of leaf sheaths and are borne in thick clumps. Petioles are up to 2.5 cm long, lamina 1 m x 15 cm, lanceolate, acuminate, highly pubescent or glabrous below, ligule up to 1 cm long, entire. The root system is shallow. Inflorescence arises from swollen base of leafy shoots and are 60-120 cm long. They are erect, recumbent or decumbent slender panicles. It is referred as a panicle of cincinni. Panicles may be single or branched. Flowers are hermaphrodite. Bracts lanceolate, acute, glabrous, rather persistent, but becoming fimbriate with age. Cincinni many flower, bracteols up to 2.5 cm long, tubular, mucronate and glabrous. Calyx is up to 2 cm long, two or three lobed, lobes mucronate and corolla tube almost as long as calyx. Labellum conspicuous, white streaked with violet, obovate, three lobed and narrowed at base. Lateral staminodes inconspicuous and subulate. Anther sessile, ovary 2-3 mm long, inferior and glabrous. Fruit is a capsule which is oblong or more or less globose. Stigma is small and in close contact with distal end of pollen sacs. Capsule contains 15-20 seeds with mucilaginous aril.

Varieties

Cultivated varieties are grouped into three main groups primarily based on nature of panicle. They are 'Malabar', 'Mysore' and 'Vazhukka'. Vazhukka is a natural hybrid between Malabar and Mysore. Distinguishing features of these varieties are summarized below:

Parameters	Malabar	Mysore	Vazhukka
Plant stature	medium (2-3 m ht)	tall (3-5 m ht)	tall (3-5 m ht)
Leaf	dorsal side may be pubescent or glabrous, short petiole	lanceolate or oblong-lanceolate, glabrous on both sides, petiole long	deep green oblong - lanceolate long petiole
Panicle	Prostrate	Erect	semi erect
Capsule	globose-oblong shaped	ovoid, bold dark green	bold, globose or ovoid
Bearing nature	early, short span of flowering	late, long flowering span	late, long flowering span
Adaptability	lower elevation (600-1000 m MSL)	higher elevation (900-1200 m MSL)	higher elevation (900-1200 m MSL)
Tolerance to drought	thrives under low rainfall	thrives well under well distributed rainfall	thrives well under well distributed rainfall

Growth and development

Investigations on growth and development revealed that the Vazhukka variety possessed more tillering ability than Malabar and Mysore. Growth pattern of an individual tiller exhibited biennial habit, wherein the first years growth was characterized by a distinct vegetative phase, reproductive phase dominating during 2nd year. Mean life span of tiller was 30 months in Malabar, 26 months in Mysore and 25 months in Vazhukka. Tillers emerge from axils of under ground stem. Growth of tiller follows a linear pattern in all cultivars. Vegetative shoots require a period of 10-12 months to attain maturity for producing reproductive buds and newly emerging panicles take a period of 7-8 months for complete growth. Histological studies showed that differentiation of panicle primordium was more during November-March in variety Malabar. Flowering commences with onset of monsoon. Flowers appear on panicles after four months and flowering continues for a period of six months. Each panicle possesses a long cane like peduncle having nodes and internodes. Each node has a scale leaf in axil of which flowers are borne on a modified helicoid cyme (cincinnus). It takes about 90-110 days for first flower in a fresh panicle to open irrespective of variety. Flower opening commences from 3.30 am and continues till 7.30 am. Anther dehiscence is between 7.30-8.30 am. Pollen fertility in Malabar, Mysore and Vazhukka cultivars ranged from 63.07-99.4%, 75.32-98.93% & 66.91-99.78% respectively. Stigma receptivity was highest between 7.00 and 8.00 h, 10.00 h and 8.00 h respectively in these cultivars. Panicles emerge from swollen base of tillers.

Flowering is spread over a period of six months from May-October. Almost 75 % of flowers are produced during June-August. Time required to reach full bloom stage from flower bud initiation ranges from 25-35 days and capsules mature in about 120 days from full bloom stage.

Pollination biology

Cardamom plants have bisexual flowers. The showy labellum, presence of nectar, position of stigma and anthers are such that it favours cross pollination by pollinating agents. Pollen fertility is maximum, at the full bloom stage. Honey bees are chief pollinating agents. About 90 % of pollination is effected by honey bees (*Apis cerana, A. indica* and *A. dorsata*).

Belavadi *et al.* (1993) and Belavadi and Parvathi (1998) carried out detailed studies on pollination ecology and biology of cardamom in Karnataka. Peak pollinator activity was between 11.00-13.00 h. Based on pollinator activity, they concluded that a minimum of three honey bee colonies/ha is required for effective pollination in a cardamom plantation with a density of 3000 plants/ha. They also suggested an isolation distance of 15 m for seed gardens. Because of close association between honey bees and cardamom pollination, it is always recommended either to avoid application of plant protection chemicals during morning hours or to use honey bee friendly chemicals.

Fruit set

Percentage fruit set/clump varies with variety. It was found high in Vazhukka (60.98%). Vazhukka and Malabar cultivars stood at par for yield of capsules per clump or fresh weight as well as dry weight basis. An early crop bearing habit was exhibited by Malabar whereas Mysore possessed late bearing habit. Extent of fruit set is the highest when atmospheric humidity is very high and setting is scanty in summer months even under irrigated condition (Ravindran and Madhusoodanan, 2002). Seeds are white in immature capsule, turn brown on ageing and become black at maturity.

Application of growth regulators like NAA (40 ppm); 2,4-D (4 ppm); GA (200 ppm) decreased the capsule drop and increased yield.

Crop improvement

Blessed with twin advantages of vegetative propagation and viable sexual reproduction, cardamom offers immense scope for crop improvement. Thrust area in the improvement programmes identified include isolation of elite clones/varieties suited to different agroecological conditions and development of hybrids/lines having higher yield, superior quality capsules and resistance/tolerance towards abiotic/biotic stresses.

Germplasm

The crop improvement programme in cardamom began with exploration for genetic variability available in its centre of origin, the Western Ghats. Being cross pollinated in nature, natural variability is high in cardamom. Germplasm collection and evaluation were started at Cardamom Research Station, Pampadumpara (Kerala Agricultural University) as early as 1956. In 1957, the Regional Research Station, Mudigiri under University of Agricultural Sciences, Bangalore, Karnataka was started. The programme was intensified with the launch of All India Co-ordinated Research Project on Spices in 1971. Thereafter explorations for germplasm were made by various research institutes in South India. Field germplasm is being maintained in following centres.

Centres	Cultivated germplasm	Wild and related taxa
Indian Cardamom Research Institute, Myladumpara, Idukky, Kerala	570	12
IISR Regional Station, Appangala, Coorg, Karnataka	314	13
Regional Research Station, Mudigire, Chikamangalore, Karnataka	236	7
Cardamom Research Station, Pampadumpara, Idukky, Kerala	78	14

Accessions having multiple branching panicles, compound panicles, terminal panicles, pink coloured tillers, narrow leaves, and clones characterized with asexuality, female sterility and clistogamy are a few of the variants noticed in the germplasm maintained at various centres. The germplasm is being subjected to characterisation and preliminary evaluation.

Protocols are developed for *in vitro* conservation of germplasm by slow growth (Nirmalbabu *et al.,* 1999). They found that ½ MS medium without growth regulators and with 10 mg/l each of sucrose and mannitol was the best for *in vitro* storage of cardamom for slow growth. By using this medium, sub culture intervals could be extended to one year or more, when incubated at 22±2°C at 2500 lux of light and at 10 h photoperiod. Choudhary and Chandel (1995) successfully cryo-conserved cardamom seeds having 7.7-14.3 % moisture content and obtained 80 % germination when tested after one year storage in vapour phase of liquid nitrogen (at -150°C). Sajina *et al.* (1997a) standardized synthetic seed production in cardamom.

Variability studies

Intra and intervarietal variability was assessed in accessions of cardamom at Cardamom Research Station, Pampadumpara (Nybe, 2001). Analysis of intra varietal variance revealed that there was no significant variation within varieties (Table 11). Investigations on intervarietal variability showed significant differences for eight characters. Panicles/plant, fresh and dry weight of capsules and capsules/panicle showed moderate to high heritability and high genetic gain.

Table 11. Inter-varietal variability in cardamom varieties

Character/varieties	Malabar	Mysore	Vazhukka	PV-1	CD (0.05)
Height of tillers (cm)	274.22	298.31	263.78	246.97	17.01
No. of productive tillers/plant	13.14	8.77	6.96	4.98	1.52
No. of leaves/tiller	15.36	14.96	14.28	14.28	0.93^{NS}
No. of panicles/plant	37.11	22.70	28.16	13.17	4.99
Length of panicles (cm)	35.01	36.03	35.83	26.07	3.49^{NS}
No. of nodes/panicle	16.51	16.36	16.07	13.31	1.24^{NS}
Internodal length in the panicle (cm)	1.57	1.80	1.67	1.43	0.01
No. of capsules/panicle	38.38	28.62	32.34	25.38	4.40
100 capsule weight (g)	74.35	75.20	61.62	78.34	4.76^{NS}
Capsule volume (cc)	75.45	74.66	63.32	83.39	4.96
No. of seeds/capsule	13.34	13.44	12.85	12.44	0.86^{NS}
Fresh weight of capsules (g/plant)	890.67	452.96	618.97	304.44	566.76
Dry weight of capsules (g/plant)	191.64	107.88	137.31	66.41	36.81

These characters which show high response for selection can be utilized in further breeding programmes.

Padmini *et al.* (1999) evaluated cardamom germplasm available at IISR, Kozhikode. Characters such as panicle number, nodes/panicle, internodal length and capsule length exhibited high coefficient of variation. Padmini *et al.* (2000) studied variability among compound panicle types and found that proximal branching is more prevalent than distal or entire branching. The contribution of such branching towards yield varied from 12-41 %.

Patel *et al.* (1998) reported that characters like panicles/tiller, panicles/clump, recovery ratio and capsules/panicle could be taken as criteria for selection for yield in cardamom.

Seventy two accessions were evaluated for five consecutive years at Pampadumpara, Kerala for vegetative and economic characters and susceptibility to insect pests (Miniraj *et al.*, 2000) (Table 12). Nine accessions recorded a dry capsule yield of above 150 g/clump.

Among these nine accessions, S-1 is the best for further breeding programmes in view of its higher yield, moderate incidence of insect pests and higher quality parameters. Accessions MBP (Multiple branching panicle) and CP (Compound panicle) possessed higher yield potential. Accessions PS-12 and PS-5 offer scope in breeding for quality.

Table 12. Evaluation of cardamom germplasm

Type	WW	DW	CV%	Vl	Ol	CB	Thrips %	NSC	NC
MBP	1050.1	243.1	79.0	6.2	10.0	30.04[ab]	1.32	11.8	OB
CP	884.9	194.8	100.6	7.3	10.0	1.40[a]	1.62	14.7	RB
Veeraputhran	613.1	192.5	710	7.3	10.5	2.76[ab]	0.71	14.8	OB
S-1	517.0	212.7	37.0	7.8	11.6	2.43[ab]	1.00	17.3	OB
PS-9	443.4	166.9	58.0	7.0	10.8	2.32[ab]	1.40	15.6	OM
PS-4	375.2	152.1	52.0	7.3	10.2	2.02[ab]	0.71	14.8	OB
PS-5	372.9	165.1	50.0	8.4	11.4	2.78[ab]	0.88	16.4	OB
PS-12	282.7	157.7	39.0	8.4	11.6	2.89[ab]	1.17	20.3	RB
Clone-57	250.8	159.2	30.0	7.0	10.3	1.84[ab]	1.35	14.9	RM
CD (P<0.05)	-	-	-	-	-	1.69	NS	-	-

*Means followed by the same letter are not significantly different in DMRT

WW - Wet weight (g plant^{-1}), DW - Dry weight (g plant^{-1}), CB - Capsule borer (%)*, NSC - No. of seeds capsule^{-1}, NC - Nature of capsule, Vl - Volatile oil % (V/W), Ol - Oleoresin % (V/W), OB - Oblong bold, RB - Round bold, OM - Oblong medium, RM - Round medium.

Selection

Breeding in cardamom is currently based on selection recombinants that are superior with regard to yield and quality of capsules and tolerance to biotic and abiotic stresses. Systematic evaluation of available germplasm, followed by comparative yield trials and multilocation trials under various agro-ecological conditions have resulted in release of several improved selections in cardamom (Table 13).

Performance of these selections are much superior to conventionally used clones with regard to yield and capsule characters. One of the major factors that contributed to a steep increase in production and productivity of cardamom in India was the wide coverage of these improved varieties of cardamom. Apart from these varieties, progressive farmers in cardamom tracts have come out with superior types like Njallani, Vallay Green Gold, Vanderkunnel, Palakkudi, PNS Vaigai etc. Of the farmer's selections "Njallani Green Gold" actually boosted cardamom production in the Cardamom Hill Reserve (CHR) in Idukky. This wonder variety yield up to 2475 kg/ha of dry cardamom under good management selection for yield and quality parameters

Table 13. High yielding cardamom varieties and their characteristics

Variety/ year of release	Pedigree / parentage and plant type	Institution/ University	Av. Yield kg/ha (dry)	Salient features
Mudigere 1 1984	Clonal selection from Malabar type	Regional Research Station, UAS Mudigere- 577 132, Chikmagalur (Dist.) Karnataka	275 (Potl. yield 1000)	Erect and compact plant, short panicle, pale green, oval bold capsules, suitable for high density planting, moderately tolerant to thrips, hairy caterpillar and white grubs, pubescent leaves. Contains 8.0% oil, 36.0% 1, 8 cineol, 42.0% α-terpenyl acetate, dry recovery 20.0%
Mudigere 2 1996	Clonal selection from open pollination in Malabar type	-do-	475 (Potl. yield 1000)	Early maturing, suitable for high density planting, round/oval and bold capsules, oil 8.0%, 1,8 cineol 45.0%, α -terpenyl acetate 38.0%
PV 1 1991	A selection from Walayar collection, a Malabar type	Cardamom Research Station, KAU, Pampadumpara, Idukki, Kerala PIN 685 553	260 (Potl. yield 500)	An early maturing type, short panicle, elongated slightly ribbed light green capsules, essential oil 6.8%, 1, 8 cineol 33.0%, α - terpenyl acetate 46.0%, dry recovery 19.9%
PV 2 2001	A selection from OP seedlings of PV-1, Vazhukka type	-do-	982 (Potl. Yield 1250)	Early maturing, unbranched lengthy panicle, long bold capsules, high dry recovery ercentage (23.8%), essential oil 6.6%, field tolerant to stem borer and thrips. Suitable for elevation range of 1000 to 1200 meters above MSL.
ICRI 1 1992	Selection from Chakkupallam collection, a Malabar type	ICRI (Spices Board), Myladumpara, Kailasanadu, Idukki (Dist.), Kerala - 685 535	325 (656 kg under irrigated condition)	An early maturing variety, medium sized panicle with globose, round and extra bold dark green capsules contains oil 8.7%, 1,8 cineol 29.0%, α -terpenyl acetate 38.0% and dry recovery 22.9%.
ICRI 2 1992	Clonal selection from germplasm collection, Mysore type	-do-	375 (766 kg under irrigated condition)	Performs well under high altitude and irrigated condition, medium long panicles, oblong bold and parrot green capsules, tolerant to *azukkal* disease. Dry recovery 22.5%
ICRI 3 1993	Selection from Malabar type	ICRI (Spices Board), Sakleshpur, Donigalpost 573 134, Karnataka	440 (790 kg under irrigated condition)	Early maturing, non-pubsecent leaves, tolerant to rhizome rot disease, oblong, bold parrot green capsules, suitable for hill zone of Karnataka, oil 6.6%, 1,8 cineol 54.0%, α-terpenyl acetate 24.0%, dry recovery 22.0%
TKD 4 1997	Clonal selection from Vadagaraparai area of lower pulenys, a Malabar type	ICRI (Spice Board) Thdiyankudisai - 624 212 Tamil Nadu	455 (960 kg under irrigated condition)	Early maturity, medium sized panicle, globose bold capsules, oil 6.4%. Suitable for low rainfall areas, relatively tolerant to rhizome rot and capsule borer
IISR Suvasini (CCS-1) 1997	Selection from OP progeny of CL 37 from RRS Mudigere. Malabar type	Indian Institute of Spices Research, Cardamom Research Centre (ICAR), Madikeri-571 201, Appangala, Karnataka	745 (1322 kg under irrigated condition)	Early maturing, suitable for high density planting, long panicle, high % of bold parrot green capsule (89), tolerant to rhizome rot, thrips, shoot/panicle/ capsule borer, essential oil 8.7%, 1, 8 cineol 42.0%, α - terpenyl acetate 37.0%, dry recovery 22.0%

Contd...

IISR Vijetha (NKE-12) 2001	Clonal selection from field resistant plant for katte, a Malabar type	-do-	643 (Potl. Yield 979 kg)	Virus resistant selection with high percentage of bold capsules (77.0%) oil 7.9%, 1, 8 cineol 45 0%, α - terpenyl acetate 23.4%, dry recovery 22.0%. Recommended to moderate rainfall areas with moderate high shaded and mosaic infected areas, field tolerant to thrips and borer as well as mosaic
IISR Avinash (RR-1) 2001	A selection from OP seedlings of CCS-1. a Malabar type	-do-	847 (Potl. yield 1483)	Has extended flowering period, dark green capsules and retains its colour even after processing. Tolerant to rhizome rot, shoot/panicle/ capsule borer, oil 6.7%, 1, 8 cineol 30.4%, α - terpenyl acetate 35.5%, dry recovery 20.8%. Highly suitable for planting in valleys and rhizome rot prone areas and intensive cultivation
Njallani Green Gold	A clonal selection from Vazhukka type	Farmers selection from Idukki district of Kerala	1600 (Potl. yield > 3000)	A higher yielding clonal selection by a cardamom grower in Idukki district. Capsules bold, over 70% of the cured cardamom above 7 mm. Reported to be of high quality. Currently the highest yielding cultivar. Yield up to 5000 kg/ha has been recorded

are continuing and several promising selections are under multilocational trials.

Selection for biotic and abiotic stress

Efforts are made to survey and collect disease escapes from hot spot areas of 'Katte' disease and they were subjected to further evaluation by artificial inoculation. A Natural Katte Escape (NKE) line was found promising and was released as IISR Vijetha (NKE-12) for the mosaic infected areas of Karnataka and Wynad in Kerala. A preliminary screening of cardamom varieties for "azhukkal" disease showed that Malabar and Vazhukka were more susceptible compared to Mysore (Radhakrishnan *et al.*, 2004). Survey, collection and evaluation of drought escapes from low rainfall areas are also progressing and several accessions of Malabar are promising.

Hybridization

Intervarietal hybridization using genetically divergent genotypes started as early as 1976. Madhusoodanan *et al.* (1998) isolated three high yielding heterotic recombinants namely MHC-10, MHC-13 and MHC-18 having an average yield of 470 to 610 kg/ha under moderate management. Cardamom hybrids being heterozygous and heterogenous in nature, have to be multiplied either clonally or through micropropagation to produce "true to type" planting materials and to avoid segregation.

Attempts on intergeneric hybridization involving *Amomum*, *Alpinia*, *Hedychium* and *Aframomum* were sterile (Krishnamoorthy *et al.*, 1989).

Polycross breeding

Cardamom yield could be substantially improved through a polycross breeding programme (Chandrappa *et al.*, 1998). Up to 60 % increase in yield was obtained from a polycross progeny of Mudigere-1. They found that a population of more than 3000 plants is essential for carrying out effective selection.

Mutation and polyploidy breeding

Induced mutants (through Xrays, gamma rays, EMS and MH) were developed in cardamom but none was derivable. Polyploids ($2n = 4x = 96$) were also successfully induced in cardamom, but they were inferior to diploids.

Biotechnology

Variability could be noticed among the somaclones developed in cardamom and a few somaclones tolerant to 'Katte' virus were identified (Nirmalbabu *et al.*, 2006). Protoplasts could be isolated successfully from leaf mesophyll tissues of *in vitro* grown plantlets and cell suspension cultures of cardamom (Geetha *et al.*, 2000).

Crop management

Climate and soil

Cardamom thrives well in the tropical rainforests of Western Ghats, in tracts that lie between 8° 30′ and 14° 30′ N latitude and longitude of 75 ° 75′. It requires warm and humid climate and prefers an altitude of 600-1200 above MSL with temperatures ranging from 10 to 35°C, fairly well distributed annual rainfall of 1500-5000 mm and moderate shade (50-60%). In India, cardamom is cultivated mostly under natural forest canopy, except in certain areas in Karnataka and Wynad in Kerala, where it is grown in arecanut and coffee plantations. In India, at present 75-80 % of the total area under cardamom is rainfed. With denudation of forests in many areas of Western Ghats, including the CHR, cardamom plantations are exposed to extreme fluctuations in local climate. Cardamom is a sensitive plant and any serious disturbance in environment, especially climatic factors will adversely affect growth, development and production. Recurring droughts coupled with desiccating winds from Tamil Nadu plains and resultant outbreak of pests caused miserable crop failures in CHR (Miniraj and Murugan, 2000). Sloppy land facing eastern and northern directions are ideal for cultivation. Cardamom prefers forest soils rich in humus. Most of cardamom growing soils are rich in organic matter and nitrogen and also low to medium in available phosphorus and potassium. Soil should be well drained and with a pH between 4.5 to 6.5. Cardamom being a surface feeder with about 70 % roots at a depth of 15 cm, availability of moisture at root zone is important for good growth and yield.

Propagation

Cardamom is propagated both asexually (through suckers) and sexually (through seeds). Seed propagation is recommended in areas affected by 'Katte' disease (a disease transmitted through suckers) as well as when large number of planting material is to be produced. Vegetative propagation gives earlier bearing and true to type plants. Micro propagation technique has also been standardised and commercialized in cardamom.

Seed propagation

Fully ripe seeds are extracted from ripe capsules from high yielding vigorous plants preferably from 2nd and 3rd round of harvest. Best time for seed capsule collection is September-October. Immediately after picking, capsules are immersed in water and seeds are extracted by gently pressing capsules. Seeds are then washed in water to remove mucilage. Scarification of seed with 2.5 % nitric acid for 10 minutes followed by washing in water before sowing assures 80 % germination. Seeds are then smeared with wood ash and dried under shade. September sowing is ideal as seedlings are ready for transplanting during ensuring planting season.

Primary nursery

Nursery site should be preferably on moderately sloppy land, in open and well drained location near source of water. Raised seed beds (20 cm) of one meter width and of convenient length are prepared. Spread a layer of forest soil over the bed. Soil solarisation of seed beds enhances germination by 25 %, reduces weed growth by 82 % and suppress pests and diseases. Seeds can be broadcasted or live sown 1-2 cm apart in rows 8-10 cm apart. Seed rate is 10 g/m^2. On an average, one kg of seed capsules are required to get about 5000 plantable seedlings. Beds are mulched to a thickness of about 2 cm with paddy straw and watered regularly in morning and evening. Sprouting commences in about 25-30 days. After germination is completed, mulch is removed and an over head 'pandal' at a height of 2 m is erected for providing shade. Regular watering, weeding and plant protection are attended. Thinning of seedlings is done at six-seven leaf state, followed by earthing up.

Secondary nursery

Nursery beds are prepared in same manner as for primary nursery. Shade is to be provided by erecting overhead pandal. Seedlings are transplanted when they are 4-6 months old (4 to 5 leaf stage) at a distance of 20 cm x 20 cm. Beds are mulched and watered regularly. Weeding, manuring and plant protection practices have to be attended whenever needed. One month before transplanting, the overhead shade is removed to enhance tillering. In Kerala 18-22 months old seedlings are used for transplanting while in Karnataka 10 months old seedlings are used. Two year old seedlings is the best for getting more tillers and higher yields. The secondary nursery can also be raised in polybags. Polybags of 20 x 20 cm filled with a mixture of humus rich top soil, cowdung and sand (3:1:1) is an excellent medium for filling polybags. Raising seedlings in polybags has many advantages like uniform growth, more tillering and good establishment in field.

Vegetative propagation

Clonal propagation is advantageous in places where 'Katte' disease is not a menace. Also plants raised from rhizomes come to bear early compared to seedlings. Cardamom is propagated vegetatively by division of rhizomes. A portion of the rhizome with one old and one young shoot consists a planting unit. The main disadvantage with clonal propagation is limited availability of planting material.

Rapid clonal multiplication technique

A rapid multiplication technique for production of vegetative planting material was standardised at IISR research centre at Appangala, Karnataka

(Korikanthimath *et al.*,1999). Suckers are collected from high yielding plants and planted at a spacing of 1.8 m x 0.6 m, thus accommodating approximately 6800 plants/ha of clonal nursery area. Overhead shade is provided and nursery is irrigated. Fertilizers at the rate of 48: 48: 96g NPK plant^{-1} is given in two splits. After 12 months of planting, each planting unit produces on an average 32-42 suckers. In an area of 1 ha clonal nursery, 1 to 1.4 lakhs planting units can be produced after one year. A crop of 190 g/plant of dry cardamom (1759 kg/ha) also could he harvested within just 19 months of planting. This system is easy and cheap and can be adopted by farmers.

Micropropagation

First commercial level production of plantlets in spice crops through tissue culture was achieved in Cardamom. Micropropagation technique in cardamom was standardised by many workers (Kumar *et al.*, 1985; Reghunath 1989; Nadgauda *et al.*, 1983; Nirmalbabu *et al.*, 1997 and Rao *et al.*, 1998). The approaches used were (1) through callusing, (2) through adventitious bud formation and (3) through enhanced axillary branching. Cost effective commercial micro propagation technology developed at ICRI, Spices Board is given below (Rao *et al.*, 2004).

Young sprouting tillers of eight to 10 cm length were found as the best explants for tissue culture of small cardamom. Time of collection of the explants is also very important for success of tissue culture of small cardamom. Collection of explants between March to May is ideal. Various media formulations were tested such as LS (1965), Gamborg *et al.* (B5) (1968), MS (1962), KC (1946) and SH (1972). Media were supplemented with various growth regulators such as BAP, KN, IAA etc. Coconut milk was added to every combination tested. Growth factors D-Biotin and calcium pantothenate also were added @ 0.1 mg/l. Hardening of tissue-cultured plantlets was done in sand filled cups under a mist chamber with 80 % relative humidity at a temperature not exceeding 33°C.

Among various media formulations tried, SH was more effective for cardamom shoot induction supplemented with BAP, KN and coconut milk each 0.5 mg/l and IAA at 2 mg/l. Growth factors D-Biotin and calcium pantothenate also were added @ 0.1 mg/l. The buds elongated on M I medium 6-8 weeks, grew to a height of 30-40 mm within 12 weeks. (M I = SH basal + BAP 0.5 mg/l + KN 0.5 mg/l + IAA 2 mg/l + CW (5 per cent V/V) + D-biotin 0.1 mg/l + Ca. pant 0.1 mg/l).

Transfer of shoots to fresh medium could be done every five weeks, although well established cultures of cardamom appeared to hardly remain in the same medium for up to two months without any deterioration.

When primary cultures were about 30-40 mm long, they were individually transferred to shoot multiplication medium M II (M II = MS basal

+ BAP 0.5 mg/l + KN 0.5 mg/l + IAA 2 mg/l + CW 5 pr cent V/V + D-biotin 0.1 mg/l + Ca. pan 0.1 mg/l).

Rate of multiplication ranged from 1:1 to 1:13 with an average of 1:3 per month and the rapidly growing cultures were normally subcultured at an interval of 27-35 days. Although subcultures could often be continued over many months and even years without adverse effects apparent in culture, genetically variant shoots could arise through somaclonal variation. To prevent this it was found desirable to initiate fresh cultures from new explants after every 12th subculture.

White's rooting medium without any hormones was found more effective than any other media in promoting shoot elongation and subsequent rooting in cardamom. White roots appeared from 4 to 14 days after transferring to rooting medium. The shoots grew rapidly to a length of 50-60 mm during this period. The rooting percentage was 86-93 %.

Rooted shoots that were 4 to 5 cm high and having well-ramified roots were transferred for hardening and acclimatization. Topsoil or a composted soil mixture preferably sterilized or pasteurized was found satisfactory as planting medium. Dipping plantlets in fungicidal solution prior to soil transplanting reduced the risk of damage by fungal attack.

Plantlets were generally allowed to acclimatize in nursery for two to three months till they reached a height of 20-30 cm before transplanting either to polybags or directly to field. Field preparation, planting and culture management were performed according to recommended practices to achieve successful field establishment and subsequent vigorous growth.

Field planting

Field is cleared and terraces made across slopes at suitable planting distances. Trenches and furrows are always advisable for better soil and moisture conservation. Planting distance should be more in high rainfall areas or irrigated condition and less in low rainfall areas. Spacing depends on soil fertility, type of cultivars and terrain. Spacing adopted is more in Kerala while in Karnataka, closer spacing is followed. Spacing of 2 m x 2 m is recommended for Malabar, 3 m x 3 m for Mysore and Vazhukka. However, in fields of low fertility, plant density can go up to 1600/ha for Mysore and Vazhukka and up to 3000/ha for Malabar. For Karnataka, a general spacing of 1.8 m x 1.8 m is suggested. A spacing of 1.8 m x 1.8 m (3025 plants/ha) and 2.7 m x 2.7 m (1371 plants/ha) is recommended in Malabar and Mysore types respectively in Tamil Nadu. Pits of size cm x 45 cm x 30 cm are dug during April-May, filled with a mixture of surface soil and compost or well-rotten farm yard manure + 100 of rockphosphate. This has to be done two months in advance of planting of seedling for better establishment and good growth. Best season for planting is with commencement of South West monsoon before heavy rains. Cloudy days with light drizzles would be ideal for planting and result in early and maximum percentage of establishment. Avoid deep planting. A small mound may be formed inside pit to cover the rhizome and immediately after planting mulching should be done. Plants should be supported by stakes. Two-year old seedlings (20 months old in a few cases) is the best for getting more tillers and higher yields.

High density planting

Studies have proved that Malabar varieties perform well under high density planting. Planting of cardamom (Malabar type) with a high density of 5000 plants/ha under rainfed condition in trench system with a fertilizer dose of 120:120:240 kg NPK/ha is recommended for Karnataka state. Studies on high density planting with a spacing of 1.8 x 1.8 m (3025 plants/ha) indicated early and high yields in initial three-four years. However, removal of alternate plants after five to six years is inevitable for facilitating easy cultural operations. Variety Mudigere-1 (Malabar type) is the best suited for high density planting (1 m x 1 m spacing). A closer spacing of 1 m x 1 m for Malabar types gave higher yields closely followed by 1.5 m x 1.5 m spacing under the rainfall conditions of Shevroy hills in Tamil Nadu. Planting of suckers at a spacing of 1.8 m x 0.9 m gave considerable high yields in initial stage. Mysore and Vazhukka varieties are not suitable for high density planting.

Shade management

Shade regulation is an important operation in cardamom plantation. Cardamom thrives well in moderate shade (50-60% light). In general, 40-50% light intensity under rainfed and about 55-60% light intensity under irrigated condition is required. Shade trees having distributed branching pattern and small leaves are ideal for cardamom.

An ideal shade tree in cardamom plantation should have the following characteristics (George *et al.*, 1984):

1. Wide canopy so that number of shade trees in unit area is minimum.
2. No shedding of leaves during flowering phase so that pollination is not affected.
3. Medium size and evergreen in nature retaining their foliage throughout the year.
4. Should have small leaves and well spread branching system.
5. Should have deep root system to avoid competition with cardamom and to resist wind.
6. Should be fast growing to provide necessary immediate shade.

It is desirable to maintain a mixed population of medium sized trees that facilitate shade regulation and maintain more or less optimum conditions throughout the year.

Some of the good shade trees suited are *Vernonia arborea* (Karana), *Cedrella toona* (Red ceder), *Cassia fistula* (Kanikonna), *Diospyros malabaricum* (Vellakil), *Terminalia tomentosa* (Thempavu), *Hopea parviflora* (Thambakam), *Acrocarpus fraxinifolius* (Balangi), *Mesopsis eminii* (Elangi) and *Artocarpus integrifolia* (jack). Shade trees like red cedar, which shed their leaves in monsoon, provide natural shade regulation. Protection from sunlight by maintaining an overhead tree shade is essential for cardamom in the initial stages of growth to enhance tillering. In newly established plantations, where shade is inadequate, fast growing species of *Erythrina* may be planted along with other trees. After the establishment of shade trees, erytrina may be removed. Similarly, shade should be thinned during monsoon and premonsoon period. In order to provide adequate sunlight during rainy season (when light intensity is less), excess shade should be regulated before onset of monsoon. Shade regulation studies on cardamom productivity revealed that one third of incident sunlight (15.20 K Lux) in the form of filtered light is optimum for growth and production. Studies at RRS Mudigere (University of Agricultural Sciences, Bangalore) revealed that shade trees in cardamom plantation play an important role in nutrient recycling through leaf fall. *Thevetia nerifolia, Terminalia catapa* and *Artocarpus integrifolia*, produced more biomass around 950 $g/m^2/year$. Jack tree is found to recycle maximum NPK (63.9kg/10 tonnes biomass). A study conducted at Cardamom Research Station, Pampadumpara to assess organic carbon content of CHR dominated by different tree species was given by Miniraj and Murugan (2000). In the undisturbed ecosystem, organic carbon content of soil was high (Table 14). Comparatively higher organic carbon content was recorded in cardamom production systems as compared to that black pepper. This is clearly attributed to lack of native tree species and low biomass production of dominant tree - *Erythrina*. Besides, warmer soil environment experienced in pepper production system due to shadeless condition directly and indirectly affect decomposition of organic matter. Cultivation of cardamom under artificial shade (coir mat) is economical and feasible.

After care

Regular schedule of cultural practices consisting of mulching, weeding, earthing up, thrashing, gap filling, fertilizer application and irrigation are to be followed. During initial growth period sickle weeding with light forking around the plant followed by mulching in May, September and January enhanced tiller and panicle production and increased yield by 6.8% with a cost benefit ratio of 1: 5 (ICRI, 2002). Mulching helps to conserve soil moisture and protect plants during summer. The leaf litter, twigs, bushes and removed weeds can be used as mulch thus conserving moisture and improving fertility of soil. Frequent weeding is necessary in first year of planting. Subsequently, two to three weedings in a year can be done depending upon

Table 14. Percentage of soil organic carbon content of CHR dominated by various tree species

Common name (Vernacular name)	Botanic name	Production system	Organic carbon content (%)
Korangatti	*Acrocarpus fraxinifolius*	Cardamom	1.78
Perumarom	*Ailanthus triphysa*	Cardamom	1.76
Jack	*Artocarpus heterophyllus*	Cardamom	1.76
Chorakkali	*Bichofia javanica*	Cardamom	1.78
Vediplaavu	*Cullenia exarillata*	Cardamom	1.77
Mullumurikku	*Erythrina indica*	Pepper	1.61
Murukku	*Erythrina lithosperma*	Pepper	1.64
Porivatta	*Mallotus tetracoccus*	Cardamom	1.54
Kulamavu	*Persia macrantha*	Cardamom	1.61
Chandanavayambu	*Toona ciliata*	Cardamom	1.76
Pottama	*Trema orientalis*	Cardamom	1.67
Karana	*Vernonea arborea*	Cardamom	1.69
Underdisturbed mixed forest not under cultivation	Underdisturbed mixed forest not under cultivation	Cardamom	2.10

intensity of weed growth. Weeding either by using herbicides like Paraquat @ 0.4 kg ai (Gramaxone) or manually is equally economical. Trashing (removal of old and dry shoots of cardamom plants) should be carried out once in a year between Feb-March, soon after completion of final harvest. Plants should be earthed up around base of clump up to collar region, which helps to keep plants intact and cover exposed roots and rhizomes. Earthing up operation should be followed by mulching. Gap filling is a regular practice to be attended in cardamom plantation.

Fertilizer requirements

Cardamom is a surface feeder. Roots are generally confined to surface soil and nearly 70% of roots are with in 5 cm depth. Therefore fertilizer is to be applied in circular bands of 15 cm width leaving a distance of 30 cm from plant base. Nutrient uptake studies indicated the uptake of N, P, K, Ca and Mg in the ratio of 6:1:12:3:0.8 respectively. Cardamom is a heavy feeder of potash. For crop yielding 100 kg dry capsules/ha, an optimum fertilizer dose of 75:75:150 kg NPK/ha (half the dose for one year old plant) in two split doses is recommended for Karnataka. First application during May-June helps production of suckers and development of capsules and second application during September-October helps initiation of panicles and suckers. In irrigated plantation, application of fertilizers in 4 split doses at quarterly interval is beneficial. Higher levels of fertilizer viz., 150:150:225 kg NPK/ha resulted in significant higher capsule yields at Mudigere (Karnataka). Evaluation of optimum fertilizer levels for high density cropping studies conducted at Mudigere confirmed that cardamom grown under natural shade responded positively for fertilizer application and significantly influenced yield. Mudigere (Karnataka) and Pampadumpara (Kerala) have come out with independent fertilizer recommendations viz. 75:75:150 kg NPK/ha and 100:100:175 kg NPK/ha respectively. While the fertilizer recommended for Tamil Nadu is 40:80:40 kg NPK/ha during June, August and November. For low rainfall areas of Tamil Nadu, recommendation is 20:40:20 kg NPK/ha in two splits during September and November. Fertilizer dose of 120:120:240 kg NPK/ha is recommended in high density planting of (5000 plants/ha) Karnataka. At Mudigere, studies on integrated nutrient management using organic and inorganic manures on yield of cardamom confirmed that 100% inorganic manure alone recorded maximum green capsule yield (707 kg/ha). Studies also confirmed that boron and molybdenum play an important role along with other management practices in increasing yield. Application of boron in the form of disodium tetraborate @ 20 kg/ha and molybdenum in the form of sodium molybdate @ 0.25 kg/ha mixed with appropriate quantity of FYM @ 1 kg/plant applied with onset of monsoon increased cardamom yield by 20%. Organic manures like compost or cattle manure may be given @ 5 kg/clump. Neem oil cake may also be applied @ one kg/clump.

Irrigation

Cardamom is sensitive to moisture stress and needs irrigation during summer. Irrigation at an interval of 10-15 days from December till commencement of monsoon is required. Analysis of rainfall pattern of 50 years in cardamom growing tracts by ICRI, Spices Board recommended that judicious irrigation during peak dry months (February to April) ensures increased yield. This is the period when development of young tillers and elongation of panicle take place. Water stress during this period of growth result in reduced yield. Any one of the methods viz. sprinkler, drip or hose irrigation could be adopted. Taking into account the crops consumptive use and climatic condition, it is recommended that irrigation is given at an interval of 10-12 days to get the equivalent of 25 mm rain in each round. Cardamom plants are also sensitive to water logging. Hence, excessive irrigation is also detrimental. Combination of irrigation at 75 % available soil moisture level and high light intensity (65 to 70% lux) resulted in sustained increase in yield.

Application of growth regulators

Exogenous application of NAA (40 ppm) or 2,4-D (4 ppm) increased plant height and panicle production. Similarly, spraying of NAA (25 ppm) twice (June and July) reduced fruit drop and increased fruit set. Application of ethrel (250 ppm) under high light intensity (15 K lux) doubled sucker production.

Bee-keeping

Pollination in cardamom is entomophilous mainly by two species of honey bees viz. *Apis cerana* and *Apis dorsata*. Honey bees contribute over 90% pollination, lack of pollination causes fruit drop and fruit set in cardamom. Number of visits of bees is directly correlated with the number of seed formed in capsules. Five to six visits by bees are required for getting full seed set. At least four bee colonies (hives) / ha to are needed to enhance pollination and fruit set. Increase in yield up to nine % could be obtained by keeping bee hives.

Soil and moisture conservation

The economic life span of cardamom crop is 10-15 years under good growth condition without severe pests and diseases. Conservation of soil and water should be taken care in hill slopes. Measures like contour trenching, bunding, bench terracing, drain channels, check dams and afforestation are recommended. Intensive tillage of soil leads to soil erosion.

Crop protection

Cardamom is affected by several diseases and infested by many insects/ pests. Ecological upset, particularly edaphic factors have added new pest problems in cardamom plantation in recent years. Many pests considered minor are assuming alarming proportions. Root grub has emerged as a serious pest in exposed, warm and less shaded conditions. Outbreak of white flies and locusts in certain pockets is another example of ill effects of changes in microclimate. The control measures developed and recommended for adoption are given in Table 15.

Diseases/pests	Control / Management
Katte disease	Eradication and roughing of infected plants and adoption of phytosanitation to control aphid vector (*Pentalonia nigronervosa)*
Cardamom vein clearing virus (Kokke - kandu)	Eradication and roughing of infected plants and adoption of phytosanitation
Azhukal disease / capsule rot (*Phytophthora* spp.)	Integration of both cultural and fungicidal management practices. Cultural practices include phytosanitation ie., removal of and destruction of affected parts, regulating excess shade and thrashing to remove senile parts. Spraying and soil drenching with Bordeaux mixture (1%) or copper oxychloride (0.2%) at bimonthly intervals just before onset of monsoon (May / June) and again after end of monsoon (August)
Rhizome rot clump rot (*Pythium vexans, Rhizoctonia solani*)	Adopt cultural practices like removal and destruction of infected plants, regulate excess shade, good drainage and three rounds of thrashing (June, August and October). Drench basin with copper oxychloride (0.2%) or Bordeaux mixture (1.0%) @ 2-3 l/plant

Root knot nematodes (*Meloidogyne incognita, Radopholus similis*)	Use nematode free planting material, soil fumigation or soil solarization at nursery sites and rotation of nursery sites. Spot application of Phorate 10 G 30 g/clump twice a year during May-June and November-December. Application of aldicarb, carbofuran and fenamiphos @ 5 kg ai/ha is reduced root knot nematode population.
Thrips (*Sciothrips cardamomi*)	Remove the dried shoot peels and leaves (thrashing) at the beginning of monsoon (April-May). Spray with monocrotophos (0.05%) and phosalone (0.05%) alternatively at bimonthly intervals during summer and winter (skipping the heavy monsoon) or use of neem based insecticides. Apply of phorate @ 10 g/clump followed by two sprays of phosalone (0.05%), Fenthion (0.05%) during May and August.
Shoot/panicle/ capsule borer (*Conogethes punctiferalis*)	Remove the borer-affected tillers during September-October and spray fenthion, monocrotophos, quinalphos or endosulphan (0.075%) twice in January/ February and September /October or nuvacrone 35EC (1.25 ml/l) during May and June followed by zolone (2 ml/l) after 40 days.
Bottle borer or scolytid borer (*Thamurgides cardamomi*)	Regulate shade, spray contact insecticides like quinalphos (0.025%)/methyl parathion (0.05%) in March-April and September-October after harvest
Hairy caterpillars (*Eupterote* spp., *Lenodera vittata*)	Collect and destroy catterpillers colonized at base of shade trees, eliminate shade trees which favour for egg laying. Spray methyl parathion, quinalphos (0.1%) or dust with BHC 10% or Zolone (4%).
Shoot fly (*Formosina flavipes*)	Remove the affected shoots and destroy, spray ekalux quinalphos (0.03%)/phosolone (0.05%), monocrotophos (0.05%) or apply carbofuran 0.3 kg ai/ha
White fly (*Dialeurodes cardamomi*)	Spray insecticides like methyl parathion or phosphamidon (0.05%) on foliage
Rhizome weevil (*Prodioctes haematicus*)	Destroy affected plants/seedlings. Drench base of clumps with BHC (0.2%) or apply phorate @ 2-4 g ai/clump
Root grubs (*Basilepta fulvicorne*)	Soil incorporation of granular insecticides like carbofuran 3G @ 8-10 kg/ha or thimet 10% G at 2.75 kg ai/ha or phorate 2-4 g ai/plant in soil (June / July or November-December).
Mealy bug, scales and lacewing bugs	Spraying monocrotophos (36 ml in 10 l of water) or ekalux (0.1%) at 15 days interval
Mites	Spray dicofol (kelthane at 20 ml in 10 l of water)

Harvesting and processing

Cardamom plants start bearing fruits (capsules) from second year after planting and satisfactory yield is obtained from 3rd year. It takes 110 days from flowering to fruit maturity. Harvesting season is in August-December; peak harvest is during October-November. Capsules are harvested at an interval of 15-20 days at Karnataka and 30 days in Kerala and harvesting will be completed in seven-eight rounds. Fruits that are just ripened or physiologically mature (at the black seeded stage) are picked by experienced workers. If capsules are immature, they will have a wrinkled appearance on curing and if they are fully ripe they will split on curing and will develop an yellow color. Essential oil content will be more at physiologically mature stage and dry recovery more at ripened stage. At physiologically mature stage, recovery is 24 %.

Two types of pickings are adopted - light picking and hard picking. In light picking, only mature capsules are harvested whereas in hard picking semi-mature crop is also removed. Choice depend on availability of labour. Cardamom harvesting is mainly done by women labourers.

Quality parameters of cardamom

Boldness and colour of cured capsules, content and organoleptic character of volatile oil are important quality parameters of cardamom. They are primarily dependent on cultivar and time of harvest. Malabar is superior with regard to flavour and quality whereas Mysore dominated in production of bold green capsules.

Retention of green colour

Colour of cured cardamom is an important factor in consumer market. The highest quality cardamom traditionally was Aleppy Green, which is still regarded as the best. Different chemical treatments were tried to retain green colour of harvested capsules. Among such treatments, soaking green (wet) capsules immediately after harvest in two per cent sodium carbonate solution for 10 minutes fixes green colour during subsequent drying and storage. Except sundrying, all other methods are found to protect the green colour.

Post harvest operations

Post harvest operations consist of washing, curing (drying), cleaning, polishing, sorting, grading and packing. The capsules immediately after harvest are washed in water to remove adhering soil. Sometimes, capsules are treated with 2% washing soda (sodium carbonate) for 10 minutes to stabilize chlorophyll and this treatment imparts better green colour.

Curing

Cardamom curing may be defined as process in which moisture content of freshly harvested cardamom capsules is reduced from 80 % to 11-12 % at an optimum temperature of 45-55°C to retain its green colour and volatile oil to a maximum extent (Vadiraj, 2004). There are mainly two types of drying in cardamom viz. natural (sun drying) and artificial (firewood, electricity, kerosene or LPG as fuel).

Sun drying

Harvested capsules are directly dried under sun for a period of five to six days. Sun dried capsules will be bleached in appearance. Splitting of capsules will be more. This method is prevalent in some parts of Karnataka.

Flue pipe curing

It is one of the best methods of drying by which high quality green cardamom can be obtained. Hand picked fruits are dried in special curing chambers under controlled temperature to retain delicate flavour and green colour. Heat required for drying is produced by burning fire wood in iron Kiln, the heat thus produced is passed through pipes made of galvanized iron sheet. Capsules left for drying are spread thinly on wire net trays arranged one above the other on racks in the upper chamber of the curing house. They are stirred frequently to ensure uniform drying and also help to retain green colour of capsule. Drying temperature is kept at 50°C for first 4 hours and subsequently reduced to 45-50°C by opening ventilators and using exhaust fan. Finally temperature is raised to 60°C for one hour. The process of drying takes about 36-48 h. Dried capsules are rubbed with coir mat/gunny cloth/ steel mesh and sieved to remove other plant debris. This process is called polishing. Polishing is also done in machines. The polished capsules are then graded according to size by passing through sieves of size 7, 6 and 5 mm. They are sorted according to size and colour. The percentage recovery is 24. The graded produce is packed in polythene lined gunny bags.

Bin drier designed by UAS, Bangalore (electricity based), Melccard drier used in Bodinaikkenur region in Tamil Nadu (firewood operated), Solar drier developed by CPCRI, Kasaragod (solar energy as fuel), Mechanical cardamom drier developed by RRL, Trivandrum (electricity as energy), and thorough flow drier fabricated by CFTRI (electricity as energy source) are some of the new generation driers available. Kerosene. LPG gas and biogas were successfully used as an alternate fuel source in cardamom curing especially in the fire wood deficient areas and also by small and marginal growers. The kerosene and LPG based system were standardised for small quantity of 50-100 kg capacity (Vadiraj, 2004). These systems are not only ecofriendly but economical also.

Processed products

Bleached cardamom

Bleached cardamom is manufactured in some parts of Karnataka. It is creamy white or golden yellow in colour. Bleaching can be done either with dried capsules or fresh capsules. Fresh capsules are soaked in 20 % potassium metabisulphite solution containing one % hydrogen peroxide for one hour. Drying of these capsules yield golden yellow colour. For dry capsules, bleaching agents used are bleaching powder, sulphur dioxide, potassium metabisulphite and hydrogen peroxide. In the conventional method prevalent still in Karnataka, bleaching of cardamom is carried out by steeping dried capsules in soap nut water (*Sapindus saponaria*). Bleached cardamom is inferior in quality (low volatile oil content) but its keeping quality is slightly better.

Cardamom seed and powder

Cardamom seeds are obtained by decorticating capsules. Cardamom powder is obtained by grinding seeds.

Cardamom oil

Produced commercially by steam distillation of crushed fruits or powdered seeds of cardamom. Oil percentage is 8-11. Main flavour components of the oil are 1,8-cineole, a-terpinylacetate, linalyl acetate and linalool. Relative proportions of these constituents have a direct bearing on cardamom quality. Cineole is responsible for camphoraceous odour while other esters and alcohols give pleasant fruity odour.

Cardamom oleoresin

Oleoresin is obtained by solvent extraction. Of late, super critical fluid extraction is also adopted (SCFE). Oleoresin is produced in a relatively small scale only.

Fixed oil of cardamom seeds

Obtained from seeds. Major constituents of oil are oleic and palmitic acids.

Grading of cardamom

Specifications for Indian cardamoms are given below:

Grade	Description	Size (mm)	Weight (g/l) min	Colour	General characteristics
AG Alleppey Green					
AGB	Extra bold	7	435	Green	Kiln dried
AGS	Superior	5	385	Green	3 cornered and with ribbed appearance
AGS1	Shipment	4	320-350	Light green	
AGL	Light	3.5	260		
CG Coorg Green					
CGEB	Extra bold	8	450	Golden to light green	
CGB	Bold	7.5	435		Round, ribbed or smooth

CG-1	Superior	6.5	415	Light Green	
				Green	
CG-2	Mota, green	6	385	Cream	
CG-3	Shipment	5	350	Brown	
CG-4	Light	3.5	280		
Bleached (half bleached)					
BL-1		8.5	340	Pale	Fully developed
BL-2		7	340	Creamy	round/3 cornered ribbed
				or smooth skin	
BL-3		5	300	Dull white	

Agmark grades are also available for Coorg Cardamom and Bleached Cardamom.

Uses

Major use of cardamom is as a spice for domestic culinary purpose in the whole or ground form. It is also used in bakery and confectionery. It is an important constituent in ready to use spice blends, garam masala etc. Cardamom flavoured beverages viz. tea, coffee, chocolates and cardamom tincture are popular in market. Cardamom flavour is encapsulated using an innovative technology by CFTRI, Mysore. Essential oil is used in perfumery, for flavouring liquors and other beverages. Oleoresin is used as a soluble spice. In medicine, it is used as an aromatic stimulant, carminative and as an antioxidant. Cardamom is an essential ingredient of digestive stimulants and is used in medicinal preparations for indigestion and flatulence. Occurrence of trypsin like protease in fresh cardamom seeds was reported by Josephrajkumar (2005).

Cardamom (Large)
(*Amomum subulatum* Roxb., F. Zingiberaceae)

Large cardamom or Nepal cardamom is a spice cultivated in the sub-Himalaya State of Sikkim and adjoining areas of Darjeeling district of West Bengal. In the past, aboriginal inhabitants of Sikkim, the 'Lepchas' collected capsules of large cardamom from forests, but later on these forests passed on to village ownership and villagers started cultivation of large cardamom. Present area under this perennial crop is 26,000 ha and annual production is estimated to 4500 to 5000 tonnes (Biswas *et al.*, 2004). The commodity is mainly sold in the domestic markets of North India and a small quantity is also exported.

Botany

Large cardamom is a perennial herb like small cardamom. Inflorescence is a condensed spike on a short peduncle bearing 40 – 50 flower buds in an acropetal sequence. Fruits have almost the size of nutmeg and are cheap and efficient substitute for true cardamom. The dark red-brown globose capsules contain 40 - 50 seeds. There are mainly five cultivars of large cardamom viz., Ramsey, Sawney, Golsey, Varlangy and Bebo. Another cultivar Seremna or Lephrakey is also popular in certain areas.

Crop improvement

A germplasm conservatory of large cardamom with 179 accessions is maintained at Pangthang in East Sikkim and Kabi at North Sikkim. Accessions are being subjected to characterization and evaluation. Two high yielding varieties ICRI Sikkim-1 and ICRI Sikkim-2 (belonging to Sawney cultivar) were released from Regional Research Station, Sikkim for large cardamom. Yield of these two varieties are 800-900 kg/ha under rainfed condition (Biswas *et al.*, 2004). Hybridisation was attempted between Varlangy and Sawney to obtain higher yield (Varlangy) with wide adaptability (Sawney) and hybrids are under evaluation.

Crop management

Crop management practices are almost like that for small cardamom. Propagation is done through seeds and rhizomes. Protocol for micropropagation is also standardised (Sajina *et al.*, 1997b). Crop comes to harvest about 2-3 years after planting. Harvesting is done when seeds of top capsules in spike attain dark grey color. Stalk of spike is cut at base, individual capsules are harvested manually. Capsules are cured either by traditional curing system called Bhati system (direct heat drying) or through flue pipe curing (indirect heat drying). Driage is 28 %. Seeds on steam distillation yield 1-3.5 % essential oil which is rich in 1,8-cineole and almost devoid of a-terpenyl acetate. Large cardamom is used as an ingredient as well as flavouring agent with masala and curry powders, in flavouring sweet dishes, pastries; as a masticatory and for medicinal purposes. Ripe fruits are eaten raw by people of Sikkim. There are two viral diseases - 'foorkey' and 'chirke' which cause considerable damage to crop.

Ginger
(*Zingiber officinale* Rosc., F. Zingiberaceae)

Ginger, one of the oldest known spices, is esteemed for its aroma and pungency. Ginger of commerce constitutes the rhizome of *Zingiber officinale.* It originated in South-East Asia, but was under cultivation in India as well as in China from ancient times. Because of easiness of carrying the rhizomes, it spread throughout tropical and subtropical regions in both hemispheres. Main ginger growing countries are India, China, Jamaica, Taiwan, Sierra Leone, Nigeria, Figi, Mauritius, Indonesia, Brazil, Costa Rica, Ghana, Malaysia,

Bangladesh, Philippines, Sri Lanka, Thailand, Trinidad, Uganda, Hawaii, Guatemala and many Pacific ocean islands. In ancient India, ginger was not significant as a spice, but as a great medicine and it finds a prime place in ancient ayurvedic texts.

Ginger was grown on west coast of India from time immemorial and later on its cultivation spread to various other parts mainly to Bengal and North Eastern India. Ginger was exported from ancient Malabar coast of Peninsular India. It was the Arabs, Portuguese and the Dutch who took it to western world. 'Cochin ginger' was the most soughtafter commodity in spice trade between India and Europe.

India is the largest producer of ginger; the annual production is about 2.63 lakh tonnes from an area of about 0.77 lakh ha contributing approximately 30 to 40 % of world production. The crop occupies the largest area in Kerala (19%) followed by Orissa (17%), Meghalaya (12%), West Bengal (12%) and Arunachal Pradesh (6%). Kerala and Meghalaya together account for nearly 40 % of the country's production. In terms of productivity, Arunachal Pradesh stands first with 7.16 tonnes/ha, followed by Meghalaya (5.14 tonnes/ha), Mizoram (5.14 tonnes/ha) and Kerala (3.44 tonnes/ha). During 1999-2000, India exported 8.773 tonnes of ginger valued at Rs.306 million, out of which dry ginger contributed Rs.199.2 million (Ravindran and Nirmalbabu, 2005).

Apart from India, South East Asia is a major ginger producing region. Ginger production in this region comes mainly from China, Thailand, Taiwan, Korea and Vietnam, China being the largest producer.

Botany

Ginger is a herbaceous perennial grown as an annual crop. The plant is erect with fibrous roots, aerial shoots (pseudostem) with leaves and underground stem (rhizome). As ginger plant grows further, several fleshy roots are also produced from rhizomes. During initial growth, apical bud of rhizome bit grows out as mother tiller and as it grows the mother rhizome develops at the base. From mother rhizome, branches arise laterally producing finger rhizomes which inturn give rise to primary and secondary tillers. Aerial shoots have many narrow leaves with long narrow leaf sheaths. A pair of ligules is present at the junction of leaves and sheath. Leaves are arranged in a distichous manner. Leaves, petiole, rhizomes and roots contain oil cells. Inflorescence is a spike which springs directly from rhizomes, oval or conical in shape. It has got overlapping bracts, from the axils of which flowers arise, each bract producing a single flower. Flowers are fragile, bisexual, short-lived and surrounded by a scariose, glabrous bracteole, with a thin tubular corolla, colourful labellum and petalloid stamen. Androecium consists of six stamens of which outer three are reduced to staminodes and are seen at base of corolla tube. The inner lateral stamens are united to form a deep purple coloured labellum. Fertile anther is cream coloured with two prominent anther lobes containing pollen grains, ovary is inferior, globose, trilocular with several ovules, style is long and filiform and stigma is hairy. Flowering is not common in ginger. Cultivar differences exist with respect to flowering. Maintaining plant as a biennial using larger rhizome bits results in flowering. Flower bud development takes 20 to 25 days from bud initiation to flowering. Anthesis is between 1.30 am and 3.30 am under west coast conditions of Kerala. Anther dehiscence almost coincides with flower opening. Flowers fade and fall on next day. Natural seed set was not reported so far in ginger. Reasons for absence of seed set under natural conditions were investigated by many workers. Studies revealed presence of spiny stigmatic surface which prevents proper adherence and germination of pollen grains (Valsala, 1994). Pollen fertility in ginger reported by various workers ranged from 0.9 to 40 % (Valsala, 1994, Bindu 1997 and Adaniya, 1998). Germination of pollen grains was very slow and they get coiled during advanced stages of pollen germination. Under *in vitro* conditions the pollen tube attained maximum size of 108 µm after 24 h on the style which was very large (39,000 µm). Flowers abscise within 12 h and so the possibility of pollen tube reaching ovule is remote. Self incompatibility was reported by Das *et al.* (1999). Damayanthi *et al.* (2003) reported that heterostyly may be a contributing factor to sterility in ginger. Different techniques tried for breaking incompatibility such as bud pollination, artificial rubbing, artificial cross pollination, mixed pollination, mentor pollination and use of growth regulators failed to produce fruit and seed set (Nybe, 2001).

Chemical composition

Ginger rhizomes contain starch (40.4 to 59%), crude fibre (4.8 to 9.8%), essential oil (1 to 2.7%), fixed oil, pungent compounds, resins, proteins and mineral elements. Of these, fibre and volatile oil contents and pungency level are the most important criteria in assessing suitability of ginger rhizomes for particular processing purposes. These constituents vary in fresh and dried forms of ginger and also at different maturity stages. In immature rhizomes, about 5 to 7 months after planting, fibre content is negligible and pungency is mild. At about 9 months after planting, volatile and pungent principles reach a maximum and thereafter their relative abundance fall as fibre content continues to increase. Drying of ginger usually leads to loss, by evaporation, of a few of volatile oil (up to 20% during sun drying) and volatile oil content of commercial dried ginger is reported to be 1 to 3 %. Ginger oleoresin is extracted from dried ginger by extraction with organic solvents (3.5-10 % is the oleoresin yield which in turn yields 15-30 % volatile oil). Pungency of ginger is attributed to gingerol, shogaol and zingerone whose relative abundance vary. If carefully prepared, fresh oleoresins of Indian ginger contains about 22 % gingerol and 3 % shogaol.

Ginger oil is prepared by steam distillation of ginger rhizomes both fresh and dry. Natural oil of fresh (green) rhizomes varies from that of dried rhizomes in composition and aroma. Chief constituent of ginger oil from dried ginger is zingiberine (38.6%), a sesquiterpene hydrocarbon. Essential oil profile of ginger from various producing countries was well established and studies revealed that there exist variation in constituents with change in geographical region, variety, climatic conditions and processing methods adopted.

Crop improvement

Genetic resources

Largest variability of ginger is reported from China, followed by India. This crop is known under cultivation and use in India and China for the last 2000 years or even more. Because of prevalence of sterility and asexual method of propagation, only local cultivars are in vogue and most of the major ginger growing regions have cultivars specially suited to that region. These local cultivars are all selected, domesticated and planted for a long time under local natural conditions; they generally have greater adaptability, higher yield, better quality and unique use value. The local types are mainly named after name of places, colour of rhizomes or other morphological features. Ravindran *et al.* (1994) opined that geographical spread accompanied by genetic differentiation into locally adapted populations caused by mutations could be the main factor responsible for variations encountered in cultivated ginger. Conscious selection for different parameters like high fresh ginger yield, good dry recovery and less fibre content over years augmented the cultivar diversity of ginger. In India, concerted efforts are being made for conservation of ginger germplasm (Table 16). The diversity is more in Kerala and in North East region of India.

Table 16. Ginger germplasm centres in India

S. No.	Institute / University
1	Indian Institute of Spices Research, Kozhikode
2	National Bureau of Plant Genetic Resources Regional Station, Thrissur, Kerala
3	Orissa University of Agriculture and Technology, Pottangi, Orissa
4	Dr.Y.S.Parmar University of Horticulture and Forestry, Solan, Himachal Pradesh
5	Rajendra Agriculture University, Dholi, Bihar
6	Uttara Bangala Krishi Viswa Vidyalaya, Pundi Bari, West Bengal
7	Narendra Dev University of Agriculture and Technology, Faizabad, Uttar Pradesh
8	Indira Gandhi Krishi Viswa Vidyalaya Regional Station, Raigarh
9	Department of Horticulture, Sikkim
10	Central Agricultural Research Institute, Port Blair, Andamans

Ginger germplasm at IISR consists of 645 accessions including exotic and indigenous collections (IISR, 2002). Over 10 accessions are conserved in the *in vitro* gene bank as medium term storage (Geetha, 2002).

Characterization and evaluation of germplasm

Good variability was reported among ginger cultivars in growing regions. In an evaluation trial with 100 accessions, Ravindran *et al.* (1994) observed moderate variability for many yield and quality traits. Tillers/plant had the highest variability followed by rhizome yield/plant. Among quality traits, shogaol content recorded the highest variability followed by crude fibre and oleoresin. Correlation and path analysis studies conducted by Sasikumar *et al.* (1992) revealed positive correlation between yield and characters like plant height, tiller and leaf number as well as length and breadth of leaf. Dry recovery had a negative direct effect on yield. Das *et al.* (1999) recommended leaf number as a parameter for selection for improving rhizome yield in ginger. Rai *et al.* (1999) reported that higher rhizome yields were strongly associated with chlorophyll A, carbohydrate content and lower polyphenol levels in the leaf. KAU (2003) reported that morphological characters were not reliable to classify ginger types, although some of them could be identified by rhizome characters. Length and girth of primary and secondary fingers were positively correlated with yield.

Correlation studies

Crop improvement work in ginger is limited to clonal selection, mutation breeding and polyploidy breeding. Conventional recombination breeding programmes were handicapped by relatively shy flowering behaviour and absence of seed set. Objectives of breeding are high yield, higher driage, wider adaptability, high quality (high oil, high oleoresin, low fibre etc.) and resistance to fungal and bacterial pathogens (bacterial wilt and rhizome rot).

Correlation studies showed significant positive relation of rhizome weight with plant height, number of tillers and leaf number. Saikia and Shadeque (1992) observed leaves/clump, tillers/clump and shoot height to be positively correlated with yield. Significant genotype differences for pseudostem length, rhizome length and breadth and yield/plant were observed by Korla and Tiwari (1999). Das *et al.* (2000) reported that linear components was significant for tiller number, plant height and yield while non-linear component was significant for plant height. Correlation studies revealed that rhizome yield/plant was positively and significantly correlated with plant height, leaves/plant, tillers/plant, leaf length and width as reported by Singh (2001). Path coefficient analysis for rhizome yield and its related traits revealed very high positive direct effects of stomatal number as reported by Das (1999). Rai *et al.* (1999) observed that higher rhizome yield was positively associated with greater chlorophyll-A and carbohydrate content and lower polyphenol levels in leaves. Significant positive association of oil content with rhizome yield/plant and oleoresin content was observed by Singh *et al.* (2000).

Selection

Both indigenous and introduced cultivars were screened under various geographic situations and selections made for yield, quality and resistance attributes. Yadav (1999) reported high heritability coupled with high genetic advance for plant height, leaf length, tillers/plant, number of primary and secondary rhizomes, weight of primary rhizomes and rhizome yield/plant indicating that desirable improvement in these traits can be brought about through selection. For higher yield of dry ginger, Nadia, Bajpai and Maran were recommended for cultivation in plains of Kerala (Nybe, 2001). Maran also recorded higher quality in terms of oil and oleoresin. In another evaluation, Rio-de-Janeiro was the most suitable for yield of fresh rhizome followed by cv. Kuruppampady. But Rio-de-Janeiro was very much susceptible to soft rot though it was superior in content of oleoresin (10.53%). Oil content was maximum in cv. Karakal. Fibre content was minimum in China and maximum in Kuruppampady. Saikia and Shadeque (1992) reported that Maran was suitable for extraction of oil and oleoresin, but it had high fibre content. Nizam and Jayachandran (1997) observed maximum nonvolatile ether extract in Kuruppampady. Korla (1999) observed that SG-61 had high oil and oleoresin but fibre content was high and yield was low. More fibre content in Khasi local and less in Nadia was reported by Chandra and Govind (1999). Singh *et al.* (2000) observed the highest rhizome yield and oil content in Thinladium and Nadia respectively.

Mutation breeding

Because of existence of sterility, alternative to induce variability is through physical and chemical mutagens. Being a vegetatively propagated crop, induced variability can be fixed immediately and true to type can be propagated. Induction of variability through mutations were tried by many workers and desirable mutants identified. Ginger buds are sensitive to irradiation and LD_{50} was reported below 2 Krad. LD_{50} for sprouting and survival was reported between 1.5 and 2.0 Krad (Nybe, 2001) for gamma rays and below 8 mM for Ethyl Methyl Sulphonate. Lower doses of gamma rays (0.5 and 0.75 Krad) and EMS (2 to 4 mM) were more effective inducing wider variations (Jayachandran, 1989). He observed effects of mutagen treatment in subsequent generations, indicating operation of strong diplontic selection. Mohanty and Panda (1991) tried EMS, Sodium azide, Colchicine and gamma rays as mutagenic agents in five cultivars and studied VM_1, VM_2 and VM_3 generations. A high yielding mutant was subsequently released for cultivation. Two more mutants were identified at Pottangi, Orissa.

Polyploidy breeding

Because of chromosomal sterility, polyploidy breeding offers scope to explore possibilities of inducing variability in ginger. There are reports of

Table 17. High yielding ginger varieties released in India

Variety/ Year of release	Pedigree/ parentage & plant type	Institution /University	Av. yield t/ha (fresh)	Salient features	Recommended state/region
Suprabha 1988	Clonal selection from Kunduli local	High Altitude Research Station, OUA & T, Pottangi-764 039, Koraput (Dist), Orissa	16.6 (Potl. yield 22.80)	Plumpy rhizome, less fibre, wide adaptability, suitable for both early and late sowing, duration 229 days, 8.9% oleoresin, 4.4% crude fibre, 1.9% essential oil and 20.5% dry recovery	Orissa and adjoining states
Suruchi 1990	Clonal selection from Kunduli local	-do-	11.6 (Potl. yield 23.50)	Profuse tillering, bold rhizome, suitable for rainfed / irrigated conditions duration 218 days, 10.9% oleoresin, 2.0% essential oil, 3.8% crude fibre, 23.5% dry recovery	Orissa, Central and South India
Surabhi 1991	Induced mutant of Rudrapur local	-do-	17.5 (Potl. yield 23.00)	Plumpy rhizome, dark skinned yellow fleshed, suitable for both irrigated/rainfed, duration 225 days, 10.2% Oleoresin, 2.1% essential oil, 4.0% crude fibre, 23.5% dry recovery.	Orissa
V_3S_1-8	Sodium azide mutant	-do-	29.0	A mutant line moderately tolerant to diseases and pests. Having an oleoresin content of 10.8%, essential oil 1.8%, crude fibre 3.2%, dry recovery 22.2%, suitable for green and dry ginger wide ecological adaptability, suitable for both hills and plains	Orissa, Andhra Pradesh, West Bengal, Madhya Pradesh, Uttar Pradesh, Bihar
V_1E_8-2	An EMS mutant	-do-	32.9	A high yielding mutant with moderate tolerance to disease and pests. Contains 10.8% oleoresin, 1.8% essential oil, 3.5% crude fibre, 21.4% dry recovery, suitable for green ginger, late planting under rainfed condition in hills and plains	Orissa, Andhra Pradesh, West Bengal, Madhya Pradesh, Uttar Pradesh, Bihar
Himgiri 1996	Clonal selection from Himachal collection	Department of Vegetable Crops, YSPUH & F, Nauni, Solan-173 230, Himachal Pradesh	13.5 (Pot yield 20.6t)	Best for green ginger, less susceptible to rhizome rot disease, suitable for rainfed condition, 4.29% oleoresin, 1.6% essential oil, 6.05% crude fibre, 20.2% dry recovery, 230 days duration	Himachal Pradesh
IISR Varada 1996	Selection from germplasm	Indian Institute of Spices Research, ICAR, Calicut-673 012, Kerala	22.6	High yielder, high quality bold low fibre content (3.29% to 4.50%), essential oil 1.7%, 6.7% oleoresin and 19.5% dry recovery, tolerant to diseases, maturity 200 days.	All over India
IISR Rejatha 2004	Selection from germplasm	-do-	22.4	High yielder, plumpy and bold rhizome, 2.36% essential oil, 6.3% oleoresin, 4.0% crude fibre, 20.8% dry recovery, maturity 200 days.	Kerala and Karnataka
ISSR Mahima 2004	Selection from germplasm	-do-	23.2	High yielder, plumpy bold rhizomes, 1.72% essential oil, 4.5% oleoresin, 3.26% crude fibre, dry recovery 23.0%, maturity 200 days. The variety is resistant to nematodes (*M. incognita* and *M. javanica*)	Kerala and adjoining States

tetraploids flowering in second year of planting and not flowering at all. Ramachandran and Nair (1992) reported tetraploids with larger rhizome and higher yield, but oil content was lowered by 0.5 %. Sheeba (1996) observed eutetraploids in ginger to have larger epidermal cells and stomata with reduced number/mm^2 of leaf and with increased pollen fertility. Flowering was also observed in autotetraploids. Shankar (2003) evaluated seven autotetraploids (from Rio-de-Janeiro, Himachal Pradesh and Maran) along with their diploid counter parts. The auto tetraploid from Rio-de-Janeiro was notable for low fibre content, high oil and oleoresin content, increased driage and thick cuticle compared to its diploid counter part. The autotetraploid from Himachal Pradesh exhibited high driage, oleoresin content and cuticle thickness. Fresh and dry rhizome yields of induced variants were also on par with diploids. However, autotetraploids were susceptible to soft rot and bacterial wilt. Both autotetraploids exhibited high pollen fertility and viability.

Biotechnological interventions

Protocol for rapid multiplication of ginger was standardised at Kerala Agricultural University (Nybe, 2001). Sprouting buds were identified as the best source of explants for initiating *in vitro* culture of ginger. MS medium was found ideal. BA (2.5 mg/l) was the most effective in getting maximum number of multiple shoots. Evaluation of TC plants indicated that rhizome yield was less compared to conventional crop.

In vitro pollination and fertilization

In vitro pollination and fertilization technique were standardised in ginger (Valsala, 1994). Placental pollination registered maximum seeds per culture with minimum effort and it was the best. Flower buds collected on day of anthesis and one day after anthesis were suitable for in vitro pollination. The in vitro produced seeds and fruits of ginger grew rapidly in initial stages and then slowly. The in vitro produced seeds germinated when 80 days old seeds were incubated initially in medium of ½ MS with 2,4-Dmg l^{-1} for two months and then in hormone combination of BAP 9 mg l^{-1} and 2,4-D 0.1 mg l^{-1}. However, the in vitro produced seeds showed erratic response to germination. Bindu (1997) reported embryo rescue and germination studies in ginger. Embryo rescue involving transfer of embryo along with endosperm to culture media with varying combinations of auxins and cytokinin did not promote development of embryo to seedling.

Crop management

Climate and soil

Ginger requires a warm and humid climate. The plant thrives well from sea level to an altitude of 1500 m in the Himalayas, optimum elevation being between 300 and 900 m (Pruthy, 1993). A well distributed rainfall (150 – 300 cm) during growing season and dry spells during land preparation as well as before harvest are required for large scale cultivation of crop. In areas receiving less rainfall, the crop needs regular irrigation.

Ginger can be grown in a wide range of well drained soils of at least 30 cm depth, ranging from heavy laterite loams to clayey loam. Laterite loams containing not more than 30 % sand or 20 % clay and free from gravel are however preferred as they give higher yields (Pruthy, 1993). Sahu and Mitra (1992) got the highest yield in sandy loam soil having the minimum bulk density (1.20 g/cc), moderate acidic reaction (pH 5.7), and high organic matter (organic carbon 1.1 %) and available potassium (351 kg/ha). The yield decreased with increase in soil clay content and decrease in pH.

Time of planting

In South India, crop is mainly grown as a monsoon crop from April-May to December and in North Central India, it is grown as an irrigated crop. As rainfed crop, the first week of April was the best time of planting for getting maximum yield under Kerala (India) conditions registering 200 % increase in yield compared to planting in first week of June. When June and July plantings were compared, planting in June recorded higher yield and low incidence of soft rot. For irrigated ginger, the best time for planting was middle of February.

Seeds and seed rate

In ginger, rhizomes are used for planting. Rate of seed rhizome varies from 900-1500 kg ha^{-1}. For selection and preservation of seeds, following method is recommended in Kerala, India (KAU, 2003). Mark healthy and disease free plants in field when crop is 6-8 months old and still green. Select the best rhizomes free from pests and diseases from marked plants. Handle seed rhizomes carefully to avoid damage to buds. Soak selected rhizomes for 30 minutes in a solution of Mancozeb and Malathion to give terminal concentration of 0.3 % for the former and 0.1 % for the latter. Dry treated rhizomes in shade by spreading on floor which is lined with sand or saw dust. It is advisable to spread layers of leaves of *Glycosmis pentaphylla.* Pits are covered with coconut fronds. Examine stored rhizomes at monthly intervals and remove the rhizomes which show signs of rotting. This helps to keep inoculum level low. Also treat seed rhizomes in same manner before planting.

Under Indian conditions, optimum size of seed bit was found to be 15 g with one or two viable buds and seed rate recommended is 1500 kg ha^{-1}. Soaking seed rhizomes for 24 hours, 10 days prior to planting results in good sprouting. Effect of storing seed rhizomes after cutting them into pieces was studied. Early and fairly uniform germination were observed in plots planted with seed rhizomes cut and stored for 45 and 30 days before planting (Nybe, 2001).

Under open and intercropping conditions, rhizome bits weighing 15 g recorded the highest sprouting (Table 18). Green ginger yield increased with increasing rhizome size both under open and intercropped condition. Plants raised from rhizomes weighing 5 g gave the lowest yield, which was inferior to other treatments. Plants from rhizomes weighing 15 g recorded the highest green ginger yield (Table 19). Under open conditions, plants raised from rhizome bits weighing 10 and 15 g gave higher dry ginger compared to plants from 5 g. Rhizome bits of cv. Kuruppampady performed better, however when intercropped in a coconut garden. Cv. Maran gave higher yield in all three sizes of seed materials used. Size of rhizomes did not cause differences in quality components like volatile oil and starch in any variety, but it induced small variations in nonvolatile ether extract (NVEE) and crude fibre. The cost benefit analysis indicated that use of mini-seed rhizomes, weighing 10g, is more profitable (Nizam and Jayachandran 2001).

In a few places, farmers sow whole rhizomes and unearth them when crop reaches about 30 to 35 cm in height (Singh, 1982). The recovery is about 94 % at three months after planting. This practice helps farmers to recover 60 to 70 % seed cost.

Table 18. Effect of size of seed rhizome and variety on green ginger yield (tonnes ha^{-1}) under open condition

Variety	Rhizome size (g)			Mean
	5	10	15	(g)
Kuruppampady	20.400	21.488	25.200	22.360
Maran	11.584	13.128	19.488	14.736
Nedumangad	11.360	16.000	13.328	13.560
Rio-de-Janeiro	11.416	15.896	15.912	14.408
Mean	13.688	16.632	18.480	
CD (0.05)	S - 2.800*	V x S - NS		V - 3.422*

Table 19. Effect of size of seed rhizome and variety on green ginger yield (tonnes ha^{-1}) as intercrop in coconut garden

Variety	Rhizome size (g)			Mean
	5	10	15	
Kuruppampady	11.476	16.317	17.354	15.049
Maran	19.024	22.999	22.875	21.633
Nedumangad	15.745	27.546	26.448	23.246
Rio-de-Janeiro	7.656	13.812	15.405	12.291
Mean	13.475	20.169	20.520	
CD (0.05)	S - 2.563*	V x S - 5.127*		V- 6.716*

Spacing and method of planting

Spacing varies with soil fertility, cultivar, climate and management practices. Earlier reports indicated that closer spacing gave better yield. Based on trials, planting of ginger is recommended on raised beds (in order to facilitate drainage) at a spacing of 20 x 20 cm or 25 x 25 cm and a depth of 4-5 cm with the viable bud facing upwards. Pandey (1999) reported that among different spacings (40 x 20, 30 x 20, 40 x 30 and 50 x 20 cm) the highest yield was observed under the closest spacing. Planting of irrigated ginger in raised beds gave maximum yield when compared to planting in ridges, furrows and flat ground (Nybe, 2001). Seed rhizome is placed 3.5-5.0 cm deep in pit and soil is pressed over it, followed by light irrigation. Mulching beds, twice with green leaves is an important practice. In general, planting depth varies with size of planting unit, soil type, and soil moisture content (Kandiannan *et al.*, 1996).Bolder seed rhizomes are planted deeper and smaller rhizome bits at shallow depths. Commonly adopted practice is to place rhizome pieces at 4 to 10 cm depth (Mohanty *et al.*, 1990; Wilson and Ovid, 1993).

Shade

The crop prefers light shade for good growth, but shade is not absolutely necessary. Jayachandran *et al.* (1991) investigated effect of shade on yield of ginger cv. Rio-de-Janeiro by growing plants under no shade (open), 25 %, 50 % and 75 % shade. Shade was provided by coconut leaves spread on a pandal [shelter]. At harvest (8 months after planting), fresh rhizome yield was the highest under 25 % shade and lowest under 75 % shade (20093 kg/ha and 10778 kg/ha, respectively). The yield under open conditions was similar to that under 50 % shade. Dry ginger recovery was the highest under 25 per cent shade (2733 kg/ha).

Screening of ginger cultivars for shade tolerance was done with six cultivars (Maran, Kuruppampadi, Himachal, Rio-de-Janeiro, Nedumangad and Amballore Local) under four shade levels (0,25,50 and 75 %) (George, 1992). The study confirmed the shade loving nature of ginger registering significantly higher yield under different shade levels than under open, with 25 % shade recording the highest value. The quality of ginger rhizomes improved when grown under shade. Based on rhizome yield, cultivars adapted to each of shade levels were identified as Kuruppampadi and Himachal (0 % shade), Nedumangad, Himachal and Kuruppampadi (75 % shade). Himachal was found adapted to all situations. Under natural shade of coconut, Amballore Local showed better performance.

Response of ginger to different shade levels (0, 25, 50 and 75 %) was studied (Babu, 1993) (Table 20) Maximum yield (22.22 t ha^{-1}) and dry matter production were recorded in low shade condition of 25 %. As performance of crop was better under shade than in open, the crop was considered as shade loving and highly suitable as intercrop in coconut gardens. Uptake studies showed that fertilizer requirement for intercropped ginger in low shade level will be 10-50 % higher than that of pure crop.

Table 20. Effect of shade and mulch on mean green ginger yield (tonnes ha^{-1})

Shade (%)	Mulch (green leaves (t ha^{-1})				
	7.5	15	225	30.0	Mean
0	13.164	14.145	17.398	19.517	16.056
25	18.029	21.439	26.030	26.036	22.883
50	15.535	20.492	21.123	22.007	19.789
75	16.292	17.303	20.839	21.786	19.055
Mean	15.755	18.345	21.347	23.337	
CD (0.05)	M – 0. 291* S x M – 0.569*			S – 0.476*	

Ancy and Jayachandran (1993) studied Rio de Janeiro plants, grown under 0, 25, 50 or 75 %t shade (provided by dry coconut fronds) and supplied with fertilizers at 75, 100, 125 or 150 % of the recommended rate (75, 50 and 50 kg of N, P_2O_5 and K_2O, respectively) at Thiruvananthapuram, India. Volatile oil content was the highest (2.19% on dry weight basis) under 25 % shade and non-volatile ether extract (NVEE) content was significantly higher under 25 or 50 % shade than under 0 or 100 % shade. Rhizome fibre content was not affected by shade and fertilizer treatments.

Nath (1993) evaluated effect of shade (provided by a pigeon pea crop) and rhizome treatment (1% formaldehyde) on rhizome rot of *Zingiber officinale* cv. Turda, caused by *Pythium myriotylum*. The highest rhizome yield was recorded in rhizome treated plants grown in shade for 150 days. Increased yield was mainly due to reduction of disease as disease incidence was 45.8 and 41.3% in control and rhizome treated plants grown without shade, respectively and 19.4% in rhizome treated plants grown in shade. It is suggested that temperature around the crop got reduced by shading and this could ultimately suppress growth of fungus. Wilson and Ovid (1993) studied effect of shade on growth and yield of ginger and observed that average yield of ginger was the highest (68.4 tonnes/ha) under okra shade (66.2 tonnes/ha).

Ancy and Jayachandran (1998) noted that nutrient uptake increased with increasing fertilizer rate under all shade intensities but response in terms of rhizome yield/unit increase in fertilizer rate was the highest under 25 % shade followed by 50, 0 and 75 % shade. Dry rhizome yield was the highest (3415.25 kg/ha) under 25 % shade with NPK at 150 % of the recommended rate.

Table 21. Effect of shade on volatile oil content of ginger (v/w %)

Shade (%)	Period (DAP)				
	120	150	180	210	240
0	1.46	2.35	2.22	2.06	2.18
20	1.94	2.79	2.57	2.20	2.26
40	2.24	2.51	2.65	2.26	2.31
60	2.44	3.35	2.78	2.40	2.42
80	2.47	3.57	2.54	2.43	2.45
CD (0.05)	0.069*	0.097*	0.039*	0.089*	0.098*

Jayachandran *et al.* (1998) recorded that under artificial shade (25 %), ginger yield was 11-27 % higher than in open fields and even under 50 % artificial shade, yield was better than under open conditions. Under natural shade in coconut plantation, there was a 32 % increase in rhizome yield.

Another experiment (Sreekala, 1999) to study effect of shade on biomass production and partitioning of photosynthates in ginger cv. Rio-de-Janeiro confirmed preference of crop to low shade levels registering better growth and yield. Different shade levels (0,20,40,60 and 80%) influenced quality of ginger rhizomes. While volatile oil was maximum under higher shade levels in general (60 and 80 %), non-volatile ether extract was higher under 20 % shade. Starch as well as crude fibre content were more in plants grown under open condition (Table 21).

Photosynthetic rate and related parameters of ginger measured at six months after planting using leaf chamber analyzer indicated that photosynthetically active radiation on leaf surface as well as stomatal conductance were high under open condition. But leaf internal carbon dioxide concentration as well as stomatal resistance were high under heavier shade levels (60 and 80 %). Photosynthetic rate as well as transpiration rate were maximum in plants grown in open. Although, at 20 % shade, photosynthetic rate was less, yield was high. This might be because of the photo-oxidation that has taken place at high light intensities or due to the inefficient translocation of photosynthates in open condition compared to 20 % shade (Ancy and Jayachandran, 2000). Another screening trial conducted in Kerala for shade tolerance with 13 culltivars of ginger and same set of shade levels affirmed shade loving nature of ginger producing the highest rhizome yield and quality rhizomes at 25 % shade. Cultivar, Valluvanad was selected as the best single variety for all situations. Cultivars selected as suitable for each of shade levels were Jamaica, Valluvanad and Kuruppampady (0 % shade), PS-667 and Jamaica (50% shade), Valluvanad, Jamaica and Jorhat (75% shade) (Sreekala and Jayachandran, 2001). The cultivar Rio-de-Janeiro raised as pure crop and as intercrop, recorded maximum dry matter production as intercrop in six year old coconut plantation when compared to two year old plantation.

It appears that relatively low temperature combined with low light intensity contributes to development of more chlorophyll in ginger plants grown in shade, leading to better synthesis of photosynthates (Sreekala *et al.*, 2001). It can also be claimed that better utilization of carbohydrates takes place as degradation of carbohydrate is slowed in decreased respiration at lower temperature. Ajithkumar *et al.* (2002) studied the effect of shade regimes on photosynthetic rate and stomatal characters using cv. Rio de Janerio. They found that the highest photosynthetic rate was in the open conditions, followed by plants grown under 20 and 40 % shade. The photosynthetic rate, stomatal conductance, transpiration rate, stomatal index and stomatal frequency were significantly reduced linearly with increasing levels of shade. Yield increase under 20 and 40 % shade compared to open condition is attributed to higher leaf area under these shade levels.

Sreekala and Jayachandran (2001) worked out influence of shade on physiological parameters of ginger. They reported higher dry matter accumulation, leaf area index, net assimilation rate and crop growth rate under low (20 %) shade levels. Radio tracer analysis done using labelled 14^C showed that under low light intensity, photosynthates produced translocated efficiently to lower portion while in open condition efficient translocation did not take place. Studies showed that crop can tolerate shade upto 40-50 per cent. Thus, the partially shaded coconut garden can be exploited for increasing area under ginger.

Nutrient uptake and removal

Growth of ginger can be classified into three distinct places ie. a phase of active vegetative growth (90 to 120 days after planting); a phase of slow vegetative growth (120 to 180 days after planting); and a phase of senescence (above 180 days up to harvest. The pattern of rhizome development also followed same trend except that development of rhizome continued up to harvest.

Organic matter

In olden days, ginger was cultivated in forest cleared virgin soils and as such there was no need to apply fertilizers. But the situation changed and now it has become impossible to realise satisfactory yield without adequate supply of fertilizers along with heavy dose of organic matter. In Kerala, major ginger producing state of India, the recommended dose of organic matter is 25 to 30 t/ha of FYM and 30 t/ha of green leaves as mulch applied in three splits, 15 t at the time of planting and 7.5 t each at 60 days and 120 days after planting (Nybe, 2001).

Khandakar and Nigam (1996) reported that rhizome yield of ginger increased with increased ratio of FYM application (33 t/ha) compared with

22 t/ha. Plant height was greater with FYM. Effect of organic manure and *Azospirillum* on the growth and yield of ginger was studied in Kerala by Chengat (1997). Application of FYM @ 48 t/ha resulted in the highest returns and benefit:cost ratio (2.32) giving an additional profit of 32.04 % over control. Sadananandan *et al.* (1998) reported that among 6 organic fertilizers (FYM, neem cake, brasicca cake, groundnut cake and gingelly cake) ground nut cake gave the highest soil organic matter. Neem cake gave the highest benefit: cost ratio followed by ground nut cake.

Nutrient requirements

Requirement of N is the most critical among the major nutrients in ginger. Though nutrient is directly available to the plant in nitrate form, it is easily lost by leaching. Under tropical conditions, loss by leaching and denitrification is very high. At same time, nitrate N moves upwards with capillary rise of water during drought. Ammonium ions perform better than nitrates under heavy leaching situations. Unlike N, P is highly immobile in soil because of its reaction with iron and aluminium hydroxides.

Therefore, amount of phosphatic fertilizer needed for the crop is relatively high. For a short duration, quick growing crop like ginger, fertilizer containing high proportion of water soluble P_2O_5 is needed for better yield (Sushama and Jose. 1994). Ginger, when grown as a homestead crop, K nutrition plays an important role. Only under high rates of K application, the crop can be grown successfully under shade (Jayaraj, 1990).

Mohanty *et al.* (1993) found that yield of ginger rhizomes increased with increasing rate of fertilizer application and maximum yield (10.16 tonnes/ ha) was recorded in treatment with the highest fertilizer dose of N at 125 kg, P_2O_5 at 70 kg and K_2O at 150 kg/ha.

Govind *et al.* (1995) reported that 90 kg of P_2O_5/ ha showed maximum plant height, tillers, leaves/ plant, secondary rhizomes/ plant and maximum fresh and dry yields of rhizome in ginger cv. Nadia in Meghalaya region. Sixty kg P_2O_5 / ha showed maximum primary rhizome/plant. Different P_2O_5 rates (104.6, 83.7 and 90.8 kg/ha) gave almost similar yield (1.47 to 1.49 tonnes / ha). Joseph and Jayachandran (1996) emphasized the necessity of increasing the quantity of fertilizer to 150 % of the recommended dose (75: 50: 50 kg NPK/ha) where ginger is grown under low to medium shade (25 and 50%). Under intense shade (75%) there was no response to fertilizer treatment.

Thakur and Sharma (1997) revealed that N and P upto 100 and 60 kg/ ha respectively increased rhizome yield significantly. N, P and K uptake by ginger increased with increase in application of N and P rate upto 150 and 90 kg/ ha. Chengat (1997) studied influence of organic manures and *Azospirillum* on uptake of N, P and K and found that uptake was maximum in plots incorporated with FYM (48 t/ ha) followed by *Azospirillum* + 75 % N. Soil nutrient status in terms of available N and K were found high in plots inoculated with *Azospirullum* + 75 % N whereas available P was the highest in plots receiving recommended NPK.Per plant yield was better under farmyard manure treatment.

Need for increased fertilizer requirement under shade were confirmed in another trial where in application of 150 kg N, 100 kg P_2O_5 and 100 kg K_2O ha^{-1} gave maximum net profit for ginger raised as intercrop in coconut (Kumar, 1999). Gowda *et al.* (1999) revealed that yield of ginger cv. Rio–de-janeiro could be increased by application of 150:75:50 kg NPK/ ha under Bangalore conditions. Uptake of nitrogen, phosphorous and potassium in leaf and pseudostem progressively increased upto 180th day and then decreased while their uptake in rhizome increased steadily till harvest. Graded doses of nitrogen, phosphorous and potassium failed to influence oleoresin content of ginger.

An investigation to determine fertilizer requirement of ginger (Cv. Rio-de-Janeiro) under varying shade intensity was carried out with four shade levels (0,25,50 and 75 %) and four fertilizer levels (75,100,125 and 150 % of the recommended dose of 75:50:50 N, P_2O_5 and K_2O ha^{-1}). Under open conditions, significant increase in dry ginger yield was obtained upto 125 % of recommended dose and maximum ginger yield was obtained upto 125 % of recommended dose and maximum profit was also realized from this treatment. Growth and yield of ginger were the highest under 25 % shade. Response to fertilizer treatments in terms of growth characters and yield were the highest under this shade level. Significant increase in green and dry ginger yield was obtained with each increment in fertilizer dose upto the highest level (Jayachandran *et al*, 1991).

Kerala Agricultural University recommends 75 kg N, 50 kg P_2O_5 and 50 kg K_2O ha^{-1}. The full dose of P and half dose of K are to be given as basal and ½ N applied 60 days after planting. Remaining half dose of N and half dose of K are to be applied 120 days after planting (KAU, 2003).

Secondary and micronutrients

Since ginger is cultivated with very high dose of organic manure and green leaf mulch, additional supply of secondary and micro-nutrients are not usually required. However, Roy *et al.* (1999), observed that in local cultivars of ginger, the highest yield of 48.8 t/ha was obtained with a combination of 0.3 % zinc, 0.2 % Fe and 0.2 % B. Wang *et al.* (1993) reported that Zn affects protein synthesis by RNA metabolism and hence amino acids accumulate under zinc deficiency. Srinivasan *et al.* (2003) carried out a study on zinc nutrition of ginger. They found that zinc deficiency exists in 49 percent of soil samples analysed from various ginger-growing regions of India. Their study indicated that zinc application at 5 kg/ha increased rhizome yield significantly. The culture model, they developed indicated that 6 kg/ha zinc is the best for maximizing yield.

Biofertilizers

Ginger responds well to application of biofertilizers. Studies conducted by Vilasini (1996) revealed that soil solarised for 30 days and incorporated with *Trichoderma* (125g/ m^2) and amended with neem cake (500 g/m^2) could control disease effectively and increase yield considerably. Sharma *et al.* (1997), opined that inoculation with *Glomus mosseae* at 21 spores (10 x 102)/g soil gave the tallest ginger plants, with the highest yield (46.5 g/pot) and the highest number of tillers/plant than other treatments. *Gigaspora margarita* soil application (2.5 g/rhizome) at time of planting increased plant height, number of leaves and tillers, root weight and yield of ginger, as did pine needle organic amendment and *Trichoderma harzianum* seed treatment.

Treating rhizomes of cultivars Mahi with *Azetobacter* and *Azospirillum* followed by application of N at 50 kg/ ha produced maximum green ginger yield of 20.34 tonnes/ ha against uninnoculated field (receiving 75 kg N / ha) which yielded 17.44 t/ ha (Konde *et al.*, 1988).

Mulching

Mulching of beds with green leaves is an important and essential operation in ginger. Effect of mulching on growth and yield of ginger was already established from various studies. Ginger being grown mainly as a rainfed crop, mulching is essential for weed control, moisture conservation and to protect beds from beating action of rain. Heavy mulch changes the physical and chemical environment of soil underneath, resulting in increased availability of phosphorous and potassium. Mulching increased germination and growth of plants in terms of height and number of tillers. Weed growth in control plots was much higher than plots mulched.

Applications of leaf mulch immediately after planting and later six weeks after planting using a total of 20 tonnes ha^{-1} of green leaves resulted in 200 % increase in yield over non-mulched crop and this was found sufficient in higher elevations of Kerala. In plains, mulching crop with 30 tonnes ha^{-1} of green leaves is recommended. Immediately after planting, beds are to be mulched with 15 t ha^{-1} of green leaves, which is repeated with 7.5 t ha^{-1} each at 2 months and 4 months after planting. Mulching is done coinciding with weeding, top dressing and earthing up. Among different mulch materials, leaves of *Glycosmis pentaphylla, Glyricidia maculata* and *Artocarpus altilis* were found good.

Valsala *et al.* (1990) studied possibility of growing daincha (*Sesbania aculeata*) in bed interspaces of ginger and utilizing green leaf material for mulching at KAU, India. Ginger crop, cv. Maran, was grown on raised beds (3 x 1 m with 50-cm interspaces). A first mulch was given using locally available leaves. *S. aculeata* was sown in the bed interspaces immediately after planting of ginger rhizomes, It was up rooted 60 days later and used as a second mulch. It provided 4 kg green leaves/1.5 m^2, which was 50 % of normal requirement of 8 kg/1.5 m^2. Daincha roots were also considered as good source of organic matter.

Pawar (1990) observed that in a comparison of no mulch, polyethylene (PE) film mulch or sugarcane trash mulch (5 t ha^{-1}) combined with different irrigation regimes (60, 80, 100 and 120 mm CPE), the highest yields of green ginger were obtained with mulches and irrigation at 60 or 80 mm CPE. Under semi-arid conditions of trials, it was concluded that, with a PE mulch, 50 % less irrigation water was required to give same yield as when no mulch was applied.

Korla *et al.* (1990) conducted trials with local cultivar, Dharja and found that mean yield was the highest (205-206 g $plant^{-1}$) in plots mulched with grass clippings + farmyard manure (FYM) or pine needles + FYM. In Kerala, the major ginger producing state of India, the recommended dose of mulch is 30 t/ha of green leaves applied in three splits, 15 tonnes at time of planting and 7.5 t each at 60 days and 120 days after planting (KAU, 2003).

Kurian *et al.* (1997) noted that green manure crops of *Sesbania rostrata, S. aculeata, S. speciosa, Crotalaria juncea* or fodder cowpea (*Vigna unguiculata*), when grown amongst a ginger crop and used as second mulch, or sown a second time two months after ginger was planted and used as third mulch, reduced weed problems in ginger crop. Ginger rhizome yields were high with *S. aculeata* used as second mulch.

Babu and Jayachandran (1994) studied effects of shade (0, 25, 50 or 75%) and mulching with leaves of *Swietenia mahagoni* on yield of *Z. officinale* cv. Rio de Janeiro. Yield increased with increasing rate of mulch. Interactions between mulching rate and shade were observed. Under 25 % shade, similar yields (5.25 and 5.24 t/ha) were obtained from treatments with 100 % (30 t/ha) and 75 % (22.5 t) of recommended mulching rate, respectively. Gupta and Awasthi (1997) observed that in trials at Jagdalpur, India, in *kharif* (monsoon) seasons of 1994 and 1995, ginger cv. Suprabha, grown with mulches of palas leaves *(Butea monosperma)*, sal leaves (*Shorea robusta*), rice straw, dry grass, leaf mould or soil, the best treatment was palas leaf mulch which produced average fresh yields of 21.90 t/ha compared with 16.75 t/ha in unmulched control.

Soil solarization

Solarization has a profound suppressive effect in ginger cultivation, especially on weed population and it lasted till harvest. Eventhough, solarization substantially reduced weed population, its effect was less in sedges. *Bulbostylis barbata, Cynadon dactylon* and *Cyperus rotundus* survived solarization. Vilasini (1996) observed that soil solarization for 30 days combined with *Trichoderma* application could control rhizome rot disease and increase yield.

Increased growth response of ginger plants was observed as a result of solarization. Growth parameters like height, number of leaves/plant, number of tillers, number of roots, leaf length, leaf breadth, fresh weight of shoots and rhizomes were influenced by solarization. Significant increase in yield was obtained through solarization. *Trichoderma* incorporated and neem cake amended 30 days solarized treatment gave the maximum yield/plant (623.23 g) and also/plot (0.01t/ha), which was 53.61% more than that of control. Availability of nitrogen, phosphorous and potassium was improved by solarization. Initial cost of solarization is comparatively high; an amount of Rs.52,500 is required for solarizing one hectare of ginger field. An additional profit generated from this technique was Rs. 40,136/ha/year for 30 days solarization (Anith *et al.*, 2000).

Water management

Ginger is grown both as a rain-fed and irrigated crop. Korla and Tiwari (1999) observed that significant effects of rainfed and irrigated conditions were observed on pseudostem length, tillers/plant, leaf length, leaf breadth and yield/plot. Significant genotypic differences were observed for pseudostem length, rhizome length, rhizome breadth and yield/plant. In general, commercially released cultivar Himgiri performed well and was consistent under both environments for most of characters. Performance of SC 646 was similar to Himgiri for tillers/plant, leaves/plant, leaf length, leaf breadth, rhizome length, rhizome breadth, yield of individual plants and total yield/plot.

Intercropping and rotation

Ginger is grown both as a pure crop and as an intercrop or in rotation with other crops. In Kerala, it is grown as an intercrop with coconut, arecanut and coffee and also in rice fallows. In irrigated areas, ginger is grown in rotation with chillies, vegetables, ground nut, ragi and maize. In Kerala as well as in Sri Lanka, ginger forms a component of homestead farming, and is grown mixed with a variety of crops. Ginger is a very successful crop component in intercropping and multicropping systems. It is intercropped with vegetables (such as cabbage, beans, cucumber, and lady's finger), pulses (such as pigeon pea and black gram), cereals (maize and finger millet), oil seeds (castor,soyabean, and sunflower) and with crops such as tobacco, pineapple, tapioca, taro, *Dioscorea* and *Amorphophallus*. It can also be grown as a mixed crop with castor, finger millet, maize and red gram. Chillies-ginger-mixed cropping is prevalent in many areas.

Sankar and Swamy (1988) studied ginger cultivar Rio-de-Janeiro grown in (a) monoculture, or as an intercrop in (b) 2-year-old and (c) 6-year-old arecanut plantations. Average day temperatures in (a), (b) and (c) were 30, 28.3 and 27.9°C, respectively. Average light intensities were 104.81, 33.09 and 15.61 klx, respectively, for ginger. Leaf area was the highest in (c) and the lowest in (a). Chlorophyll-a and -b content increased up to 150th day after planting in (b) and (c), and up to 180th day in (a); thereafter it declined in all cases. Rhizome yields in (a), (b) and (c) were 2666, 9191 and 13 487 kg ha^{-1}, respectively.

Jasural *et al.* (1993) investigated performance of ginger, under rainfed conditions in pure stands and as intercrops with 5-year-old poplars planted at three spacings (5 x 5, 5 x 4 and 5 x 3 m) at Himachal Pradesh. Average illumination below canopies was 53, 46 and 38 % of incident radiation, respectively. Crop performed better as intercrop than as pure stand as measured by growth (height, tillers and leaves/plant, and leaf length and breadth), yield (rhizome length and breadth, yield/plant and hectare, and dry matter content) and survival. However, all parameters decreased as poplar spacing became closer than for pure crop. For quality parameters, only oil content in ginger showed significant differences. Among poplar spacings, 5x4 m was delineated as the ideal spacing.

Sujatha *et al.* (1994) noted that in a trial carried out during 1993-94, the highest sprouting percentage (100 %) was recorded in cultivars Earanadan and Valluvanad when 26 ginger cultivars were grown in 3 x 1 m plots to be screened as intercrops for coconuts at Regional Agricultural Research Station, Pilicode, Kasargod District, India. The highest fresh yields were recorded for cv. Kuruppampady (356.5 g/plant, 17.11 kg/3 m^2 plot and 57.05 t/ha), followed by Wynad-Mananthody and Earandan. The highest dry yield was recorded in PGS-667 (4.53 kg/plot, 15.5 t/ha), followed by Kuruppampady and SG-551.

In Sikkim, India, ginger is intercropped with mandarin orange (*Citrus reticulata* Blanco) and mixed with maize. Many farmers believe that raising maize and ginger together gives more yield than sole crops (Patiram *et al.*, 1995). Partial shade provided by mandarin orange provides a congenial atmosphere for ginger. After harvesting of maize, ginger plants become exposed to more sunlight and that favours development of bolder rhizomes.

Manjunath *et al.* (1998) evaluated ginger as intercrop for coconut in trials carried out between 1993 and 1996. Ginger (cv. Sangli Local) was a suitable annual intercrop for intensive management during May-June sowing, resulting in good yields and the highest net returns. Intercropping increased coconut yields from an average of 13 nuts/palm before intercropping to 39 nuts/palm after 3 years of intercropping.

Sharma and Bajaj (1998) undertook studies to improve ginger yield against *Pratylenchus penetrans, Meloidogyne incognita, Helicotylenchus dihystera* and *Tylenchorhynchus mashhoodi* in Himachal Pradesh, India, by intercropping with bell pepper (*Capsicum annum*) in 8 different treatments. Intercropping of one rhizome of ginger with one plant of bell-pepper gave the highest ginger weight (600 g/rhizome).

Kandiannan *et al.* (1999) investigated the effect of intercropping ginger with maize under rainfed conditions. The crops were planted in June on raised beds (3 x 1 m), ginger at 25 x 30 cm and maize at each corner of raised bed. The maize was harvested 90-100 days after planting and ginger 8 months after planting. There were no significant differences in percentage sprouting, total number of tillers/bed and fresh rhizome yield, between ginger grown alone and with maize indicating that intercropping had no deleterious effect.

Organic farming in ginger

Approaches to organic cultivation of ginger involve following aspects (Potty and Krishnakumar, 1999; Arya, 2001).

1. Nutritional management through:
    - The application of FYM, oil cakes, vermi-compost, and bio-fertilizers
    - Raising leguminous cover and intercrops
    - Use of permitted fertilizers such as powdered rock phosphate and sulfate of potash
2. Insect-pest management through:
    - Cultural practices
    - Mechanical (manual) collection and removal of root grubs, beetles, and others
    - Use of biopesticides such as neem products, extracts of *Lantana* and incorporation of *Trichoderma* in soil
    - Use of bioagents such as *Bacillus thuringiensis* strain kurusakthy, into borehole of pseudostems to kill the larvae of the stem borer; *Trichoderma,* for control of other soilborne pathogens
3. Use of tolerant varieties
4. Disease management through practices such as crop rotation and adjusting planting time

5. Adoption of proper cultural practices such as providing drainage, shade, and timely removal of affected plants.

Pests and diseases

Diseases

Ginger is prone to many diseases of which rhizome rot, bacterial wilt, phyllosticta leaf spot and storage rots are major diseases that cause economic losses. In Kerala, the major ginger growing state, bacterial wilt and rhizome rot affect the crop and cause considerable yield losses. Major diseases, their causal organisms and control measures are given below:

Disease	Causal organism	Control measures
Soft rot	*Pythium sp*	Healthy rhizome selection, mulching beds with mahaneem (*Melia azadirachta*) (Das, 1999), soil solarisation, biological control using *Trichoderma* and chemical control by seed treatment, soil drenching with fungicides viz. Dithane M-45, Difolatan, metalaxyl, Bordeaux mixture etc.
Bacterial wilt	*Ralstonia solanacearum*	Selection of disease free rhizomes, soil solarization. Rhizome solarization (Kumar *et al.*, 2003). Crop rotation, biological control using *Pseudomonas fluorescens.*
Phyllosticta leaf spot	*Phyllosticta zingiberi*	Crop sanitation, providing shade, chemical control using Bordeaux mixture.

Pests

Shoot borer (*Conogethes punctiferalis*) and rhizome scale (*Aspidiella harti*) are major insect pests of ginger of which only shoot borer incidence is common in India. In Kerala up to 25 % damage was reported (Nybe, 2001). Pruning of freshly infested shoots during July to August at fortnightly intervals and chemical control using Malathion (0.1%) during September to October (at monthly intervals were effective for management of shoot borer.

Harvesting and processing

Harvest maturity varies according to end use. If the rhizomes are for vegetable use or for preparation of such products as ginger preserves and candy, harvesting should be done 4 to 5 months after planting. For dry ginger production and for distillation of oil or solvent extraction of oleoresin, harvesting is done 8 to 9 months after planting. Harvesting is delayed at higher elevations, in cooler climates and under irrigation.

Pawar and Patil (1990) observed that dry weight of rhizomes/ plant and green ginger yield were the highest when crop was harvested eight months after planting.

Patiram *et al.* (1995) reported that ginger is harvested twice in Sikkim. In the first stage during May and June, the mother or seed-rhizomes are harvested. This harvested product(*mau* in the local language) is of inferior quality. Main harvesting comes 7 to 8 months after planting and continues up to January. Rai and Anita (1997) recorded mother rhizome extraction practiced by local farmers for many years in the Hills of Sikkim and Darjeeling, India. By extracting mother rhizome, farmers get back their investment on seed rhizome even if there is severe out break of rhizome rot disease. If the disease do not appear, mother rhizomes yield 5 to 15 times higher than seed rhizomes, and hence are popular among low income group of farmers. There is 8-12 % loss in weight as compared with seed rhizomes sown in general. The mother rhizome has equal market value as freshly harvested ginger because of the large size of rhizomes (100-500 g) planted. The wound created while detaching mother rhizome may serve as an entry point for pathogens. Air-drying, packing in 250 gauge LDPE bags followed by irradiation at 60 Gy. kept rhizomes in good marketable condition for up to 2 months at ambient temperature without sprouting or significant loss of quality and <5 % weight loss. Singh *et al.* (1999), Nagaland, India, observed that Thinglaidum, Nadia and Khasi Local had the highest rhizome yields (more than 30 t/ha). The lowest rhizome yield was recorded in Tura and HP 666 (less than 20 t/ha). Thinglaidum, Nadia and Rio-de-Janeiro had the highest fibre and oil contents.

Storage of rhizomes

Ginger rhizomes being bulky and perishable, storage of seed rhizomes for 3-4 months from harvesting to next planting season is faced which many problems. Problems like rotting, sprouting, rooting and shrivelling occur resulting in huge loss. Adopting efficient storage technique in ginger will go a long way in minimising storage loss of the valuable planting material.

Trials conducted in Kerala showed that storing ginger in 200 gauge thick polythene covers of size 35 x 25 cm with 125 punch holes (each hole with 4 mm diameter) was an effective method (Jayachandran *et al.*, 1992).

In Kerala, traditionally ginger seed rhizome is stored with leaves of *Glycosmis pentaphylla* in wooden pole racks or sleeves and kept under shade. Rhizomes are also stored in pits dug under shade, floor of which is lined with sand or sawdust (KAU, 2003).

Rai and Hossain (1998) reported that three traditional methods of seed rhizome storage, namely, storage in soil pits, storage in a dry shady place, and storage in field involving delayed harvesting. The first method is the best for small scale growers, but expensive and laborious for large scale growers. Storage in a dry, shady place is economic for larger growers but there is a problem of rhizome drying. Storage in field by delayed harvesting is not to be encouraged as it harbours rhizome rot causing organisms and pests.

Processed products

Ginger is commercially available in various forms such as green ginger, dry ginger, ginger powder, ginger oil, ginger oleoresin and preserved ginger..

Dry ginger

Although there is substantial consumption of fresh ginger world wide, most of production is converted into dry ginger and traded. For dry ginger production, the crop is harvested at full maturity. Rhizomes are soaked in water over night and rubbed well for thorough cleaning. After cleaning, they are removed from water, outer skin is then scrapped off with a bamboo splinter or wooden knife having pointed ends. During peeling, it should be ensured that cortical parenchyma, which is rich in essential oil cells is not removed or cut, as it would cause loss of volatile oil. Scraping of ginger is a laborious and skilled work. Hand peeling is superior to mechanical peeling in getting a dried product of uniform appearance, size and color. Abrasive peeling using machines was tried by Ali *et al.* (1991) and Radha *et al.* (1993) and they got up to 80 % peeling efficiency. The peeled rhizomes are washed and dried uniformly under sun for one week. Drying is to be carried out on clean concrete floors or plain uncoated bamboo mats. Improved drying methods like mechanical and solar drier are also available (Bhuyan *et al.*, 1990). Rhizomes are dried to a moisture content of 8 to 10 % and stored properly to avoid infestation by storage pests. The driage is 16 to 25 % depending on cultivar. Ginger that is cleaned, peeled and dried with no other processing is commercially known as rough or unbleached ginger. This serves as starting material for preparation of ginger powder, oil and oleoresin.

Ginger oleoresin

Bleached ginger is produced by dropping scraped fresh ginger in a slurry of slaked lime, $Ca(OH)_2$, followed by sun drying. The process of dipping in slurry is repeated until rhizomes become uniformly white in colour. Dry ginger can also be bleached by this method. Liming gives ginger a better appearance and less susceptibility to attack or storage pests. However, major importing countries prefer unbleached ginger, since bleached ginger usually contains calcium beyond permissible limits (Spices Board, 1995).

Ginger powder

Ginger powder is made by pulverizing dry ginger to a mesh size of 50 to 60.

Ginger oil

On steam distillation, dried, cracked and comminuted ginger yields 1.0 to 3.0 % oil. The oil possesses aromatic odour but not pungent flavour of spice. Ginger oil can also be distilled from fresh ginger and from ginger scrapings.

Oleoresin ginger

Ginger oleoresin is obtained by extraction of powdered dried ginger with suitable organic solvents like alcohol, acetone or any other efficient solvent. It is commercially known as gingerin. Oleoresin is also obtained from fresh ginger, known as green ginger oleoresin. Micro encapsulated oil and oleoresin are also available.

Grading

Once ginger is dry, it is sorted and graded based on size of rhizome, its colour, shape, extraneous matter, presence of light pieces and extent of residual lime (in bleached ginger). Major grades are Garbled Non bleached Calicut (NGK), Ungarbled Non bleached Calicut (NUGK), Garbled Non bleached Cochin (NGC), Garbled Bleached Cochin (BGC), Ungarbled Bleached Calicut (BUGK) etc. In International trade, Jamaican ginger is the first grade, followed by Cochin ginger.

Uses

Ginger is perhaps the most widely used spice both for flavouring and for medicinal purpose. It is used in food preparations, confectionery, beverages, making ginger candy/preserves. In medicine it is considered as a carminative and stimulant. Ginger oil is also used as a flavourant, in pharmaceuticals and also in perfumery.

Turmeric
(*Curcuma longa* L., F. Zingiberaceae)

Turmeric, the sacred spice of Asian countries, is the rhizome of *Curcuma longa* L., a herbaceous perennial native to Indo Malayan region. Turmeric is essentially a tropical crop grown in India, Pakistan, Malaysia, Thailand, Philippines, Japan, China, Srilanka, Africa and Central America. It is the third important spice crop of India, next to chillies and black pepper. India is the largest producer and exporter of turmeric contributing about 80 % of production and 45 % of export. The crop occupies major share of area in Andhra Pradesh followed by Orissa, Tamil Nadu, West Bengal and Maharashtra. Kerala which produces the world renowned Alleppy finger turmeric contributes a negligible share of Indian turmeric production.

Turmeric is very much identified with human civilization, religion, customs and it finds use both in developed and under developed countries. Curcumin - the yellow colouring pigment present in rhizome is gaining importance with ban on artificial colours in food industry. In pharmaceuticals, it is valued for the anticancerous, anti-inflamatory and antiseptic properties. Demand growth rate of turmeric is around 10 % pointing to future prospects of turmeric cultivation in the country.

Morphology

Nearly 40 species of *Curcuma* are known to exist in India, while only one species *C. longa* contributes to commercial production of turmeric of commerce (96%). *C. aromatica* is cultivated to a lesser extent for 'Kasturi turmeric' used in cosmetics. Turmeric is a herbaceous perennial with underground rhizomes. The pseudostem is tall, robust with oblong/elliptic leaves narrowed at base. Flowering is stray and the species is a sterile triploid (3n = 63) and do not normally set viable seeds.

Flowering

Flowering in turmeric depends on cultivars and climatic conditions. It was observed that out of 19 types studied, 15 types flowered under Vellanikkara (Kerala) conditions. Percentage of flowering ranged from 1.2 to 95.28 per cent with maximum in Dindrigam Ca-69 and minimum in Rajpuri. Flowering was high in Amalapuram also. It was observed that all the seven types studied except Suvarna flowered within a period of 109 to 155 days. VK 70 was the earliest flowering cultivar while Kanthi and Sobha were late flowering types.

Flowering season extended from July to October. Turmeric inflorescence took 7.6 to 11.6 days for blooming after emergence of inflorescence. Duration of flower opening within an inflorescence lasted for 8 to 11 days. Anthesis started from 7 am and continued up to 9 am, maximum flowers opened before 8 am. Anther dehiscence took place between 7.15 and 7.45 am.

Inflorescence

Inflorescence in turmeric is a compound spike which is terminal on leafy shoot. Bracts are large sized, greenish white and the upper most tinged with pink and were spirally arranged. Rest of bracts carry one to three flowers and very rarely four in their axils. Mean number of flowers/inflorescence ranged from 26 to 35.2 (Nazeem and Menon, 1994; Renjith, 1999).

Flowers are yellowish white in colour, zygomorphic, bisexual, bracteolate and epigynous. Perianth is arranged in two whorls, consisting of six segments, making a gamosepalous calyx and a gamopetalous corolla. Calyx is tubular and short dividing above into three ovate/oblong lobes. Androecium consists of six stamens which are epipetalous arising from throat of corolla tube in two whorls. Outer whorl of three stamens is represented by petalloid trilobed labellum, the most conspicuous part of the flower. In the inner whorl, the posterior stamen is fertile and had a prominent two lobed anther having broad connective projected beyond the anther. The other two were reduced to staminodes and seen at base of flower. Style is elongated and slender lying in the groove formed by anther lobes and ends in stigma.

Ovary is inferior, syncarpous, trilocular and with axile placentation. Ovary is 0.26 cm in length and 0.24 cm diameter. Mean number of ovules/ovary is 29.31 with 611.44 μm length and 436.65 μm diameter. Fruit is a thick walled trilocular capsule with numerous arillate seeds. Arillate seeds have two seed coats, the outer thick and inner thin. Seeds are filled with massive endosperm and embryo is seen towards upper side of ovule.

Pollen grains and fertility

Pollen grains are ovoid to spherical in shape, light yellow and slightly sticky and measure to a mean length of 76.68 μm and breadth of 57.39 μm. Pollen grains show heterogeneity in size among cultivars. In turmeric, pollen grains have high degree of fertility. Stain test indicated pollen fertility ranging from 71.1 (Kodur) to 84.66 % (Kuchipudi). Mean pollen fertility in seven cultivars studied was 78.5 % with no significant variation among cultivars. Pollen fertility and viability varied depending on position of flowers in inflorescence. It was high in flowers at lower portion and low at middle and upper portions. Cultivar difference was noted with respect to pollen germination. Germination percentage in short duration cultivars ranged from 23.75 to 46.08 while in medium duration cultivars it ranged from 8.22 to 11.5. Among the various media tried for pollen germination *in vitro*, modified ME_3 medium at pH 6 was the best (Renjith, 1999).

Pollination and seed set

In turmeric, pollination is brought about by insects. Fruit set and seed set are not generally observed in turmeric. Nazeem and Menon (1994) observed seed set in crosses involving *C. domestica* and *C. aromatica* types. Renjith (1999) and Vijayasree (2001) reported seed set in crosses involving short duration cultivars of *C. domestica* but failed to obtain success in crosses involving short and medium duration cultivars. Fruits matured within a period of 20-22 days. Seeds/fruit ranged from 4 to 6. Pollen pistil interaction studies after *in vivo* cross pollination of short and medium duration cultivars showed pollen germination on stigma and pollen tube passing through style and reaching ovules. Lack of seed set could be due to other factors (Vijayasree, 2001). Lad (1993) observed seed setting in turmeric cultivars Ca-92/11 and Ca 17/1.

In vitro pollination and fertilization

Seed set was obtained in turmeric through *in vitro* pollination (Renjith, 1999, Vijayasree, 2001) using flowers on day of anthesis and one day prior to anthesis. Placental pollination was the best method. Modified ME_3 medium was used for pollen germination. The culture medium ½ MS + BAP 1 mg^{-1} + Kinetin 1 mg l^{-1} + NAA 0.5 mg^{-1} + 3 % sucrose along with organic supplements CW (15% v/v) and CH (200 mg l^{-1}) supported ovule development after *in vitro* pollination. In *in vitro* developed seed, endosperm development was not complete.

Crop improvement

Genetic resources

The genus *Curcuma* consists of about 117 species whereas from India, around 40 species are reported (Velayudhan *et al.*, 1999). Apart from *C. longa*, the common culinary turmeric, there are several other species of *Curcuma* which are mainly used as colouring agents, for production of arrow root, cosmetics and for medicinal purposes. Species such as *C. aromatica*, (Kasthuri manjal), *C. caesia* (Black turmeric), *C. amada, C. zedoaria, C. purpurescens, C. mangga, C. heyneana, C. xanthorrhiza, C. aeruginosa, C . phaeocaulis* and *C. petiolata* are also cultivated in different parts of India.

Varieties

In India, a number of cultivars are distinguished by names of localities where they are grown (Purseglove *et al.*, 1981). Menon (1975) found that turmeric produced in different localities vary in quality, Alleppey, Rajpuri, Guntur and Madras being popular trade names of Indian turmeric.

Turmeric types can be grouped into three, based on maturity period as short, medium and long duration types. Short duration types are known as Kasturi. They mature in seven months, rhizomes possess pleasant aroma, good yielders of dried turmeric and rich in volatile oil content but low in curcumin and used in culinary preparation. Flowering is common in these types and seeds produce gametic seedlings. Medium duration Kesari types (Bontha) which mature in eight months are referred as intermediary types and are high yielders of fresh rhizomes than Kasturi types and rich in curcumin and volatile oil. Long duration types mature in nine months and are moderately good both for rhizome yield and other quality constituents. Flowering and seed set are rare in medium and long duration types (Rao *et al.*, 1975 and Rao and Rao, 1988).

All short duration types are not in conformity with above statements. Released short duration varieties, Suguna and Sudharsana are low in curing percentage (12 to 14) but some what high in curcumin content (>5%) (Edison *et al.*, 1991).

Long duration types are economically more important accounting for 96.4 % of total area under turmeric cultivation and the remaining 3.6 being under short duration types (Pruthy, 1976).

Some of the popular Kasturi types of Andhra Pradesh are Kasturi Kothapeta, Kasturi Tanuka, Kasturi Amalapuram, Chayapasupu, Gummalakshipuram etc. They have an yield potential of 10 to 15 t of raw turmeric/ha. The curing percentage of Kasturi Kothapeta, Kasturi Tanuka and Kasturi Amalapuram is about 24 to 25 while in Chayapasupu it is about 19 and 14 in Gummalakshmipuram. Gummalakshmipuram has a high curcumin content of 5 % (Rao and Rao, 1994).

Popular Kesari types of Andhra Pradesh are Kesari Deevur and Amrutapani Kothapeta (Rao and Rao, 1994). Popular long duration types are Duggirala, Tekurpet, Mydukur, Armoor, Sugandham, Vontimitta, Nandyal, Avanigadda. They have an yield potential ranging from 15 to 30 t ha^{-1} and curing percentage 15 to 22.4 (Rao and Rao, 1994).

Erode and Salem are popular cultivars of Tamil Nadu with curcumin 3.9 and 4.5 % respectively (Muthuswamy and Ahmed-Shah, 1982).

Shillong and Tall Karbi are popular cultivars of Assam (Rathiah, 1986). Shillong is high yielder of fresh rhizomes (40 t ha^{-1}) and low in curing percentage (14) but good in curcumin (4.1%). Tall Karbi is also a good yielder (30 to 40 t ha^{-1}) but low in curing percentage (17.6) and curcumin content (3.06%).

Rajpuri and Evaigon are popular cultivars in Maharashtra and Gujarat. Rajpuri is a moderate yielder (20 t ha^{-1}) with 3.4 % curcumin content, while Evaigon is a high green yielder of (45 t ha^{-1}) with low curing percentage (17) (Pujari *et al.*, 1987).

Popular cultivars in Orissa are Duhgi, Jabedi and Katingia. Duhgi is a low yielder (10 t ha^{-1}) but has good curing percentage (26%) and poor in curcumin content (1.2%). Jabedi is low in curcumin content (1.2%), while Katingia has high curing percentage (36) but poor in yield (8 tonnes ha^{-1}) and curcumin content (1%) (Rao and Rao, 1994).

Gorakhpur, the popular cultivar of Uttar Pradesh has a green yield of 15 tonnes ha^{-1}, moderate curing outturn (18%) and curcumin content (3.5%) (Rao and Rao, 1994).

Alleppey turmeric has an yield potential of 25 t ha^{-1} and high curcumin content of 6.5 % but with slightly low curing percentage of 19 (Lewis, 1973 and Purseglove *et al.*, 1981).

Mannuthy Local from Central Kerala is a good yielder of rhizome with 24 t ha^{-1} and rich in curcumin content (7.58 %) (Philip, 1978). Superiority of Mannuthy Local cultivar was also reported by Kurian and Valsala (1995). They found that the local cultivars from Kuruppampadi (Ernakulam District) and Mupliyam (Thrissur District) are rich in curcumin content with 7.82 to 7.69 % respectively.

Turmeric produced in different localities varies in quality and price. Alleppey, Salem, Erode, Rajpuri and Nizamabad fingers are popular trade names of Indian turmeric. Velayudhan *et al.* (1994) classified 568 collections of *Cucurma longa* into 21 morphotypes based on aboveground and underground characters. Menon *et al.* (1992) evaluated 39 open pollinated progenies of turmeric variety, Nandyal and found significant differences among progenies with respect to all plant characters except tillers/plant, rhizome characters, yield, curing percentage and curcumin content. The list of released varieties with details are furnished in Table 23.

Genetic variability and correlation studies

Path analysis studies of 30 turmeric genotypes indicated that plant height, weight of primary rhizome, weight of secondary rhizome and number of secondary rhizomes / plant had direct positive effect on rhizome yield at genotypic level (Pandey *et al.*, 2003). They suggested that plant height, number of secondary rhizomes/plant, weight of primary and secondary rhizome/ plant and number of tillers/clump can be used as criteria for selection of better genotypes. Dash and Jana (2004) observed wide genetic diversity among turmeric genotypes. Pandey *et al.* (2003) in a multivariate analysis for divergence among turmeric germplasm for seventeen characters grouped the genotypes into seven clusters. No relationship was found between genetic divergence and geographical distribution of accessions.

Existence of wide variability in existing cultivars of turmeric in respect of yield components, yield and quality was reported by several workers (Philip, 1983; Philip and Nair, 1986, 1987; Philip *et al.*, 1982; Mohanty, 1979; Rao *et al.*, 1975; Pujari *et al.*, 1987). Mohanty (1979) studied genetic variability, heritability and expected genetic advance in 40 turmeric accessions in the germplasm under Pottangi conditions of Orissa. The fingers, rhizome yield, tiller production and plant height exhibited a greater proportion of apparent variability which was genetic and heritable. Degree of association between rhizome yield and yield components was also studied and it was found that plant height, number of leaves and leaf area were positively correlated with yield of rhizomes. Philip and Nair (1986) reported a significant and positive correlation of rhizome yield with plant height, petiole length and roots/plant and suggested that these components could be used as selection indices. Later, Philip and Nair (1986) carried out a similar study on 19 South Indian cultivars. In their studies, plant height, rhizome yield, curing percentage, curcumin content, oleoresin content and resistance to leaf blotch and leaf spot diseases recorded higher percentage of genetic advance indicating scope for selection. Number of tillers, number of nodes, internodal distance and resistance to shoot borer were weak indices for selection.

In an analysis of intercorrelations of coefficients among morphological characters and yield, Muralidharan *et al.* (1980) observed that number of tillers, plant height and number of fingers had significant correlation with yield. Number of tillers, number of leaves, plant height and leaf length and breadth recorded highly significant inter-correlation among themselves. Path coefficient analysis indicated that wherever significant positive correlations between yield and morphological characters were established, it was mainly due to substantial positive contribution by plant height and number of fingers either directly or indirectly. Based on this, it was concluded that plant height in turmeric is a single important morphological character for which selection for yield could be made. Estimation of oil and curcumin contents in different cultivars of *C. longa* and *C. aromatica* indicated that variability for oil and curcumin content was high in *C.longa* compared to *C. aromatica* and for identifying cultivars with high curcumin and oil content, it may be worthwhile to carry out selection in *C. longa* alone (Nambiar, 1979).

Mehta and Patel (1983) studied certain stability parameters for rhizome yield under Gujarat conditions. Genotypes (g), environment (e) and g x e interaction components were found significant based on phenotypic stability analysis. Among eleven cultivars studied, the highest mean yield of rhizome/ ha was obtained in Kesar (10.86 tonnes/ha) showing a stability parameter value greater than one. It was concluded that cultivars, Kesar, T.R. and Jugijan gave higher mean values than average with bivalue (varietal regression on environmental indices) greater than unity indicating that these cultivars will be able to perform well under better management conditions. Path coefficient analysis by Subramanian (1986) revealed that diameter of mother rhizome is the main determinant of yield in turmeric. It recorded significant and positive association with rhizome yield. Genetic analysis of yield components in two varieties revealed that diameter of mother rhizome had the highest percentage of habitability, genotypic and phenotypic coefficient of variations and high genetic advance. Geetha and Prabhakaran (1987) studied association of eighteen biometric characters with yield in 19 turmeric varieties under Kerala conditions. Their study revealed that the highest GCV was recorded by yield followed by nodes/mother rhizome. Yield also recoded the highest PCV. The highest heritability was noticed for girth of mother rhizome followed by plant height in which the latter recorded high genetic advance. Correlation studies indicated that all characters were positively related to yield. Path coefficient analysis showed plant height and length of secondary fingers to be major contributing characters as could be seen from Table 22.

Table 22. Path coefficient anlaysis in turmeric showing direct and indirect effects of characters on yield

Characters	LSF	GM	NPF	LPF	LFL	PL	HP	NLT	GCY
LSF	*1.6664*	0.0829	-1.0312	-0.0961	-0.3587	-0.3349	0.4420	0.1809	0.5513
GM	0.2538	*0.5444*	-0.7625	-0.1025	-0.4606	-0.8125	1.5607	0.4575	0.6783
NPF	1.4130	0.3414	*-1.2162*	-0.1166	-0.4177	-0.6979	1.1073	0.3389	0.8122
LPF	0.9199	0.3205	-0.8141	*-0.1742*	-0.5020	-0.2017	0.7834	0.3698	0.7016
LFL	0.9459	0.3968	-0.8038	-0.1384	*-0.6320*	-0.7715	1.3342	0.5378	0.8690
PL	0.4684	0.3613	-0.7123	-0.0295	-0.4092	*-1.1915*	1.7864	0.3836	0.6672
HP	0.4017	0.4635	-0.7743	-0.0744	-0.4599	-1.1610	*1.8333*	0.4708	0.5997
NLT	0.5314	0.4392	-0.7265	-0.1136	-0.5991	-0.8058	1.5214	*0.5673*	0.8143

Source: Geetha and Prabhakaran (1987)

Residual effect h = 0.6492

Diagonal entries in italics indicate direct effect of the relevant factors

LSF - Length of secondary Finger, GM - Girth of mother rhizome at centre,

NPF - Nodes/primary finger, LPF – Length of primary finger,

LFL - Length of first fully opened leaf, PL - Petiole length, HP - Height of plant,

NLT - No. of leaves/ tiller, GCY - Genotypic correlation with yield

From studies conducted by Jalgaonkar *et al.* (1986) under Konkan conditions of Maharashtra, it was found that significant variations existed among different varieties of turmeric in respect of growth characters, plant height, leaves/ plant, length of leaves, breadth of leaves and leaf area. Sugandham was the tallest among varieties studied while Waigon recorded the lowest height. No significant variation was noticed with regard to tillers/ plant. Leaves/plant was the highest in Erode variety and considerable variation among different varieties of turmeric were noted for leaf length, breadth and area.

Hybridization

Hybridization work is limited in turmeric as commercial cultivars are sterile triploids of *C. longa*. Cultivars of *C. aromatica* are tetraploids and set seed. Hybridisation between short duration and long duration (*C.longa*) turmeric cultivars was attempted to evolve short duration hybrid varieties with high curcumin content. In the 18 crosses, seed set ranged from 4.0 to 60.7 % (Nazeem and Menon, 1994). Seed set was noticed in eight of the 11 crosses tried. The crosses failed to set seeds in combinations involving short and medium duration cultivars.

Mutation breeding

Efforts were made to evolve improved varieties through mutation breeding which resulted in release of a few varieties, such as BSR-1, BSR-2, CO 1 and Suroma.

Biotechnology

Organogenesis and plantlet formation were achieved from the callus cultures of turmeric. Variations with high curcumin content were isolated from tissue cultured plantlets (Nadgouda *et al.*, 1982). *In vitro* microrhizome formation was reported in turmeric by Nirmalbabu *et al.* (2003). Protoplasts could be isolated successfully from leaf tissues as well as from cell suspension cultures of turmeric (Nirmalbabu *et al.*, 2006).

Crop management

Turmeric is successfully cultivated from sea level to an altitude of 1200 m above Mean Sea Level with an optimum range of 450-900 m. Well drained loamy or alluvial soils with good organic status are well suited. Moderate rainfall at sowing, fairly heavy and well distributed rainfall during growing period and dry weather about one month before harvest are more suitable. The crop cannot withstand water logging.

Planting materials

The crop is commercially propagated by vegetative means. Mother rhizomes and finger rhizomes are used for propagation. Mother rhizomes, when used as seed material, gave high yield over finger rhizomes. Comparing between 20 g, 40 g and 60 g seed weight, it was found that 60 g gave high yield but it has the disadvantage of increased seed rate. Whole mother rhizomes, cut mother rhizomes and fingers weighing 25 to 30 g of two varieties, Sugandham and Wynad Local were compared. Results indicated that use of whole mother rhizomes as planting material contributed towards maximum yield in both varieties. Split mother rhizomes were superior to fingers and gave more vigorous sprouts. For planting one hectare, 2500 kg rhizome is needed.

Type of rhizomes used for planting varies from region to region. In Andhra Pradesh, planting of mother rhizome is mostly preferred while fingers are used for planting in Tamil Nadu. Mother rhizomes, cut mother rhizomes and fingers are generally used for propagation. Research results favour use of mother rhizomes (Valsala *et al.*, 1988; Rao and Rao, 1988). Shanmugam *et al.* (2000) reported that sprouted primary rhizomes (14 days old) produced more vigorous plants and produced significantly higher yield than the unsprouted primary rhizomes and rhizomes at other stages of sprouting.

Micropropagation

Micro propagation of *Curcuma* sp. was first reported by Nadgauda *et al.* (1978) and later modified by Kuruvinashetty *et al.* (1982), Keshavachandran and Khader (1989), Balachandran *et al.* (1990), Rajan (1997) and Nirmalbabu *et al.* (1997).

Season

The land is prepared with receipt of early monsoon showers. Early planting results in more yield. In Kerala, planting in middle of May is ideal (KAU, 2003). Soil is brought to fine tilth by giving 3-4 ploughings. Beds of 1.0-1.5 m width, 15 cm height and of convenient length are prepared with spacing of 50 cm between beds. Planting can be done on ridges and furrows also. Trials confirmed that a plant to plant spacing of 25 x 25 cm was optimum for realizing maximum yield/unit area, rhizome bits are planted on small pits of 10 cm depth and covered with soil or dry powdered cattle manure. In ridge and furrow method, spacing recommended is 45-60 cm x 25 cm. In bed planting, mulching is done immediately after sowing with 15 t of green leaves/ha, which is repeated two months after planting after the earthing up.

Nutrition

For analysing nutritional status of crop with respect to N, P and K, third leaf from top was identified as diagnostic leaf. Regarding optimum age of plant for sampling, period between 90 and 120 days after planting was ideal (KAU, 2003). Being an exhaustive crop, turmeric requires heavy manuring

Table 23. Improved varieties of turmeric and their features.

Variety and Year of release	Pedigree/ parentage & Plant type	Institution/ University	Av. yield t/ha (fresh)	Salient features
Co - 1 1983	Vegetative mutant by x-ray irradiation of Erode Local	Dept. of Spices and Plantation Crops, HC & RI, TNAU, Coimbatore-641 003, Tamil Nadu	30.5 (Potl. yield 35.0)	Bold and bright orange yellow rhizomes, curcumin 3.2%, oleoresin 6.7%, essential oil 3.7%, dry recovery 19.5%, suitable for drought prone, hilly, saline and alkaline areas. Crop duration 270 days. Plants are robust, vigorous and taller (40-60 cm) with more no. of leaves (25-32) and tiller production 3.7 to 5.0
BSR – 1 1986	Clonal selection from Erode Local irradiated with x-rays	-do-	30.7 (Potl. yield 39.6)	Bright yellow rhizome, curcumin 4.2%, oleoresin 4.0%, essential oil 3.7%, dry recovery 20.5%, crop duration 285 days, suitable for drought prone areas of Tamil Nadu
BSR – 2 1994	Induced mutant from Erode Local	-do-	32.7	A high yielding short duration variety (245 days) with bigger rhizomes, resistant to scale insects
Krishna 1983	Clonal selection from Tekurpeta collection	Maharashtra Agrl. University, Kasba, Digraj, Maharashtra-416 305	9.2 (Potl. yield 11.8)	Plumpy rhizomes, curcumin 2.8%, oleoresin 3.8%, essential oil 2.0%, dry recovery 16.4%, duration 240 days. Moderately resistant to pests and diseases
Sugandham 1984	Clonal selection from germplasm	Spices Research Station GAU, Jagudan-382701 Mehsana, Gujarat	15.0 (Potl. yield 20.0)	Thick, round rhizomes with short internodes, Curcumin 3.1%, Oleoresin 11.0%, essential oil 2.7%, dry recovery 23.3%, duration 210 days. Moderately tolerant to pests & diseases
Roma 1988	Clonal selection from T. Sunder	High Altitude Research Station, OUA&T, Pottangi-764 039, Korapurt (Dist.), Orissa	20.7 (Potl. yield 40)	Suitable for both rainfed and irrigated condition, suitable for hilly areas and late season planting. Curcumin 6.1%, oleoresin 13.2%, essential oil 4.2% and dry recovery 31.0%, duration 250 days
Suroma 1989	Clonal selection from T. Sunder	-do-	20.0 (Potl. yield 44.9)	Round and plumpy rhizome, field tolerance to leaf blotch, leaf spot and rhizome scale, curcumin 6.1%, oleoresin 13.1%, essential oil 4.4% and dry recovery 26.0%, duration 253 days
Ranga 1992	Clonal selection from Rajpuri Local	-do-	31.3 (Potl. yield 37.1)	Bold and spindle shaped mother rhizome, suitable for late planting and low lying areas. Moderately resistant to leaf blotch and scales, curcumin 6.3%, oleoresin 13.5%, essential oil 4.4% and dry recovery 24.8%, duration 250 days
Rasmi 1992	Clonal selection from Rajpuri Local	-do-	31.3 (Potl. yield 37.5)	Bold rhizomes, suitable for both rainfed and irrigated condition, early and late sown season, curcumin 6.4%, oleoresin 13.4%, essential oil 4.4% and dry recovery 23.0%, duration 240 days
Rajendra Sonia 1989	Selection from Local germplasm	Dept.. of Hort., Tirhut College of Agriculture, RAU, Dholi-843121, Bihar	23.0 (Potl. yield 42.0)	Bold and plumpy rhizome, grows widely under all north Indian conditions. Curcumin 8.4%, essential oil 5.0% and dry recovery 18.0%, duration 225 days
Megha Turmeric 1996	Selection from Lakadong type	ICAR R.C. NEH Region, Shillong, Meghalaya	20.0 (Potl. yield 27.5)	Suitable for North East hill and North West Bengal. Bold rhizomes, high curcumin content 6.8% and dry recovery 16.37%, duration 300-315 days
Pant Peethabh 2001	Clonal selection from local types	Department of vegetable Science GB Pant University of Agriculture and Technology, Pantnagar-263 145, Dist. Udam Singh Nagar, Uttaranchal	20.0 (Potl. yield 29.0)	Long attractive fingers, curcumin 7.5%, essential oil 1.0%, dry recovery 18.5%, resistant to rhizome rot.

Contd...

Suranjana (TCP-2) 2000	Clonal selection from local types of North Bengal	Uttar Bengal Krishi Viswa Vidyalaya, North Bengal Pundibari (PO)-736 165, Dist. Cooch Behar, West Bengal	Potl. yield (29.0)	Suitable for open and shaded conditions, sole or intercrop, suitable for rainfed as well as high rain fall areas. Curcumin 5.7%, oleoresin 10.9%, essential oil 4.1%, dry recovery 21.2%, duration 235 days, tolerant to leaf blotch and rhizome rot. Resistant to rhizome scales and moderately resistant to shoot borer.
Suguna 1991	Selection from germplasm collection from Assam	Indian Institute of Spices Research, ICAR, Calicut-673 012, Kerala	29.3 (Potl. yield 60.3)	Short duration type (190 days), curcumin 4.9%, oleoresin 13.5%, essential oil 6.9% and dry recovery 20.4%, field tolerance to rhizome rot
Suvarna 1991	Selection from germplasm collection from Assam	-do-	17.4 (Potl. yield 43.5)	Bright orange coloured rhizome with slender fingers. Maturity 200 days, field tolerant to pest and diseases. Curcumin 4.3%, oleoresin 13.5%, essential oil 7.0% and dry recovery 20.0%.
Sudharsana 1991	Selection from germplasm collection from Singhat, Manipur	-do-	28.8 (Potl. yield 54.9)	High yielding variety, short duration type (190 days). Field tolerant to rhizome rot. Curcumin 5.3%, oleoresin 15.0%, essential oil 7.0% and dry recovery 20.6%.
IISR Prabha 1996	Open pollinated progeny selection	-do-	37.47	High yielding variety, curcumin content 6.5%, oleoresin 15.0%, essential oil 6.5% and dry recovery 19.5%, crop duration 205 days.
IISR Prathiba 1996	Open pollinated progeny selection	-do-	39.12	High quality line, 6.2% curcumin content with high yield, 16.2% oleoresin, 6.2% essential oil, 18.5% dry recovery, crop duration 225 days.
IISR Alleppy Supreme 2004	A clonal selection from Alleppy turmeric	-do-	35.45 5.58 (dry)	Shows tolerance to leaf blotch disease. Rhizomes contain 5.55% curcumin, 16.0% oleoresin, 19.0% dry recovery, crop duration 210 days.
IISR Kedaram 2004	Clonal selection from germplasm	-do-	34.5 5.28 (dry)	Tolerant to leaf blotch disease, Rhizomes contain 5.5% curcumin, 13.6% oleoresin, maturity 210 days and 18.9% driage.
Kanthi 1996	Clonal selection from Mydukur variety of Andhra Pradesh	Dept. of Plantation Crops and Spices, College of Hort., KAU, Vellanikkara-680 656, Trichur, Kerala	37.65	Erect leaf with broad lamina, big mother rhizomes with medium bold fingers and closer internodes. Medium duration. Curcumin content (7.18%), oleoresin 8.25%, essential oil 5.15%, dry recovery 20.15%, duration 240-270 days.
Sobha 1996	Clonal selection from local type	-do-	35.88	Mother rhizome big with medium bold and closer internodes. Inner core of rhizomes is dark orange like Alleppey. More tertiary rhizomes. Dryage 19.38%, curcumin content (7.39%), oleoresin (9.65%), essential oil (4.25%), medium duration 240-270 days.
Sona 2002	Clonal selection from local germplasm	-do-	4.02 (dry)	Orange yellow rhizome, medium bold with no tertiary fingers. Best suited for central zone of Kerala. Rhizome medium bold, field tolerant to leaf blotch. Curcumin 7.12%, essential oil 4.4%, oleoresin 10.25%, 18.88% dry recovery, medium duration. 240-270 days.
Varna 2002	Clonal selection from local germplasm	-do-	4.16 (dry) (Potl. yield 6.37)	Bright orange yellow rhizome, medium bold with closer inter nodes, tertiary fingers present. Suited to central zone of Kerala. Field tolerant to leaf blotch, curcumin 7.87%, essential oil 4.56%, oleoresin 10.8%, 19.05% dry recovery, medium duration 240-270 days.

Source: Johny and Ravindran (2005)

for higher yield. Manurial and fertilizer requirements vary according to agroecological situations prevailing in a particular locality.

Saifudeen (1981) observed that crop growth period of turmeric can be divided into four phases viz. a phase of moderate vegetative growth, a phase of active vegetative growth, a period of slow vegetative growth and a phase approaching senescence. Generally, as dry matter production increases, uptake also increases and phase of active vegetative growth is also the period during which maximum uptake of nutrients takes place. It was observed that uptake of nutrients was higher upto 3rd month for K, upto 4th month for N and upto 5th month for P with subsequent decrease. The crop attains maximum vegetative growth during 4th to 5th month suggesting need for earlier application of N, P and K for increasing plant growth (Rao and Rao, 1988). Among major nutrients, nitrogen was essential for boosting up vegetative growth and thereby yield. Phosphorus was found necessary for root development while K was involved in transportation of carbohydrates to rhizomes (Muralidharan and Balakrishnan, 1972). In addition, K may also improve curcumin content (Mohanbabu and Muthuswamy, 1984). However, Ahmedshah *et al.* (1988) reported that there was little response to incremental doses of potassium upto 180 kg/ha in respect of morphological and rhizome characters of turmeric. Among micronutrients, iron deficiency was observed in turmeric grown on calcareous and alkaline soils. In that case application of $FeSO_4$ at 30 kg/ha recorded the highest yield of rhizomes followed by foliar spray of $FeSO_4$ (0.5%) during 3rd, 4th and 5th month (Balashanmugam *et al.*, 1990). In the case of zinc deficiency, application of $ZnSO_4$ at 15 kg/ha is recommended.

Organics

Turmeric responds to heavy dressings of organic matter and many experimental evidences are available on beneficial effects of organic matter either alone or in combination with inorganic fertilizers on growth and productivity of turmeric. Usually large quantities of organic manures in the form of FYM, oilcakes and green leaves (as mulch) are applied in different turmeric growing states. Recommendations differ according to variety, soil type, soil fertility and climatic conditions. Use of biofertilizers like Azospirillum was reported to reduce, 50 % of recommended dose of nitrogenous fertilizers besides increasing fresh rhizome weight by 20.68 % over recommended dose of fertilizer (Subramanyam *et al.*, 2004).

Nutrient management

Response to applied nutrients in turmeric is governed by a number of factors viz. soil type, nature of planting material used, variety, spacing, cropping system adopted etc.

Tayde and Deshmukh (1986) observed significant interaction between planting materials and N levels. When finger rhizomes were used, the highest level of nitrogen (200 kg/ha) was required to get maximum yield. But when mother rhizomes were used, application of 100 kg N/ha was sufficient. Nitrogen application could be reduced by 25 to 50 % when mother rhizomes were used. A three year study in Andhra Pradesh revealed positive interaction between varieties and nutrients (Sankaraiah and Reddy, 1987). A fertilizer trial conducted in Karnataka revealed that total uptake of nutrients was maximum (193:39:257 NPK kg ha^{-1}) in Suguna followed by Sudarshana and Accession 295 (187:43:231 NPK kg ha^{-1}) (Mohan *et al.*, 2004). Similarly, in turmeric based cropping systems it is necessary to apply extra doses of nutrients to meet requirements of component crops (Sivaraman, 1992). Response to fertilizer application also varied with shade cast by the taller component crops. Jayaraj (1990) studied the response of turmeric to different levels of K (60, 75, 90 and 120 kg/ha) under four shade intensities (0, 25, 50 and 75%) and time of application (complete as basal and split application). Turmeric performed well under 25 % shade with 90 kg K_2O/ha when entire quantity of K was applied in a single dose. Curcumin content was also found maximum at 25 % shade and showed an increase with increase in K level.

Fertilizer recommendations

Muralidharan and Balakrishnan (1972) recommended a fertilizer dose of 100-100-120 kg/ha of NP&K under Wynad conditions of Kerala for variety Armoor under rainfed condition (pH of soil 4.8-5.1). In Meghalaya, 60:40:60 kg/ha of NP&K was recommended by Govind *et al.* (1990). Sankaraiah and Reddy (1987) recommended 250-80-200 kg/ha of NP&K for variety Duggirala.

Kerala Agricultural University recommends 30-40 t of FYM/ha supplemented with N, P_2O_5 and K_2O at 30:30:60 kg ha^{-1}. Full dose of P_2O_5 and half dose of K_2O are applied as basal dose. Two third of N is applied one month after planting and the rest of N and K_2O are applied 60 days after planting. Sreenivasan and Rema (2002) recommended fertilizers @ 60:50:120 ha^{-1} for more yield.

Intercropping

Turmeric could be grown both as a sole crop and as inter / mixed crop. However crop rotation is essential as it depletes more nutrients from soil. Turmeric is rotated with chillies,sesame, little millet etc. in rainfed conditions and finger millet, ground nut, maize and vegetables under irrigated condition. As an annual spice crop, turmeric is ideal for intercropping in coconut gardens. Varietal adaptability to shade levels was the prime factor governing yield as an intercrop. Under rainfed conditions, cultivars Sugandham, Wynad Local and T. Sundar were suitable for cultivation in coconut garden (Paul, 1992). Economics of an intercropping experiment in Tamil Nadu (Kumar *et al.*, 2002) revealed that coconut + turmeric gave a net profit of Rs.24, 071/ha whereas sole crop gave a net profit of Rs.22,075/ha. High curcumin varieties from Kerala Agricultural University viz. Kanthi, Shobha, Sona and Varna performed well as intercrops in coconut plantations (Nybe and Miniraj, 2005).

Shade response

Response of turmeric to different shade levels (0, 25, 50 and 75%) was studied. The highest yield was recorded at 50 % shade. The crop is classified as shade loving and it can be successfully intercropped in coconut garden. Uptake studies showed that fertilizer requirement of turmeric shaded to 50 % was about 20 % more than the general recommendation for sole crop in open.

Varietal adaptability to shade levels was the prime factor governing yield of turmeric as intercrop. Sugandham, Wynad Local and T. Sunder were adapted for cultivation in coconut garden under rainfed condition. Performance of different cultivars (PCT 5, BSR-1, Ethamukulam, PTS 9, PCT 8, PTS 38) under varying shade levels (0, 25, 50 and 75%) and fitness of these cultivars as intercrop in coconut garden were studied. Though no significant difference was observed among cultivars at different shade levels, all cultivars gave the highest rhizome yield at 50 % level reaffirming its shade loving nature. Yield parameters such as dry matter production and harvest index were also more at 50 and 75 % shade, respectively. PCT-8 and PTS-9 performed better both under shade and in the open. BSR 1 showed the best performance as intercrop in coconut garden.

Screening of twelve turmeric varieties for shade tolerance under varying shade levels (0, 25, 50 and 75%) was done. All the varieties tested recorded the highest yield at no shade, suggesting that response to shade depends on varieties. Yield parameters were also the highest in open whereas harvest index was not altered by shading. Significant interaction was noticed between shade levels and varieties on rhizome yield. Six varieties were selected as superior for all situations. Varieties PCT 5, PTS 24, BSR 1 and Mydukur were selected for 0, 25, 50 and 75 % shade, respectively.

Use of growth regulator

Effect of three growth regulators MH, CCC and Kinetin were studied and it showed significant influence in reducing leaf area and increasing number of primary fingers. At 50 ppm, MH decreased leaf area. Number of primary fingers was significantly increased by MH at 25 and 50 ppm and kinetin at 125 ppm whereas CCC at 1000 and 1500 ppm showed a decreasing trend. Yield of rhizomes was not influenced by growth regulators while kinetin 75 and 100 ppm, CCC at 1000 and 1500 ppm and MH at 50 ppm increased curcumin content of rhizomes.

Pests and diseases

Shoot borer (*Conogethus punctiferalis*) is the most serious pest of turmeric. Spraying malathion (0.1%) or monocrotophos (0.075 %) at 21 days interval during July-October is effective in controlling pest infestation. Rhizome scale (*Aspidiella hartii*) infestation is also reported on a small scale.

Leaf blotch caused by *Taphrina maculans*, leaf spot by *Colletotrichum capsici* and rhizome rot by *Pythium graminicolum* are the diseases reported. Bordeaux mixture (0.1%) or mancozeb (0.2%) application control these diseases.

Harvesting and processing

Depending upon variety, the crop becomes ready for harvest in 7-9 months after planting. Early varieties mature in 7-8 months, medium varieties in 8-9 months and late varieties after nine months. Harvesting at proper maturity results in the highest yield of cured turmeric. The land is ploughed and rhizomes are gathered by hand picking. Harvested rhizomes are cleaned off mud and other extraneous matter adhering to them. Yield/per hectare comes to 20-25 t of green turmeric. High yielding varieties yield up to 35 t/ha.

Curing

Fresh turmeric is cured to get dried commercial product. Fingers are separated from mother rhizomes. Mother rhizomes are stored for seed purpose. In the traditional method of curing, fresh rhizomes are boiled in water just enough to immerse them. Copper/ galvanised iron/ earthern vessels are used for the purpose. Boiling is stopped when froth comes out and white fumes appear giving out a typical odour. At this point (about 45-60 minutes) rhizomes turn soft. The stage at which boiling is stopped largely influences color and aroma of the final product. Over cooking spoils color of final product while under cooking renders dried product brittle. In improved method of curing, cleaned fingers (approximately 50 kg) are taken in a perforated trough of 0.9 m x 0.5 m x 0.4 m size made of GI or MS sheet with extended parallel handle. Perforated trough containing fingers are then immersed in a pan, 100 litres of water is poured into the trough to immerse turmeric fingers. Cooked fingers are taken out of the pan by lifting the trough so as to drain water into the pan. Water used for boiling can be used for curing fresh samples (Sreenivasan and Rema, 2002). Processing is to be done two or three days after harvesting. If there is delay in processing, rhizomes should be stored under shade or covered with saw dust or coir dust.

Drying

Cooked fingers are dried in sun by spreading them in 5-7 cm thick layers on bamboo mats or on drying floor. During night, rhizomes should be heaped or covered with any material that provides aeration. It may take 10-15 days for drying. Artificial drying, using cross-flow hot air at a maximum temperature of 60°C also gives a satisfactory product. In sliced turmeric, artificial drying gives a brighter coloured product than sun drying which tends to undergo surface bleaching. Yield of dry product varies from 10-30 % depending upon variety and locality.

Polishing

Dried turmeric has a poor appearance and a rough dull outer surface with scales and root bits. Appearance is improved by polishing outer surface by manual or mechanical rubbing. Turmeric is also polished in power operated drums. Yield of polished turmeric from the raw material varies from 15-25 %.

Colouring

Colour of processed turmeric influences price of produce. For an attractive product, turmeric powder may be sprinkled during last phase of polishing.

Grading

Agmark grades are prescribed for whole turmeric and turmeric powder separately for export and for internal trade. Alleppy Finger Turmeric, Rajapore Finger Turmeric, Turmeric bulbs etc. are a few of the popular grades.

Processed products

Turmeric powder, volatile oil (5-6% from dry rhizomes) and turmeric oleoresin are the processed products made from turmeric. All the three products are exported from India. Turmeric oleoresin contains 30-35% curcumin, the main colouring pigment. New generation products based on turmeric oleoresin include spice emulsions/ aqua resins and encapsulated water-dispersible powders.

Uses

Turmeric (whole) is a unique, colourful and versatile natural plant product combining properties of a spice or flavourant, a colourant, a cosmetic and a drug useful against in a number of diseases. The largest quantity of turmeric is utilized as a food adjunct and colourant. It is also used as a dye. In Indian system of medicine turmeric occupies an important place, as an ingredient in the preparation of medicinal oils and ointments. It is a stomachic, carminative, tonic, blood purifier, vermicide and an antiseptic. It is also used in cosmetic preparations. Curcumin is an antioxidant and finds application in modern drugs for cancer therapy.

Chilli
(*Capsicum annuum* Linn., F. Solanaceae)

Chillies are the dried ripe fruits of species of genus *Capsicum*. They are also called red peppers or capsicums and they constitute an important well known commercial crop used as a condiment, culinary supplement or as a vegetable. Among the spices consumed per head in India, dried chillies contribute a major share. *Capsicum annuum*, originated in the South America, is mainly cultivated in tropical and subtropical countries: Africa, India, Japan, Mexico, Turkey and the USA. Chilli was known to Indians about 400 years ago, when this crop was first introduced to India by Portuguese towards the end of 15th century. Within a span of over four centuries it has become a major crop in India covering almost all states of the country. Chillies have adapted very well to the Indian conditions so much so that India is considered as the secondary centre of origin. There are many varieties and types of chilli, differing in habit and size, shape, colour and pungency of fruit. Among these, paprikas with mild pungency cultivated mainly in European countries and chillies with higher pungency grown in tropical countries are important. Hungary is by far the most famous paprika producing country. In Indian Horticulture literature, chilli is mostly grouped under vegetables and hence only a brief description is made here.

India leads in area and production of chilli but the average yield in India is low as compared to USA, South Korea and Taiwan. The major chilli growing states are Andhra Pradesh, Karnataka, Tamil Nadu and Maharastra. Andhra pradesh has been leading both in area and production contributing on an average 25 % of the total area and over 40 to 50 % of the total production. India has emerged today as the foremost producer and exporter of chillies. India exports both whole chilli and chilli products to various countries.

Variability

The genus *Capsicum* originated in the new world tropics and subtropics. Within the genus *capsicum* five species are commonly recognized *viz.*, *C. baccatum*, *C. pubescens*, *C. annuum*, *C. frutescens* and *C. chinense*. The first two species are taxonomically clearly different, but the remaining three species together is currently treated as a species complex.

C. annuum

This is the most important species on a world scale since it includes all the commercially important types viz., red pepper, cayenne, paprika, chillies and sweet pepper.

C. frutescence

Also known as bird chilli or tabasco chilli, these are grown or exported throughout Latin America. These are extremely pungent, suitable for oleoresin extraction. In India we find them growing in coastal Kerala, Karnataka, Maharashtra, Andhra Pradesh and North Eastren states.

C. chinense

This is the most variable of the domesticated species of *Capsicum*. It is widespread within Latin America. In India we find *C. chinense* in Kerala and North Eastern states.

C. baccatum

This species is grown on a garden and field scale in South America. This species is not known in India, outside the research stations. It has very good yield potential and useful genes against powdery mildew and anthracnose (Deshpande, 2005)

C. pubescens

This is a high land pepper found throughout the Andean region and high lands of Middle America. The species is very much restricted in its distribution and the variability is very much limited.

Botany and floral biology

Capsicum is a variable sub-shrub. There are distinct morphological features which distinguish the various species. Chromosome number n=12. There are white flowered and purple flowered types. Chilli is an often cross pollinated crop. Extent of cross pollination is 63 %. Majority of the flowers open at 5 am. Stigma is receptive from a day earlier to anthesis and continous for two days after anthesis. Pollen grains are fertile a day before anthesis with maximum fertility on the day of anthesis. The extent of cross pollinaton depends on flower structure which varies from variety to variety.

Chemical composition

The main source of pungency in pepper is the chemical group of alkaloid compounds colled capsaicinoids (CAPS) which are produced in the fruit. Capsaicin is the most abundant CAPS, followed by dihydrocapsaicin and others. Varieties of chilli differ widely in CAPS content.

The pungency level of Indian chillies from different places was compiled by Spices Board (2000).

Type	Location	Capsaicin (%)
Bird eye chillies	Mizoram & parts of Manipur	0. 589
Byadgi (Kaddi)	Dharwad (Karnataka)	Less or no pungency
Elachipur Sannam (S4 type)	Amarawati (Maharashtra)	0. 200
Guntur Sannam (S4 type)	Guntur (AP)	0. 226
Hindpur (S7)	Hindpur (AP)	0. 240
Jwala	Kheda & Meshana (Gujarath)	0. 400
Kanthari (white)	Kerala & Tamil Nadu	0. 504
Kashmiri chilli	HP, J & K	0. 325

The amount of CAPS in a given variety can vary depending on the light intensity and temperature at which the plant is grown, the age of fruit and the position of the fruit on the plant. Red colour of mature pepper fruits is due to several related carotenoids pigments including capsanthin (30-60 %), capsorubin (6-18 %) etc. Intensity of red colour is primarily a function of the amount of these pigments; the Hungarian and Spanish varieties used for paprika have very high amounts of capsanthin and capsorubin compared to other varieties (Govindarajan, 1985).

CAPS in oleoresin are very stable compounds and generally do not breakdown even during processing at high temperature and during storage. CAPS in dry products (fruits, powder etc) are not as stable as in oleoresin. The temperature at which fruits are dried affects the CAPS content.

Crop improvement

Being an often cross pollinated crop, considerable variability exist in chilli. Excellent cultivars/land races are available locally in many parts of India. Quality of chilli fruit is judged by the pupose for which it is used. Smooth glossy pods with medium pungency, bright red colour, thin pericarp and tight pedicel are ideal for whole red chilli purpose (Gopalakrishnan and Indira, 2006). 'Byadgi' chilli from Karnataka gives bright red colour, mild pungency and is suitable for export. Mathur *et al.* (2000) reported Tezpur cultivar (*C . frutescens* var. naghari), as the hottest chilli known to date. It contains the highest amount of capsaicin and dehydro capsaicin (4.28 and 1.42 % w/w respectively), contributing to a pungency rating of 855000 SHU (Scoville Heat Units). Under the All India Co-ordinated Vegetable Improvement Project (AICVIP) chilli germplasm is maintained at various co-ordinating centers which is evaluated every year, leading to release of improved selections.

Genetic resources

National Bureau of Plant Gentic Resources, IARI, New Delhi, IIHR, Bangalore, Indian Institute of Vegetable Research, Varanasi and State Agricultural Universities maintain exhaustive collection of chilli. In addition, the following co-ordinating centres of AICRP (vegetables) are also entrusted with the task of collection and maintenance of chilli germplasm.

S. No.	Centre	Number of germplasm
1.	TNAU, Coimbatore	89
2.	ANGARU, Hyderabad	246
3.	KAU, Vellanikkara	16
4.	BCKV, Kalyani	40
5.	UAS, Dharwad	40
6.	AAU, Jorhat	35
7.	IIVR, Varanasi	110

Main breeding objectives in chilli are ripe fruit colour, shape, size, pungency, retention of colour on storage, pericarp thickness, seed content and resistance to diseases (virus complex, anthracnose, fruit rot, powdery mildew, cercospora leaf spot and fusarium wilt) and pests (thrips, aphids, mites). In India majority of the improved varieties are developed either through pureline selection or pedigree method of breeding. Mutation breeding has also been attempted successfully. Besides, emphasis is also laid on development of F1 hybrids. Most of the crop improvement work is carried out at Indian Agricultural Research Institute, Millet Research Station, Lam, Guntur, Chilli Research Station, Kovilpatti (Tamil Nadu) and various State Agricultural Universities. Besides varieties, there are ample numbers of F1 hybrids in chilli marketed by various private seed companies. Table 24 gives the description of improved chilli varieties developed in the public sector in India.

Table 24. High yielding varieties of chilli developed in India

S. No.	Institute	Varieties
1	Tamil Nadu Agricultural University, Coimbatore	K 1, K 2, Co 1, Co 21984, PKM-1, Co 3, Co 4, PMK-1, PLR-1, MDU-1
2	Indian Agricultural Research Institute, New Delhi	Pusa Jwala, Pusa Sadabahar, NP 46 A, Pusa Deepthi
3	University of Agricultural Sciences, Bangalore	KDC-1, HCH-9646, GPC-82
4	GB Pant University of Agriculture and Technology, Pantnagar	Pant C-1, Pant C-2, Kalyan Selection. 1
5	Acharya N G Ranga Agricultural University, Hyderabad	G 1 1962, G 2 1962, G 3 1962, Bhagya Laxmi, Andhra Jyothi, Sindhur, Apsara, Bhaskar, Prakash, LCA-305
6	Panjab Agricultural University	Punjab Lal, CH-1, CH-2, CH-3
7	Mahatma Phule Krishi Vidyapeeth, Rahuri	Musalwadi, Phule Jyothi, Phule Suryamukhi, RHRC, Shankeswhar, Agni Rekha, AKC-86-39
8	Jawaharlal Nehru Krishi Vishwa Vidyalaya, Jabalpur	Jawahar-218, JCA-154
9	Indian Institute of Horticultural Research, Bangalore	Arka Harita, Arka Lohit, Arka Meghana, Bharath
10	Orissa University of Agriculture and Technology, Bhubaneswar	BC-14-2
11	Kerala Agricultural University, Vellanikkara	Jwala Sakhi, Jwala Mukhi, Ujjwala, Anugraha
12	Haryana Agricultural University, Hissar	Hissar Vijay, Hissar Sakthi

Variability studies

Majority of the crop improvement programmes in chilli pertains to performance analysis of local germplasm, improved varieties and culture.

Heritability studies, gene action and association and inheritance of characters were studied in chilli by many workers. Direct influence of fruit length and fruit weight on yield was observed by Korla and Rastogi (1977) and Jose (2001). As a prerequisite to heterosis breeding programme, Rani (1977) and Indira (1997) assessed genetic distance between genotypes. Mathew (2004) successfully utilized molecular characterization using RAPD analysis for the estimation of genetic divergence in chilli. Cherian (2000) and Manju (2001) reported considerable morphological variability in *C. chinense* with respect to duration, growth habit, branching, flower and fruit colour, shape, pungency etc. This perennial species, known as 'mali mulaku' is gaining importance because of export potential to Maldives. Red type has high content of capsanthin (32.29 ASTA units) and yellow type has high capsaicin (0.69%) (Thomas *et al.*, 2004). Variability within *C. frutescens* is limited (Sheela, 1998). Gene action of important polygenic character and inheritance of discrete characters were studied by many workers.

Manifestation of heterosis was reported by Gopalakrishnan (1985), Phillippose (1986), Pradeepkumar (1990) and Seena (1996). Chilli hybrids CH - 1 and CH - 3 developed by PAU, Ludhiana revolutionized chilli cultivation in Punjab.

In the resistance breeding programme, resistant genes were reported in *C. baccatum* (Anthracnose, *Phytopthora* and CMV); *C. chinense* (Anthracnose, TMV, leaf curl virus and verticillium wilt) and *C.frutescens* (leaf curl virus). List of varieties reported resistant/ tolerant to pests and diseases are given below (Gopalakrishnan and Indira, 2006):

Disease/ pest	**Variety/ Culture/ Species**
Virus leaf curl/ Mosaic	LCA-305, PBC 555, PBC 830, Phule Jyothi, Phule Suryamukhi, Pusa Sadabahar, Pant-C-1, Pant-C-2, Punjab Lal, Jawahar-218, HC 1, HC 15, HC 22, HC 28, RHRC Clustering Erect, LCA 248, LCA 218, LCA 154, NP 46 A, Pusa Jwala
Bacterial wilt	Ujjwala (CA 219), Anugraha, Manjari (CA 33)
Die Back/ Anthracnose	Musalwadi, Pusa Sadabahar, Perennial, CA 87 of TNAU, Achar
Powdery mildew	Musalwadi, Phule Jyothi, Phule Suryamukhi, Arka Lohit, IIHR 1238, PMR 10/95 K
Fruit rot	Jawahar-218, PBC 36
Leaf curl complex	KDC 1, GPC 82, GPC 69, GPC 80, Jwala
Thrips/ mites	Andhra Jyothi, Bhaskar
Thrips	K 2, Chamatkar, Guchedar
Root knot nematode	K 2
Mites	LCA 235, LCA 230, LIC 13, LIC 45, Yellow Anther Mutant
Multiple reistance	Perennial, Punjab Lal, Lorai, Indonesian Selection, Tiwari, Laichi 2, LS 111 CH 1

Markose (1996) studied anatomical and biochemical basis of resistance to bacterial wilt and found that total phenol, OD phenol, protein content and enzyme activities had positive association with resistance.

Biotechnology

Micropropagation and plant regeneration in chilli was reported using various explants (Christopher and Rajan, 1994). Most of the *in vitro* culture experiments in chilli is aimed either at production of secondary metabolites in cell cultures or utilization of culture techniques for genetic manipulation (Nirmalbabu *et al.*, 2006). George and Narayanaswamy (1973) developed haploid capsicum through androgenesis. Organogenesis and plant regeneration from isolated protoplasts was reported by Saxena *et al.* (1981) and Anu *et al.* (2004) reported somaclonal variation in chilli. Sivegowda *et al.* (2002) reported *in vitro* regeneration and transformation in *C . annum*. Regeneration of transgeneic chilli plants resistant to TMV and CMV was reported by Ping *et al.* (1999). Tai and Starkawicz (2000) constructed an yeast artificial chromosome libraury of *C. annum*.

Crop management

Soil and climate

Chilli prefers well drained loamy soils in the pH range of 6.5-7.5 for optimum growth. Acidic and alkaline soils are not suitable for growth. Red loamy and black soils are suitable for cultivation. The ideal temperature required is 20-25°C. Warm humid conditions improve growth and dry condition enhances maturity. Pungency increases when night temperatures are high. Chilli is grown as rainfed as well as irrigated crop.

Seeds and sowing

Chilli is a transplanted crop. Seeds are sown in nursery and one month old seedlings are transplanted to main field. Seed rate is 0.50-1.0 kg/ha. For sowing seeds, raised seed beds of 90-100 cm width and of convenient length are prepared after incorporating well decomposed organic matter. After sowing seeds, mulch the beds with green leaves and irrigate daily. Remove mulch immediately after germination of seeds. Restrict irrigation one week before transplanting and irrigate copiously on previous day of transplanting. Seedlings are retained in the nursery for 40-50 days.

Field planting

The best season for starting rainfed crop is May-June. Irrigated crop can be started during October-November and January-February. Seedlings are transplanted in shallow trenches for irrigated crop or on ridges/level lands during rainy season. Transplanted seedlings may be given shade for three to four days during summer months. Spacing followed is 45-60 cm x 45-60 cm, depending on the variety.

Manuring

The crop responds well to manures and fertilizers. Apply well rotten FYM / compost @ 20-25 tonnes/ha at time of land preparation and mix well with soil. A fertilizer dose of 75:40:25 kg N:P_2O_5:K_2O/ha may be given. Half of nitrogen, full phosphorus and half of potash may be applied as basal dose before transplanting. One fourth of nitrogen and half of potash may be applied 20-30 days after planting. Remaining quantity may be applied two months after planting.

After cultivation

Irrigate at three to four days interval during summer. Weeding follow by fertilizer application and earthing up may be done at one and two months after planting.

Pests and diseases

Thrips, aphids and mites are main pests. Damping off, anthracnose, leaf curl, mosaic and powdery mildew are major diseases. Continous cultivation of the crop in an area also lead to serious icidences of bacterial wilt (Gopalakrishnan *et al.*, 1998).

Harvesting and processed products

Crop starts yielding green chillies 40-50 days after transplanting. Dry chillies can be harvested 70-80 days after transplanting. Fruits are hand picked. Yield of green and dry chillies is 7.5-10 t/ha and 0.75 to 1 t/ha for rainfed crop and 20-25 t/ha and 2-2.5 t/ha for the irrigated crop.

Ripe chillies are harvested and sun/machine/thermal dried to a moisture content of 10 %. Diseased and discoloured fruits are discarded for obtaining good quality. For export market, quality standards have to be strictly adhered to. The product should be free from pesticide residues, heavy metals, aflatoxins, residual solvents, yeast and mould. Paprika chilli is to be stored in a cool dry place for retention of color. Chilli powder, chilli oleoresin (6.8 to 18.0%) and paprika oleoresin are other value added products from chilli.

Paprika

Chilli fruits having deep red colour, without pungency are known as paprika. Paprika powder and paprika oleoresin are in great demand in the international market. Colour is the principle criterion in assessing the quality of paprika. The pigment content of paprika powder ranges from 0.1 to 0.8%. Best paprika contains pungent matter ranging from 0 to 0.0003 per cent only (Gopalakrishnan and Indira, 2006). The cv. Byadgi Dabbi cultivated in Dharwad district of Karnataka and 'Tomato chilli' grown in Warangal in Andhra Pradesh is similar to paprika. Three improved paprika lines namely

Kt-pl-8, Kt-pl-9 and Kt-pl-18 were identified at IARI Regional Station, Katrain (HP). Their yield varies from 4.9 to 8.6 t/ha with colour value ranging from 178 to 233 ASTA units. Arka Abhir, developed at IIHR, Bangalore has high colour value and is specifically suitable for oleoresin extraction. Augustine (2001) identified Papri King, PBC 006, Kt-pl-8, PBC 134 and Kt-pl-20 as promising paprika lines.

Paprika is commercially produced in Spain, Portugal, Central Europe, Southern Africa and the US. Once the fruit are harvested, they are loosely stacked in long narrow, cylindrical mesh bags made of red plastic and allowed to cure for 3-4 weeks. Then the peppers are washed, dried, crushed and finally ground into powder. To produce the desired level of pungency and colour, proportion of seeds to be ground with the pepper pod is decided. During grinding, the crushed pepper are heated to release the oil in the seeds which interacts with the pungent principles in the fruits to produce the intense red colour. Colour has no effect on the pungency of the paprika. Bright red paprikas may be sweet or pungent. Only undamaged fruits less than a year old are used for the top grades of Hungarian paprika.

Uses

There are many uses of pepper in food processing, as a food colourant, as a source of pungency in food, as a source of flavour, as a source of pain relief or pharmaceutical use and as a repellent. In many cases two or more of these peoperties are included in the same product, for example, paprika may be a source of colour, pungency and flavour.

Chilli is valued for its pungency and colour. In India, no dish is complete without chillies. It is used as whole dry chilli, chilli powder, chilli paste, chilli sauces, chilli oleoresin and as mixed curry powder. It is also a rich source of vitamin A and C. The crude fibre content varies from 22.9 to 31.0 %.

Vanilla
(*Vanilla planifolia* Andrews., F. Orchidaceae)

The vanilla pods or "Sticks"of commerce are the cured fruits or beans of climbing orchid *Vanilla planifolia*. It is a native of Atlantic coast from Mexico to Brazil. Cultivation of vanilla spread after discovery of America by Columbus. The aromatic qualities of vanilla beans were known long before discovery of new world. Once the vanilla ingredient was finally known in 1510, the Spanish conquerors imported vanilla beans from Mexico into Spain where factories were established for the manufacture of chocolates flavoured with vanilla. By 17th century, vanilla was introduced to France and England. Congenial conditions for growth of vanilla were available in several parts of world, but introduction of this plant was not possible because of the absence of specific pollinating agents. For this reason Mexico, West Indies and other regions, where it is indigenous, were the only source of vanilla beans for more than two centuries. Discovery of method of artificial pollination of flowers and amenability of the crop to vegetative propagation paved way for commercial cultivation of vanilla in many countries (Potty, 2003). Spread of vanilla cultivation to tropics of globe has been gradual and steady.

History of introduction of vanilla into India appears to be obscure since no authentic information on this subject is readily available. Documentary evidences reveal that this crop was experimentally introduced to West Bengal, Bihar, Tamil Nadu, Pondichery, Karnataka, Kerala and Assam only 100 years ago. The earliest record which could be traced with regard to cultivation of vanilla in India was at Kallar Burliar Fruit Research Station in the Nilgiris during 1945 (Spices Board, 1992). Later in 1960, preliminary investigations on the cultivation aspects of vanilla was also initiated at the erstwhile Horticulture Research Station and present Regional Agricultural Research Station, Ambalavayal, Wynad in Kerala. Both stations worked mainly on propagation, hand pollination, disease management and curing with a fair degree of success. The initial attempts by enthusiastic farmers in Kerala and Tamil Nadu did not succeed due to various reasons like lack of proper care, continued technical support and proper marketing arrangements for the produce. Neverthless, these plantations served as sources of planting material for vanilla development programs initiated by various agencies later.

Vanilla is one of the most expensive spices traded in the global market. It accounts for 0.75 % of the total world trade on spices. Major vanilla growing countries are Madagascar, Indonesia, India, Mexico, Comoros and Reunion. It is also grown to a lesser extent in French Polynesia, Tonga, Guadeloupe and Zimbabwe. Indonesia is the largest producer in the world and together with Madagascar, contributes nearly 90 % of world production (Potty, 2003). Vanilla produced in Madagascar is called bourbon vanilla and this fetches the highest price in international market.

Systematics

Vanilla belongs to family Orchidaceae. Apart from *V. planifolia,* two other cultivated species are *V. pompona* (West Indian vanilla) and *V. tahitensis* (Tahitian vanilla) (Madhusoodanan *et al.*, 2003). Only *V. planifolia* is cultivated in India. *V. andamanica, V. pilifera, V. walkeriae* and *V. wightiana* are native to India of which *V. wightiana* is endemic to South India. This species is easily distinguished from cultivated species by its vertical grooves on stem which are absent in latter. Another prominent feature is absence of well developed leaves. *V. tahitensis* differ from *V. planifolia* in having more slender stems, narrower leaves, longer perianth segments and a lip that is shorter, with reddish brown beans of 12-14 cm long, 9 mm width, broad in middle, and tapering towards ends.

V. pompona resembles *V. planifolia* except that the leaves are larger, flowers greenish yellow and larger and more fleshy and the lip has a tuft of imbricated scales instead of hairs in centre of disc. Beans are three angled and more fleshy. They show little or no tendency to split at maturity. Quality of tahitensis and pompona beans is inferior to planifolia.

Morphology

Vanilla is a herbaceous perennial vine climbing up trees or other supports to a height of 10-15 m by means of adventious roots. In cultivation it is trained to a height which facilitates hand pollination and harvesting. Aerial roots occur singly opposite to leaves and adhere firmly to support on which vine climbs. They can absorb moisture from atmosphere. Basal roots ramify in humus or mulch layer. Stem is round and dark green in colour, succulent with internodal length of 5 to 15 cm. Girth is about 1 to 2 cm. Leaves are thick, twigged, dark green, alternate and oblong-elliptic to lanceolate. Tip is acute to acuminate and the base somewhat rounded. Petiole is thick, short and canalized above.

Flower

Flowers are produced on stout axillary racemose inflorescence, rarely branched. They are usually borne towards top of vine and are 5-8 cm long with usually 10-26 flowers opening from base upwards. Rachis is stout and 4-10 mm in diameter. Bracts are rigid, concave and persistent. Flowers are large, waxy, pale greenish yellow, bisexual and zygomorphic. Sepals and petals look alike and it is commonly called perianth. Lower petal is short, broad and is modified into a trumpet shaped labellum. Lower part of labellum envelops a central structure called column (gynostemium) which is a little shorter than perianth and also have a tuft of scales (Bhat and Sudarshan, 2004). Column has at its apex two pollen sacs (pollinia 1.5 mm length and 1.9 mm breadth) covered by a hood like structure as their cap. Stigma is physically prevented from coming in contact with anther by a flap like structure (4.1 mm length and 3.0 mm breadth) known as rostellum. This prevents natural pollination in vanilla. Slender stalk like portion is the ovary which is 4-5 cm in length (Kuruvilla *et al.*, 1996). A cross section of ovary of an opened flower shows three carpels, three pairs of fibro vascular bundles and three pairs of placentae. The placentae extend throughout length of the ovarian cavity.

Pollination

Structure of flower is not conducive for natural self pollination. In its native habitat, Mexico and Central America, some of the flowers are pollinated by stingless bees of genus Melipona and by humming birds. Elsewhere, hand pollination is unavoidable for fruit set. Even in Mexico, only a small percentage of fruits set naturally and hand pollination is carried out in commercial gardens. Method of hand pollination being practiced now was first demonstrated by Charles Morgen in 1836 and was perfected by Edmond Albius in 1841. Pollination is performed with help of a pointed bamboo splinter, stem of a stiff grass or a sharpened tooth prick. Using one of these aids, rostellum is pushed back and overhanging anther is pressed against stigma with thumb and thus smearing pollen over it. Each flower lasts for a day only and so pollination should be carried out on same day. Ideal time for pollination is between 6 am to 1 pm. The fruit set % was maximum at 8 am and showed a decreasing trend as pollination get delayed (Bhat and Sudarshan, 1998). Studies showed that complete transfer of pollen result in maximum fruit growth and with less than 50 % pollen, fruit size gets reduced considerably (Bhat and Sudarshan, 2000). Hence, complete pollinia have to be transferred to stigmatic surface so as to get maximum growth of fruit. A skilled worker on an average pollinates about 1000 flowers a day. Flowers which are not fertilized fall off within a day or two. Normally eight to ten flowers located on lower side of inflorescence are pollinated and not more than 10-12 inflorescences in a vine. Remaining flower buds are nipped off to reduce strain on vine. Once fertilization takes place, the ovary elongates rapidly for 45 days until full length and girth of beans are attained within 75 days (Bhat and Sudarshan, 2004).

Phenology of flowering in vanilla

Phenology of flowering was studied in detail by Bhat and Sudarshan (2004) under Saklespur conditions in Karnataka. Vanilla vine comes to flower in 3rd year of planting. Flowering generally takes place on branches which are one and a half year old. Inflorescence initials are generally activated in leaf axils during December-January, as light green protuberances. These buds attained a length of 3.56 cm in first fortnight. After 30 days, length was 4.59 cm. At this stage, flower buds commence to emerge from base of inflorescence. After 45 days, flowers are fully developed and blooming commences. The inflorescence which was 2.5 cm long in beginning attained a mean length of 7.34 cm by the 60th day. In an inflorescence, the blooming period varied from 5 to 40 days depending on number of flowers in that inflorescence. In about 50 % of inflorescences, blooming period was 21-25 days. Under high elevation, Kuruvilla *et al.* (2004) reported that zynchronised flowering is lacking and blooming in an inflorescence continued for 14 to 30 days. Under high range conditions (1050 m above MSL) flowering extended for a period of 2-3 months (March to May) and one to two months ahead in mid lands and plains. At Sakleshpur (Karnataka) conditions peak blooming occurred during 2nd fortnight of March (Bhat and Sudarshan, 2004). They also observed that there was no significant correlation between weather parameters and phenology of flowering. In about 50 % of flowers, blooming was observed every day. Flowers bloom from base of inflorescence upward and last for a day.

Anthesis

Majority of flowers open at about 4 am - 6 am under high altitudes. It takes 11 hours for complete anthesis. Flower starts closing slowly at 1 pm and completely by 4 pm. The total pollen viability period is 57 hours (23 hours before flower opening and 34 hours after flower opening). Stigma receptivity commenced 40 hours before anthesis and it lasts 16 hours after anthesis.

Fruits

Fruits are commercially known as beans which are pendulous, cylindrical and three angled. Elongation of fruits continues for a period of 3-4 months and attains a length of 10-25 cm and 4.5 to 5.0 cm girth. The fertilized flower develops into a mature fruit in about 9 to 11 months under various conditions. The beans contain numerous tiny black seeds which germinate only under *in vitro* conditions.

Crop improvement

The vanilla received serious research attention only after 1990 and since then ICRI (Spices Board), ICAR institutions and State Agricultural Universities are actively involved in crop improvement programmes in vanilla.

Genetic resources

Being a vegetatively propagated crop, introduced in a limited area, genetic variability is very low. Germplasm collections are maintained at ICRI stations at Saklespur, Karnataka (56 accessions collected from Kerala and Karnataka and three related species) and at Myladumpara, Kerala (26 accessions and five related species). These accessions are subjected to characterization based on morphological and biochemical characters (Kuruvilla *et al.*, 2004).

Selection

Initial evaluation at ICRI, Myladumpara resulted in identification of four genotypes MCV-1, MCV-2, MCV-3 and MCV-4. These genotypes yielded consistently above 250 kg/ha under horizontal coiling system. However, their yield potential under vertical system of coiling with varying number of hanging shoots are yet to be assessed.

Hybridisation

Hybridisation has been initiated with an objective of transferring alien gene (s) responsible for natural bean formation and abiotic stress.

Mutation breeding

Presently, segregating progenies developed through *in vitro* seed germination is the source of variation in vanilla. This has limitations considering the pollination system. In this context, mutation breeding is the viable alternative to be adopted for inducing variability both *in vitro* and *in vivo*.

In vanilla, crop improvement studies are still in infant stage. Thrust should be to generate diversity within the crop so as to exploit it in breeding process.

Area and production of vanilla

Global cultivation of vanilla is estimated to be 40846 hectares from which production is about 5583 t. In India, it is grown over an area of 2545 hectares covering Kerala, Tamil Nadu and Karnataka producing around 100 t (Kuruvilla, *et al.*, 2004).

Crop management

Climate and soil

Vanilla requires warm and moist conditions of humid tropics and thrives well between 10°N and 20°S latitudes having a well distributed rainfall of 150-300 cm with a temperature range of 25-32°C. Ideal relative humidity is 80 %. It comes up well up to 1500 m above MSL and prefers land with gentle slope and light porous soil with adequate drainage. Forest soil rich in humus is ideal but plants survive well in sandy loam to even laterite soils. Top soil rich in humus and organic matter is essential for cultivation of vanilla. The crop requires a short dry spell for flowering. Water stagnation, strong winds and arid conditions are detrimental to the crop.

Vanilla does not grow well under intense light or heavy shade. It grows well under trees which provide filtered light of about 50 %. It can be successfully cultivated as an intercrop in coconut, arecanut, black pepper and coffee plantations.

Propagation

Vanilla is commercially propagated by stem cuttings. They can either be planted directly in field or can be initially grown in polythene bags and transplanted after rooting. Cuttings with 8 to 10 internodes would be preferable as they come to flower earlier than shorter cuttings. Any part of vine with dormant axillary vegetative buds can be selected for taking stem cuttings. However, vines of current years growth which are in vegetative phase leaving tender shoots at tip are the most ideal (Kuruvilla *et al.*, 2003). Length of shoot cutting used for propagation has a direct bearing on growth of vine and yield (Suryanarayana, 2004). Longer cuttings flower in second year itself whereas shorter cuttings take atleast three years to flower. Length of cutting is to be adjusted depending on availability of planting material and area to be planted. Cuttings shorter than 60 cm length are not recommended for *in situ* planting, instead they are rooted in polythene bags and then field planted.

A simple and rapid multiplication technique was developed at ICRI, Myladumpara for production of large number of cuttings from limited area. Rapid multiplication nursery is raised in 50% shade. One meter cuttings are planted in trenches of 60 cm from wide and deep leaving 40 cm in between. Trenches are filled with top soil, well dried and powdered farmyard manure and sand in the ratio of 3:1:1. Standards for trailing of vines are planted at a distance of 1 meter along trenches atleast two months before planting vanilla. Vines are allowed to grow on supports and later coiled around branches of supports. Mulching, manuring and irrigation are given judiciously. A well maintained plant would produce about five to seven meters of growth per year. On an average one hectare of nursery would be sufficient to produce about 40-50 thousand meters annually (Kuruvilla *et al.*, 2003).

For raising rooted cuttings, 2-3 noded cuttings are planted in polybags such that atleast one node is in potting mixture and two nodes above. The cuttings are tied on to sticks planted in polybags. Partial shade should be provided either by using agroshadenets or coir nets. Plants are maintained with regular watering and manuring. Cuttings will root in about 20 days and they will be ready for transplanting after six to seven months.

Micropropagation

Tissue culture is now acknowledged as a reliable technology for large scale multiplication of quality planting materials of vanilla in a short span of time. The protocol developed at ICRI, Mayiladumpara (Rao *et al.*, 2005) is as follows:

Protocol

Surface sterilization and preparation of explants

Nodal segments of vanilla from greenhouse maintained plants are used as explants. Surfaces sterilize with 0.08% $HgCl_2$ for 12 min preferably under vacuum. After thorough rinsing, suspend explants in presterilized streptomycin solution (0.1 %) for 15 m, trim and blot dry explants prior to inoculation.

Culture medium

Primary culture initiation and establishment: Medium: Murashige and Skoog (1962)/Knudson-C (1945) + BAP 1 mg/l., IAA 0.1 mg/l; D-Biotin 0.1 mg/l Ca pantothenate 0.1 mg/l; Sucrose 3% and agar 7.5 mg/l. at pH 5.8. Multiplication ratio was 1:2.5.

Further subcultures

At 40-45 day's intervals on same medium.

Shoot elongation and rooting

Elongation and rooting of shoots could be facilitated on MS + 0.1 mg/l KN; 0.1 mg/l NAA; Sucrose 2 % and agar 0.65 %.

Duration of culture: 50-60 days

Hardening and acclimatization

Primary hardening

- Suspend ex agar plants in 0.1 % bavistine solution for 15 min.
- Transfer to micropots containing soilrite and place under mist chamber in green house.
- Plantlets should be sprayed with a bio-agent preferably *Pseudomonas* (10 ml in 1 litre water) once in 15 days to prevent any disease incidence (Duration-6 weeks). A prophylactic spray of bavistine (2 gm/litre water) may also be given at well spaced intervals of about 15 days before *Pseudomonas* sprays.

Secondary hardening

- After primary hardening, plantlets should be transferred to potting mix and allowed to get acclimatized in secondary nursery for two months (or) when plantlets attain a length of at least 30 cm.
- Plantlets should be sprayed with *Pseudomonas* and one or two rounds of NPK at a concentration of 2 g/litre water.
- Transfer to field conditions, when plantlets attain a length of 25-30 cm and are robust. Field preparation, planting and management should be performed as per recommended package of practices, for successful establishment and growth of plantlets.

  With above stated protocol, cost of production worked out to be Rs.9/ plantlet.

Synthetic seeds

Synthetic seed production in vanilla was reported by Sajina *et al.* (1997a) and Marymathew *et al.* (1999).

In vitro seed germination

Vanilla seeds can be germinated aseptically and nonsymbiotically through *in vitro* techniques. The medium employed is Knudson C supplemented with an assortment of growth regulator combinations of BAP, NAA and coconut water. As many as 20 buds may develop from a single seed in a five week old culture passage. The shoot buds upon transfer to MS medium supplemented with KN and NAP each at 0.1 mg/l develop into plantlets with well developed leaves and root system. The fully developed plantlets are acclimatized and transferred to polybags. Plantlets from seed cultures exhibit lesser internodal length compared to those derived from nodal cultures. However, they attain normal growth during hardening.

Performance of micropropagated planting material in comparison with traditional stem cuttings is being studied in field. In initial years, they exhibit slow growth. However, once sufficient growth is achieved, plants grow normally on par with stem cuttings. Initial slow growth is usually observed upto 6 months of planting (Kuruvilla *et al.*, 2003).

Hena (2005) studied existing variability in field established vanilla plants derived from *in vitro* seed culture using morphlogical and molecular markers. The study revealed that among the 10 morphological characters selected, leaf length, width, leaf area, number of leaves and total growth showed significant variation. Twenty different accessions which were highly variable could be selected from the 360 accessions studied. Based on isozyme analysis, these 20 accessions could be grouped into nine clusters (based on esterase pattern) and into fifteen clusters (based on peroxidase pattern). Molecular studies involving RAPD analysis using 10 primers grouped the 20 accessions into three major clusters.

Crop Management

Land preparation and planting

Vanilla being a climbing vine requires support for climbing. It flourishes well in partial shade of 50 per cent sunlight. A good support tree which can provide adequate shade to the plants should be grown well before planting of vanilla cuttings in the main field. Low branching trees with rough bark and small leaves are ideal for this purpose. Some of the trees now being used as support trees include *Jatropha curcas, Plumeria alba, Casuarina equisetifolia* and a few species of *Glyricidia* and *Erythrina*. Under Indian conditions, *Glyricidia* is a very good standard for vanilla. The support trees should be planted atleast six months before planting of vanilla. Standards are planted at a spacing of 2.5-3.0 m between rows and 2 m within row making a population of 1600 to 2000 trees/hectare.

Vanilla is generally planted at a time when weather is not too moist or dry. Soil near support tree is loosened to a depth of at least 30 cm. Soil is mixed with equal portions of leaf mulch and compost and a small mound of 60 cm diameter and 25 cm height is made. Cuttings are collected well in advance and after removing three to four basal leaves, dipped in one percent Bordeaux mixture and kept under shade for withering. They are then planted at the rate of one per support. Defoliated part of vine is laid on loose soil surface and covered with a thin layer of about two to three cm soil. Basal end of cutting should be kept just above soil to prevent rotting. Growing end is gently tied to support for clamping by aerial roots of cuttings which are planted close to support. Under South Indian conditions, cuttings are planted on North Eastern side of support tree to avoid bright sunlight. When rooted cuttings are used for planting, pits of suitable size just to hold polythene bag are dug and cuttings are planted without disturbing root system. Newly planted vines are shaded properly. It takes about four to eight weeks for cuttings to strike roots and to show initial sign of growth.

Maintenance

Once established, vines have to be given constant attention. Vanilla needs lot of organic matter for proper growth. Well decomposed cowdung, compost leaf mulch etc. may be spread around plants three weeks after planting. Mulching with green or dry leaves is to be carried out twice a year. As roots are very tender confining to surface layer of soil, it should be ensured that they are not damaged by cultural operations.

Manures and fertilizers

Decomposed mulch is the main source of nutrients to vanilla. Well rotten cowdung, compost, composted and well rotten poultry manure, vermicompost, neem cake etc. can be applied to promote growth of plant. Vermicompost may be applied at the rate of 20 kg/grown up vine. No research work has been done in detail with regard to application of fertilizers to vanilla plants.

Potty and Krishnakumar (2003) reported that vanilla yield can be enhanced by soil application of 20:10:30 g NPK/vine/year and foliar application of urea, single superphosphate and muriate of potash at the rate of 1.0, 0.5 and 1.5 % respectively during January, May and September. Positive responses to application of biogas slurry and groundnut cake have been reported by Anuja (2005). She also found that foliar application of 17:17:17 complex fertilizer resulted in higher foliar potassium content in leaves (3.80%).

Irrigation

Vanilla requires moist climate for the best growth. Excessive rainfall and extended droughts are equally detrimental to the crop. The soil around the vine should be kept moist. The aerial roots and leaves also absorb moisture from the atmosphere. Hence maintaining high relative humidity in vanilla plantation by using microsprinklers and foggers will help in better growth of the plant.

Training and pruning

If the vine is permitted to grow up on a tree, it will rarely blossom, so long as it is growing upward. Vanilla is to be trained to keep the vines within the limits so as to facilitate cultural management. Vines are allowed to grow up to 1.5 m and then brought down. Vines are twisted around lower branches of supporting tree or over lattice of trellis so that they may hang down. Bending of vines appears to be an important factor in causing it to flower and fruit beyond the bend, which may be due to accumulation of carbohydrates and other flower inducing substances in that region of vine. If vegetative growth of vanilla vine are arrested by pruning, it turns to reproductive phase. In Seychelles where climatic conditions are not favourable, pruning of vines eight to nine months before expected flowering is essential to get a satisfactory crop. At pruning time special bearing branches are prepared. These are shoots which when 1-1.2 m long are bent down with tip pruned at about 45 cm from soil. Any shoot produced on these branches is cut off when 7.5-10 cm long, but shoots appearing on rest of plant before bends are allowed to grow. They will constitute bearing branches of following year. As a result of decreased sap flow towards bearing branches, they flower. After harvesting the beans, old branches are cut off. Mean while, following years bearing branches have already been prepared. Growers in Puerto Rico follow the practice of removing 10-15 cm growing shoot tip each year before flowering. This results in accumulatiion of carbohydrates, thus encouraging flower production in axils of leaves on e hanging branches.

In South India, planters follow their own techniques for pruning, the ultimate aim is to alter physiology of plant to prepare for reproductive phase. When harvesting is completed the whole branch which carries them is cut off and only new shoots of previous year are retained so that there is no waste of nutrients.

Shade management

Judicious lopping of branches of living supports is very important to adjust shade level to around 50 %. Under high rainfall and high relative humidity, vanilla can withstand more sunlight than during low humidity and drought periods. During flower bud formation and blossoming, more light is beneficial, but during dry periods support trees should maintain much of their foliage for providing shade. All pruned materials can be used as mulch.

Growth regulators

Earlier reports showed that parthenocarpic development of pod can be induced in vanilla with growth regulating chemicals though they are poor both in quantitative and qualitative characters. Anuja (2005) obtained parthenocarpic development of beans with 2,4-D and GA. Length and volume of beans were more in 2,4-D (0.20 mg/flower bud swabbing). However, maximum growth and weight of beans were observed in hand pollinated beans. Experiments to improve vanillin content revealed that growth regulators significantly influenced quality of beans. Moisture content was more in beans sprayed with IBA (100 ppm). Vanillin content fluctuated throughout curing period. After sun drying, hand pollinated beans showed a high vanillin content of 3.98 %. After slow drying period IBA (200 ppm) recorded the highest vanillin content of 4.22 % and after conditioning kinetin (200 ppm) recorded the highest vanillin content (4.20%). Phenol content after curing was more in beans which received kinetin (200 ppm) spray and the highest sugar content was recorded by NAA (100 ppm) (Anuja, 2005).

Intercropping

Vanilla is a suitable intercrop under coconut, arecanut or any other tall crop, which provide shade. In recent times, vanilla cultivation is gaining importance as one of the intercropping components under high density multispecies cropping system (HDMCS) in coconut plantations (Nybe *et al.*, 2004). Coconut gardens of age between 8-40 years old offer favourable shade requirement of 50-60 %. In between two rows of coconuts, two rows of standards are planted at 1.8 m x 1.8 m spacing and vanilla planted. In areca gardens, generally vanilla is planted on to areca palm itself and during 2nd year vines are trained by providing trellis in between areca palms. Korikanthimath *et al.* (1999b) reported economic feasibility of vanilla cultivation in coconut gardens in plains and low rainfall areas of Karnataka under irrigation.

Diseases and pests

Diseases and pests found in vanilla in India are:

Diseases		**Causal organism**
Root rot	-	*Sclerotium rolfsii*
Root wilt	-	*Fusarium oxysporum*
Stem rot	-	*Fusarium oxysporum*
Shoot tip rot	-	*Fusarium & Colletotrichum gloeosporioides*
Bean rot	-	*Phytophthora meadii, Sclerotium rolfsii*
Bean yellowing and rotting	-	*Fusarium oxysporum*
Leaf spot	-	*Colletotrichum vanillae*
Vanilla mosaic	-	Virus

Pests

No serious pests attacking vanilla are noticed. Surveys located incidence of following pests to a lesser extent (Varadarasan *et al.*, 2003).

Vanilla bug - *Halyomorpha picus*
Leaf footed bug - *Riptortus pedestris*
Vanilla vine weevil - *Sipalus sp.*
Vanilla beetle - *Saula ferruginea*
Vanilla caterpillar - *Archipes sp.*

Adverse conditions like drought, over nutrition, inadequate or heavy shade, prolonged rainy weather, excessive moisture, damage to roots by cultural practices and dense planting invite many diseases. Proper phytosanitary measures, drainage, application of biocontrol agents, shade regulation and application of fungicide/ pesticide can check spread of diseases and pests to a great extent. Most of the diseases appear during rainy season. A package of control measures for disease management is given by Thomas and Vijayan (2004) as follows:

- Regulation of excess shade and plant sanitation during the first week of May
- Spray Bordeaux mixture (1%) on the entire vines after one or two summer showers by 2nd week of May
- Apply *Trichoderma* at base of vines, two weeks after Bordeaux mixture application after removing the mulch (Mass multiplied *Trichoderma* mixed with cowdung (1:50) one kilogram/vine)
- Apply *Pseudomonas* as foliar spray at the rate of one litre/vine during 4th week of May (10 ml liquid culture/litre of water or 20 g powder formulation/litre of water)
- Repeat application of *Trichoderma* and *Pseudomonas* as above during August and October months.

Harvesting and curing

Beans should be harvested at right stage for obtaining good quality processed beans. Vine normally starts yielding from 3rd year after planting and yield stabilizes by fifth year. Yield declines after 15 years and replanting can be taken up at this stage. When beans are fully mature and ready for harvest, they will have a pale yellow colour at distal end. Immature beans will not develop aroma and will not be supple after curing. So also the ripe and split bean can not be cured properly and will be inferior in quality. The bean is harvested by holding it in hand and pressing pedicel with thumb or cutting with a sharp knife. Around 75-80 beans of 18-20 cm length give one kilo gram of fresh beans. In India, the harvesting season commences in late October in plains and stretches to end of December to early January of next year in high elevations. A good vanillery can yield 500 to 800 kg of cured beans.

Curing

It is a process of alternately sweating and drying of beans until moisture is lost by as much as 80 % and the beans undergo enzymatic reactions for production of vanillin which is responsible for the characteristic flavour of vanilla. During the process of curing, free vanillin is developed in beans as a result of a series of enzymatic actions on several glucosides. Simultaneously various aldehydes, aromatic esters, protocatechic acid, benzoic acid, vanillic acid and anisic alcohol are also formed and all these components together give the fragrance of natural vanillin, well distinguishable from synthetic vanillin. Various curing processes are developed in the vanilla growing countries to meet quality requirements of vanilla market. There are mainly four different processing techniques followed in different parts of the world, viz. Peruvian method, Guiana process, Mexican process and the Bourbon method. Bourbon method with necessary modifications is followed in India, Madagascar and most of other countries.

Basically any curing method involves following four stages.

1. Killing vegetative life of beans to allow onset of enzymatic reaction.
2. Raising temperature to promote this action and to achieve rapid drying to prevent harmful fermentation.
3. Slower drying for development of different fragrant substances.
4. Conditioning product by storing for a few months.

In India, modified Bourbon method of curing is followed to get processed beans. Curing process involves four stages as mentioned below:

Stage 1: Killing or wilting of beans

Mature beans are first cleaned and then sorted according to stage of maturity and length of beans. Split beans are separated and processed separately. About 25 to 30 kg of sorted beans are taken in a bamboo basket and immersed in a broad mouthed vessel containing hot water at 68°C. Longer beans are immersed for 5 minutes, medium sized beans for 4 minutes and small sized beans for 3 minutes. This is called scalding. Immersion in hot water stops further vegetative development in the beans and initiates enzymatic reactions responsible for production of aroma and flavour. After immersing in hot water for the required period, beans are taken out, excess water is rapidly drained and hot beans are rolled in a woolen blanket when still hot and kept in wooden chests lined with blankets for a period of 24 hours. Temperature of water is a critical factor in killing or scalding as higher temperature deactivates the enzymes responsible for conversion of vanillin and other secondary flavour inducing components. Care should, therefore, be taken to ensure that temperature of water does not increase more than 70°C.

Beans are taken out from chest after 24 hours and examined. Beans develop brown colour 24 hours after immersion in hot water. Beans which have not developed brown colour will have to be separated and immersed again in hot water for required period and steps will have to be repeated till they develop brown colour.

Stage 2. Sweating

Sweating beans involves rapid dehydration and slow fermentation to develop key flavour components. Enzymes responsible for conversion of vanillin and related compounds are most active at this stage. Beans in wooden chest are taken out, cleaned and spread out under sun on dark coloured blankets or cotton cloth. This is called sunning. Sunning area must be dry and away from roads to avoid contamination with dust. Beans are kept in sun till they attain a temperature around 55°C. This can be ascertained by holding beans in the hand. If beans are too hot to hold, then it would have attained required temperature. Normally this would require keeping beans under sun for 2 to 3 or 4 hours depending on season. It is better to expose beans to sun for sweating on a raised platform than keeping them on ground. A layer of agro shade net may be provided in plains to prevent excessive build up of heat on beans. Beans should not be suddenly exposed to sun during the hottest part of day. They may be kept under sun at about 11 A.M. and then temperature will slowly rise.

Beans are then rolled in same blanket or cloth when they are still hot and kept in sun for another half an hour and then stored in the woolen lined wooden chest. This process is repeated for 8 to 10 days. At end of this period, beans will have lost moisture by 40 to 50 % and acquire a deep chocolate brown colour. The beans will be supple and have started to give out pleasant aroma. Beans may be slowly massaged so that ß-glycosidase, responsible for conversion of glucovanillin to vanillin, located on fleshy wall of beans comes in contact with most of glucovanillin present in central segment of placental mass.

If sweating in sun is not possible due to unfavourable weather conditions, electric ovens can also be used for the purpose. Time needed for sweating beans in electric oven is less. Temperature in the oven may be maintained at 45°C and humidity may be increased for faster sweating.

Stage 3: Slow drying

Moisture content of beans is further reduced in this stage. Moisture content of beans will be about 50 to 60 % at end of Stage 2 which will be reduced further to 25 to 30 % in Stage 3. The slow drying is done under shade in well ventilated room. Beans are spread on wooden trays mounted on stands in a well ventilated and open room and subjected to slow drying. In arid places, clean wet clothes are kept in room for cooling air and maintaining relative humidity in the room at more than 70 %. The beans are dried for 15 to 20 days. They should be examined every day to ensure that moulds do not develop. The beans which have developed mould should be separated, immersed in hot water and Stages 2 and 3 are to be repeated. Beans may be periodically turned over for uniform drying.

Stage 4: Conditioning

Conditioning is an aging process necessary for flavour development. Beans are conditioned after moisture level is brought down to desired level by slow drying process. Normally, after slow drying process, the beans are bundled (50 to 100 beans in each bundle) according to their size or into bundles of known weight. They are then wrapped in butter paper or cellophane paper or polypropylene bags and stored in airtight containers under ambient temperature for two to three months. Conditioning is normally done at 35 to 45°C. During this stage, various chemical reactions like esterification, etherification, oxidative degradation, etc. take place and a range of volatile aroma compounds are formed.

Beans should be examined at regular intervals during conditioning so as to ensure that moulds do not develop or they are not infested with insects like mites. Vanilla beans are also subjected to attack by mildews caused by *Penicillin* and *Aspergillus*, especially if they are improperly cured and insufficiently dried. The mould attack starts from peduncle of beans and spreads to entire beans. Once beans become permeated with odour of mould, mildew, etc., it is practically impossible to eliminate it and value of beans is much reduced.

Although major moisture loss in beans occurs during sweating and drying stages, it also occurs during first three months of conditioning. Thereafter, moisture loss is negligible. Beans with an average moisture content of 32 % will be dark brown in colour and have a well developed suave aroma and a high degree of flexibility. They can be twisted round the finger without rupturing. They will exude pleasant aroma and exhibit characteristic luster on account of oil contained in them. Vanillin and more than 170 related secondary aromatic chemicals will be developed in the beans. Beans with high moisture content of 50 to 54 % will have slightly fermented aroma and are less suave than those conditioned at lower moisture content.

Quality attributes of vanilla

The most important quality attributes of cured vanilla beans for grading are length of beans, aroma, colour, flexibility, luster, absence of blemishes, mildew and freedom from insect infestation. The best quality pods are 17-25 cm long, highly aromatic, free from mildew, blemishes and insect infestation, dark brown, fleshy, supple and some what oily in appearance. Quality of vanilla beans should not be judged only by their vanillin content, equally important is flavour profile of beans. Proper harvest and post harvest practices determine both vanillin content and flavour profile. However, agroclimatic region where vanilla is grown influences flavour components. Processed beans contain about 2.2 to 2.4 % vanillin depending upon place where crop grows. Indian vanilla beans contain 2.2 to 2.4 % vanillin while Java vanilla contains 2.7 % vanillin. Mexico vanilla contains 1.7 % vanillin.

Grading

Harvested vanilla takes about four months for curing. Properly cured and dried beans are dark brown and supple enough to be twisted around the figure without rupturing. Processed beans are sorted and graded depending upon their length, colour, moisture content and flavour. Vanilla beans of commerce are divided for convenience into five geographic types.

Type	Place where grown
Mexican	Mexico
Bourbon	Island of Reunion, Madagascar, Comoros and Seychelles and India
South American	French West Indies
Tahiti	Tahiti
Java	Indonesia

Indian processed vanilla for international market is graded as per Bourbon grading procedure. Five main grades are

1st grade - whole beans >14 cm length, moisture content <25 %, no scratches, blemishes, good flavour.

2nd grade - whole beans >14 cm length, moisture content 25 to 28 %. Some scratches are permitted.

3rd grade - whole beans >14 cm length, moisture >30 %, more scrtches/ blemishes.

4th grade - whole beans with less flavour, length <14 cm with more scratches, moisture >30 %.

5th grade - broken, cut, short beans which do not have good flavour. Higher moisture level.

Packing and storage

Even after completion of curing, vanilla beans often continue to develop a desirable aroma and flavour, if stored under suitable conditions. After grading beans are again sorted according to their length into bundles having lengths varying from 12 to 26 cm. Each bundle will have 70 to 100 beans and will weigh 150 to 500 g. These bundles are wrapped in butter paper and stored in air tight tin boxes.

Processed products and uses

Vanilla beans consist of a broad array of flavour profiles of more than 170 components in addition to vanillin. These components are easily volatilized by heat or chemical distillation process. They enhance the flavour character of vanillin. This is done by value addition of the beans. Vanilla extract, tincture, oleoresin, powder, paste, perfumery tincture and absolute are the value added products available in international market. Vanilla today, constitutes the world's most popular natural flavour for numerous sweetened foods, viz., chocolates, diary products, soft drinks, confectionary, backery, beverages and in perfumery. About 50 % of vanilla is used in icecream industry alone. Vanillin is an antioxidant and finds use in pharmaceutical preparations also.

ŽŽŽ

Black Pepper

Cardamom (Small)

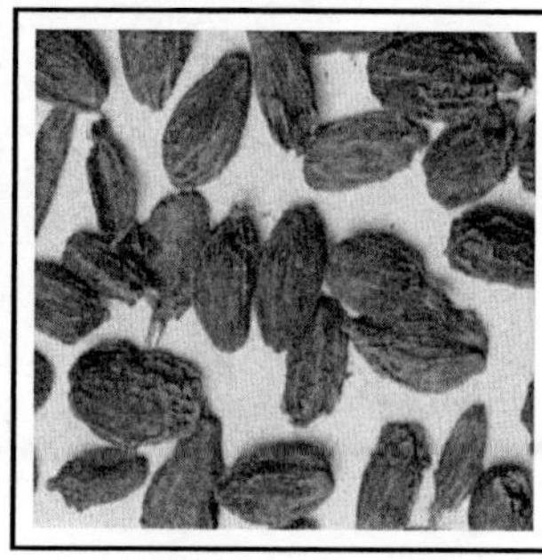

Cardamom (Large)

Ginger

Turmeric

Chilli

Vanilla

Chapter 3

Tree Spices

Nutmeg
(*Myristica fragrans* Houtt., F: Myristicaceae)

Myristica fragrans Houtt., the nutmeg tree is unique among spice plants as the donor of two distinct spices; nutmeg and mace.Native of Indonesia (Moluccas islands), nutmeg tree grows there abundantly and is now naturalised in West Indies, Srilanka, India, Phillipines, Tropical America and Pacific Islands. (Varghese, 2000). In India, the plant is grown in certain pockets of Kerala, Tamil Nadu, Karnataka, Goa, Maharashtra, North East India and Andamans (Krishnamoorthy, 2000). It is also grown in a small scale in Srilanka, Trinidad, China, Tobago, Zanzibar and Mauritius.

Chemical composition

According to Gopalan *et al*. (1971), analysis of nutmeg gave the following values: moisture 14.3 per cent; protein 7.5 per cent; ether extract 36.4 per cent; carbohydrates 28.5 per cent; fibre 11.6 per cent; mineral matter 1.7 per cent; calcium 0.12 per cent; phosphorus 0.24 per cent and 4.6 mg/100 g. Principal constituents of nutmeg are fixed oil (fat), volatile oil and starch. Flavour and therapeutic action are due to the volatile oil whose content varies from 6 to 16 per cent based on the origin and quality of nutmeg. (Lewis *et al*. 1973). Nutmeg butter is obtained by pressing of the ground and cooked or steamed kernels. The yield is 24 – 30 per cent. It is a solid, yellowish red fat possessing nutmeg odour. Nutmeg fat contains eight fatty acids, myristic acid forming the largest component.

The mace contains the following : moisture 15.9 per cent; protein 6.5 per cent; ether extract 24.4 per cent; carbohydrate 47.8 per cent; phosphorus 0.10 per cent and iron 12.6 mg/100 g (Gopalan *et al*. 1971). The oil closely resembles nutmeg oil in odour, flavour and composition. Mace yields a fat similar to that from nutmeg, but in a much smaller amount. The leaves of M.fragrans have 0.41-0.62 per cent of a light brown volatile oil with a pleasing spicy odour on water distillation (Khan and Krishnaswamy, 1953). Volatile oil can be extracted from the bark, flowers and fresh pericarp (Verghese, 2001).

Botany

The genus *Myristica* consists of about 120 species of which five are described from India. They are *M.fragrans* Houtt, *M. malabarica* Lamk, *M. magnifica* Bedd, *M. beddomei* King and *M. contoita* Warb (Santapau and Henry, 1973).

Nutmeg is a dioecious or occasionally monoecious evergreen aromatic tree usually 10-20 m in height (Shanmughavelu and Rao, 1977) with spreading branches which carry oblong-ovate leaves, acute at apex and base, 5-15 cm long and 2-7 cm wide, of feathery structure, dark green and lustrous (Verghese, 2000a).

Nazeem (1979) observed the shoot growth in nutmeg to be cyclical, a period of growth followed by quiescence. Six flushes were observed in an year. All the flushes were not seen in all the shoots which resulted in continuous growth. Two growth peaks were observed, in May-June and September.

Joshy (1946) described the inflorescence of M. fragrans as an axillary raceme, but Joseph (1980) observed that it is a branched raceme in male plant and simple cyme in female plant. Flowers are drooping, creamy yellow and fragrant. Nutmeg is usually dioecious. Krishnamoorthy (2000) identified five different types of trees viz. pure male, pure female, bisexual male, bisexual female and hermaphrodite. The hermaphrodite nature is very rarely seen whereas the bisexual types are seen in about three per cent of the flowered trees.

The flower is bracteate and bracteolate (Nair and Bahl, 1956). Sastri (1954) and Nair and Bahl (1956) described the perianth as receiving, ten vascular traces and have postulated a pentamerous origin. The androecium consists of a solid column or androphore to which is attached 14-22 bilocular anthers. The single pistil is more or less flask shaped with a short to non-existant style and bilobed stigma. The ovule is single (Joshy, 1946; Sastri, 1954 and Nair and Bahl, 1956). The fruit is pyriform and yellow in colour. The pericarp is fleshy. When the fruit matures, it splits into two, exposing the scarlet-coloured net like aril covering the dark brown seed.

The seed anatomy has been worked out by Sastri (1955). There is a massive vascular supply to the testa, tegmen and aril. The testa and tegmen are thick and multiplicative. Nucellus enlarges by cell growth without division, then it is absorbed and crushed. Endosperm is oily and starchy. Ruminations formed by outgrowths of the inner brown vascular layer of the tegmen, push the nucellar tissue into the endosperm, except at the pointed micro pylar end. The aril arises from the outer integument, but Nair and Pillai (1959) considered that it was funicular in origin.

In the pollen germination studies, Nazeem (1979) reported that the highest germination percentage of 96.9 was obtained in a nutrient medium containing 4 per cent sucrose, 25 ppm calcium nitrate and 75 ppm boric acid whereas Krishnamoorthy (1986) obtained the highest pollen germination (57.8%) with 5 per cent sucrose, 100 ppm calcium nitrate and 200 ppm magnesium sulphate.

Flowering and fruit set

Flowering pattern of male and female trees differ (Nazeem, 1979). She observed monthly variation in the extent of flowering of both male and female trees under Kerala conditions. In females, flowering was confined to seven months, whereas in male, flowering was observed throughout the year. Maximum flowering in both the cases was in July followed by October. Flower bud development in male and female trees followed specific pattern. Male flowers took only about half the period taken by the female flowers to develop. Female flowers took 154 days for complete development. In male flowers, peak anthesis was between 21.00 hrs and 03.00 hrs. Anther dehiscence occurred about 24 hours prior to anthesis. Stigmatic receptivity lasted for six days after anthesis with the maximum for the first three days.

Studies at Kallar and Burliar in Tamil Nadu showed that the pistillate and staminate flowers opened between 6 pm and 7 pm and anther dehiscence occurred 12 hrs ahead of opening of flowers. Stigma was then receptive and it retained the receptivity till 2.30 pm on the day after flower opening. Chief agent of pollination is wind. Percentage of set varied for different trees and for different aspects with maximum set on western and eastern aspects (Nazeem, 1979). Mean percentage drop after set was 74.4. Number of fruits harvested amounted to only 8.47 per cent of the total flowers produced. Period of maximum fruit drop after set coincided with the period of maximum development of the fruit. Fruits attained maturity in 206 to 237 days after fruit set. The developing fruits followed a sigmoid growth pattern.

Dioecy

Attempts to characterise dioecy in nutmeg yielded no definite methods to identify the male and female plants. Nair *et al.* (1977) postulated a method for distinguishing the sex by the shape of calcium oxalate crystals in the lower epidermal cells of the leaves of the plant of at least two years age. Male plants showed a single large rhomboidal or prismatic crystal with rectangular or square flat faces. Female plants have a large cluster of small crystals. Phadnis and Chowdhary (1971) reported colour differences in the leaf extracts of male and female nutmeg plants when treated with ammonium molybdate. But Thomas (1997) could not make out any such difference in the reaction of ammonium molybdate. Thomas (1997) observed that the average leaf width was larger in female plants compared to male plants. The length by width ratio of leaves was higher in male plants than in female plants. A slightly higher percentage of essential oil was noticed in females whereas in males the leaf phenol content was slightly on the higher side. However, the observations did not confirm to a definite pattern.

Babu *et al.* (2000) standardised the DNA isolation and RAPD protocol in nutmeg for sex determination. No polymorphism was observed among male and female plants of nutmeg.

Crop improvement

Hardly any work has been done in crop improvement. The main difficulties are the long juvenile phase, the difficulty in propagating vegetatively, the dioecy and the single ovule in female flower (Sriram, 1977).

A high amount of variability was reported in growth rate, productivity, size and shape of the leaf, flower size and shape and size of the fruit and nut (Shanmughavelu and Rao, 1977 and Sriram, 1977).

Variability and inter character association for fruit number, fruit weight, seed plus mace weight, seed weight and mace weight were studied by Krishnamoorthy *et al.* (1991). The study indicated maximum variance for fruit number per tree which had a significant negative correlation with mace weight. Seed weight also had a very high positive significant association with mace weight. So selection will be effective in improving fruit number in nutmeg and selecting nutmeg trees with optimum fruit number and moderately good seed weight.

Haldankar *et al.* (1999) observed that the fruit characters viz. weight of fresh nutmeg and aril, pericarp weight, total fruit weight, weight of dried nutmeg and aril showed positive and significant genotypic and phenotypic correlations with yield. High estimate of heritability for fruit attributes along with moderately high genetic advance suggested that these characters are under the control of additive gene action.

Varieties

Konkan Sugandha

This variety was developed by Dr. Balasaheb Sawant Konkan Krishi Vidyapeeth at Regional Fruit Research Station, Vengurle. This is the only hermaphrodite variety released. It yields 2.63 kg dry nuts (526 nuts) per tree at the age of 15 years. The nut size is 5g with 1.2 g mace.

Konkan Swad

This variety was developed by Regional Coconut Research Station, Bhatye, Ratnagiri, Maharashtra. It gives an average yield of 761.38 kg fruits/tree. Seeds contain 39.8 % essential oil and mace 10.9 %.

IISR-Vishwasree

This is a high yeilding nutmeg selection from IISR Kozhikode. It gives an average yield of 3122 kg nut/ha (dry) and 480 kg mace (dry)/ha. This variety posesses 7.1 per cent oil in nut and mace, 13.0 per cent oleoresin in mace and 2.5 per cent oleoresin in nut. Butter content is 30.9 per cent in nut.

Crop managemant

Nutmeg thrives well in places from sea level upto about 600 M height, with a warm humid climate, as in the west coast of India and the Nilgiris. Awell distributed annual rainfall of 250 cm is ideal for the crop. It can not tolerate excessively dry as well as water logged conditions. Nutmeg can be adantageously grown mixed woth coconut and arecanut.

Propagation

General method of propagation is through seeds collected from regular bearing and high yielding trees, yielding more than 10,000 fruits per tree per year and having 30 g weight per fruit, 1 g wet mace per fruit and 10 g wet weight of nuts per fruit (Bavappa and Ruettimann, 1981).

Seed propagation

Seeds from well ripe naturally split healthy fruits harvested during June-July are used. Seeds have low viability and hence are to be sown immediately after collection. They can be preserved in moist sand for 3-4 days (Mathew, 1992) in polythene bags (Madhusudanan and Babu, 1994) or other containers having suitable rooting medium (Kannan, 1971 and Krishnamoorthy, 1987). Old seeds and those in which kernels rattle inside the shell will not germinate. Seeds can be sown at a spacing of about 30 cm apart and 2.5 – 5.0 cm deep. Regular watering is necessary for good germination (Krishnamoorthy and Rema, 1988). Heavier seeds sown immediately after harvest during June recorded highest germination percentage and maximum vigorous seedlings (Mathew, 1992).

Transplanting : Germination commences from about 40th day and lasts for upto 90 days after sowing. Germinated sprouts must be transferred to polythene bags (30 cm x 15 cm) containing a mixture of soil, sand and well decomposed cowdung in the ratio 3:3:1, without much delay. One year old seedlings are transplanted to bigger polythene bags (35 x 20 cm). Seeds treated with 200 ppm gibberellic acid increased the germination percentage (75%). This treatment also showed earlier germination. Height of the seedlings was also increased by gibberellic acid treatment. Seedlings are transplanted to the main field from 18-24 months (Kannan, 1971; Shanmughavelu and Rao, 1977; Nair, 1978 and Krishnamoorthy and Rema, 1988).

Growth pattern: Nutmeg tree exhibits dimorphic growth pattern. The tree exhibits orthotropy in the trunk and plagiotropy in the branches. Attempts are made to modify the growth pattern by various workers, but with no success. Rani (1994) reported that severe pruning can induce orthotrops on otherwise plagiotropic shoots. Age of the wood was the main deciding factor in the induction of orthotrops. Physical or chemical treatments imposed could not induce orthotrops on nutmeg grafts having one year old plagiotrop as the scion part.

Vegetative propagation

Nutmeg being a dioecious crop, proper ratio of female and male plants is to be maintained in the plantation. Vegetative propagation is the practical way to achieve this. Many vegetative propagation methods were standardised to develop a suitable method for relatively rapid multiplication. Vegetative propagation techniques like stem cuttings, air layering, approach grafting, epicotyl grafting, budding and top working were tried in nutmeg with varying degrees of success (Sundararaju and Varadarajan, 1956; Kannan, 1973; Shanmughavelu and Rao, 1977; Rasalam, 1978; Mathew and Joseph 1982; Mathew, 1985 and Rethinam and Edison, 1991)

Epicotyl grafting : At Central Plantation Crops Research Institute (CPCRI), Kasaragod, Mathew and Joseph (1982) tried epicotyl grafting using M. beddomei as root stock and attained 48 per cent survival of grafts. Krishnamoorthy (1987) obtained 80 per cent successful grafts during the month of August when M. fragrans was used as the root stock.

At the Indian Institute of Spices Research, Kozhikode, epicotyl grafting was the most successful method (Krishnamoorthy and Rema, 1988). The technique of epicotyl grafting is given below :

1. Take 20 days old healthy sprout (root stock)
2. Select a good scion (of lead pencil thickness) from a high yielding female tree
3. Cut the top and make a vertical split to a length of 3-4 cm in the root stock
4. Make a wedge cut of 3-4 cm length in scion
5. Insert the scion in the slit of the root stock and tie it firmly with a polythene strip
6. Transplant the grafted sprout to the polythene bag filled with the potting mixture. Grafted portion is covered with a polythene bag to ensure high humidity which helps in graft union
7. Remove the cover bag when new leaf emerges. Also remove the polythene strip at the side of the grafting.

Maximum percentage of successful grafts (about 80%) could be obtained if grafting is carried out during August-September.

Epicotyl grafting : It was standardized at Konkan Krishi Vidya Peeth, Dapoli (Haldankar *et al.*, 1999). Medium matured (4 months) and fully matured (6 months) scion sticks recorded higher success. It is essential to provide a cover of polybag on scion stick, especially in post rainy season. The location of scion stick did not influence the success of epicotyl grafting and September was the most favourable season for epicotyl grafting.

Harvesting season of nutmeg in Konkan region is confined from June to October. Hence availability of root stock particularly for epicotyl grafting is a limitation. Soft wood grafting method developed at Regional Coconut Research Station, Bhatye, Maharashtra provides scope for grafting nutmeg for prolonged period. May was the best month for soft wood grafting with maximum success of 80 per cent. Retension of leaves on root stock did not influence the success of softwood grafting. Characteristic of scion stick and method of grafting were similar to that of epicotyl grafting (Haldankar *et al.*, 1999).

Approach grafting: In India, successful approach grafting done on root stocks of *M. fragrans, M. malabarica* and *M.beddomei* is reported by Sunderaraju and Varadarajan (1956). Those on M. beddomei and M. malabarica develop into low spreading trees whereas nutmeg seedlings generally develop into tall trees. Grafts come to flowering in 4-6 years of planting vs 6-8 years by the seedlings. Rasalam (1978) claimed 100 per cent success on wild species of nutmeg, when erect branches of female nutmeg trees were used as the scion, flowering commenced in the second year after grafting.

Trial on approach grafting at Dapoli showed that the approach grafts can be prepared throughout the year. High percentage of graft success was recorded on wild nutmeg root stocks (*M. malabarica*, 30-100 %) while on *M. fragrans*, it was 40-90 per cent. The mortality after separation of grafts was 30 per cent on *M. fragrans* and 50 per cent on M. malabarica (Haldankar *et al.*, 1999). However, approach grafting is time consuming and a tedious method of grafting.

Budding: Forkert method of budding was tried by Mathew and Joseph (1982) in *M. fragrans* and *M. beddomei* and observed that percentage of success was very low (4%) in both the cases. They noted that maximum bud take was in May which coincided with the flushing season in nutmeg. Buds remained alive even after one year without sprouting. Forkert budding on sprouts of the topworked trees also gave only initial success, which failed to sprout even with physical and chemical treatments (Beena, 1994).

Commercial coverage through vegetative propagules is very much limited due to the unavailability of orthotropic shoots for use as scion. As a result, plagiotropic shoots are collected as the scion material. But such plants have a peculiar spreading habit resulting in squatty plants with stunted growth compared to seedlings of the same age. Therefore, the fruiting area is comparatively less resulting in low yield. Because of these factors, seeds continue to be the major propagating material even now.

Top working: The sex of nutmeg tree can be identified only 7-8 years after planting, when they begin to flower. Generally, male and female trees are produced in 1:1 ratio. Since a single male tree is sufficient for every 20 female trees for pollination, the rest of the unproductive male trees can be made productive by converting them into female trees by topworking.

Trials by Beena (1994) on topping of male trees indicated that heading the trees above the first tire during August was the best with regard to sprout production and reducing the time for sprouting. Successful graft union by cleft grafting was obtained during March with scion shoots having mature leaf and full green stem and stock having two months growth.

The top working techniques adopted at Indian Institute of Spices Research (IISR), Kozhikode by (Rema and Krishnamoorthy, 2000) is as follows : Male trees are beheaded about 30 cm above the ground level in May, the cut end is smeared with Bordeaux paste. Decapitated trees are irrigated regularly till the onset of monsoon. New shoots develop from the main trunk within 45-60 days. The newly emerged shoots are ready for grafting when they attain about 20 cm length. A longitudinal cut of about 3 cm is made from the top downwards on the retained stock. This slit gives a 'v' shape when split opens. Orthotropic scions of the same thickness as that of the stock are collected in water from high yielding female trees of nutmeg on the same day of grafting. Scions with two leaves are used for grafting. The lower end of the scion is shaped into a wedge by removing the bark with a little wood from both the sides of the scion with a sharp knife. The scion is inserted into 'v' shaped slit of the stock and secured in place firmly with a 1.5 cm wide polythene strip. Grafted shoot is covered with a polythene cover to maintain humidity and to prevent the scion from drying. A temporary shelter is provided above the tree to prevent the harmful effects of rain or scorching sun. The polythene cover is removed after 30-45 days when perfect union takes place. Development of new buds from the grafted scion is an indication of perfect union. After successful union, all other shoots are removed from the main trunk and Bordeaux paste is applied on the cut surface. Overhead shade can be removed after ensuring successful union of the graft. If the graft fail, regrafting can be done on either the same shoot or on a different shoot. This technique assures cent per cent success. Top working can be used for conversion of unproductive male to productive female trees. The top worked trees yield from the 3rd year onwards. Aftercare of top worked trees include removal of new shoots that sprout from the main trunk and regular management practices especially irrigation.

In situ budding : In situ budding is now the most popular method followed by farmers of Kerala to solve the two major problems that restrict large scale cultivation of nutmeg namely doiecy and long juvenile phase.

Budding can be done on plants within 2-5 years of planting. Orthotropic scion shoots from elite mother trees are budded on seedlings two years after planting in the main field. The root stock and scion should be at the bark slipping stage. Budding is done just above the first whorl of lower leaves.

Studies at Kerala Agricultural University by Beena (1994) proved that *in situ* budding on hard trunk was the most successful. Forkert method with brown bud wood having fallen leaves and the buds inserted by leaving space on all four sides gave maximum success. In order to achieve a quicker bud burst, stumping the plant two months after budding was the most effective. Budding on hard trunk could be done with maximum success in three year old plants and July was found the best season.

Rani (1994) obtained maximum success with patch budding on *M. fragrans* and forkert method on *M. beddomei*. August was the best season for in situ budding and July for budding in polybag plants. Partial shade was ideal for better bud take than the mist chamber conditions. Stumping the root stock above bud union retaining the whorl of lower leaves gave better sprouting and late survival.

Planting and aftercare

Nutmeg requires a hot humid climate with no pronounced dry season. Soil should be rich in organic matter and well drained. The tree prefers partial shade. Sheltered valleys are the best suited. It can be grown upto about 900 m above MSL (KAU, 2003)

In India, seedlings are planted during south-west monsoon. Various spacings and pit size recommended are furnished hereunder :

Spacing (m)	Pit size (cm^3)	Reference
8.0 x 9.0	60	Shanmugavelu and Rao, 1977
9.0 x 9.0	75	Krishnamoorthy, 1987
6.0 x 6.0	80	Bavappa and Reuttimann, 1981
8.0 x 8.0	80	KAU, 2003

Ten per cent of males may be retained for pollination and the remaining male trees may be either removed or converted to females by top working.

Nutmeg is a shade loving plant (Kannan, 1971), young as well as grown up plants requiring certain amount of shade. Locations with permanent natural shade will be the optimum. For open space, artificial shade is provided by growing *Glyricidia* sp, Dadap (Bavappa and Ruettimann, 1981), banana, *Acacia* sp and Subabool (Krishnamoorthy, 1987). Lopping of branches may be done at later stages to regulate shade.

As an intercrop in coconut plantations, nutmeg is planted in between two rows of coconut so as to accommodate 125-150 plants per ha (Varghese *et al.*, 1990). In the Andaman and Nicobar islands, combinations of coconut or arecanut + nutmeg + robusta coffee + or forest trees + nutmeg + robusta coffee + pepper were successful (Rao, 1991). Spectacular yield increase of coconut palm and thereby total yield by intercropping nutmeg in coconut plantation were reported by Patil *et al.* (1991).

Regular weeding and mulching keep the field clean and conserve miosture (Shanmugavelu and Rao, 1997). Cover crops like *Mimosa* sp and *Stylosanthes* sp may also be cultivated for suppressing weed growth (Bavappa and Ruettimann, 1981). Application of herbicide mixture (gramaxone and fernoxone) controlled weeds up to six months (Anandaraj *et al.*, 1989). Seedlings may be irrigated periodically during summer (Kannan, 1971 and Shanmughavelu and Rao, 1977). For four year old plants, 20 l of water per plant thrice a week are given, the quantity of water is to be increased at later stages of growth (Krishnamoorthy, 1987).

Nutrition

Mineral nutrition of nutmeg was studied in relation to deficiency symptoms and flowering by Philip (1986). Deficiency symptoms of macro and micronutrients and the influence of macro and micronutrient deficiencies on the foliar level of other nutrients are furnished hereunder:

Table 25. Nutrient deficiencies and foliar nutrient content in nutmeg

Nutrient deficiency	Elements increased	Elements decreased
Nitrogen	P, B, Zn	Mg, Fe, Mn
Phosphorous	N, Zn	Mg, Mn
Potassium	Ca, Mg	Nil
Calcium	K, Mg, N	B
Magnesium	Ca, K, Cu	N
Sulphur	N, P	Nil
Iron	P, Zn, Mn	K, Ca
Manganeze	Fe, K, N	Nil
Copper	Fe, Mg	Ca
Zinc	P, fe	Mg
Boron	N, P	Ca, K

Flowering shoots registered significantly higher C/N ratio in leaves than that of non-flowering shoots. Flowered shoots always registered a relatively higher P/S, Ca/S, N/P and N/K ratios and a lower foliar level of Ca/Mg, N/P and N/K ratios.

Shanmugavelu and Rao (1977) recommended a fertilizer shedule of 1 kg each of ammonium sulphate, super phosphate and muriate of potash along with 50 kg of compost per year to mature plants. Seedlings and young plants are to be supplied with one tenth and half the dose respectively. The fertilizers are to be applied during May-June and September-October. The Kerala Agricultural University (KAU, 2003) recommended a fertilizer schedule of 20:18:50 g of NPK along with 15 kg of compost per year during the first year of planting which is to be gradually increased to 500:250:1000 g NPK and 50 kg of compost per year from 15th year onwards. Rao (1991) reccommended 400:320:400 g of NPK respectively for an adult tree per year in Andaman and Nicobar islands.

Whole quantity of organic manures may be applied in one dose just at the commencement of monsoon in May-June. Fertilizers should be applied at least in two doses, one in May-June along with the organic manures and again in September-October. Manures should be applied in shallow trenches dug around the plant. As a large number of roots are seen almost on the surface of the soil, trenches should not be deep or very near to the plant (Kannan, 1971). About 8 to 10 cm deep and 15-20 cm wide trenches may be taken about 10-20 cm away from one year old plant. Distance from the base of the plant to the trench should be increased every year and for a 15 year old plant it should be about 1 to 1½ metres. Trench should be covered after application of manures.

Pests and diseases

Die back caused by *Diplodia natalensis* causes drying of branches from tip downwards. Cutting the infected branches and smearing Bordeaux paste in the cut ends controls the disease. Leaf spot caused by *Cladosporium oxysporum* was reported from Dapoli (Lawate *et al.,* 1986). Bavistin (0.2 %) application controls the disease.

Harvesting and processing

Fruiting commences from fifth or sixth year but may take even eight or nine years (Verghese, 2000a). Plants which begin to fruit early are short lived. Delayed fruiting in the 9th year is the best. Optimum productivity is attained in about 15 years. Adequate yield is obtained even from 45 years old trees. Fruits ripe in 6-9 months after flowering (Nazeem, 1979). Trees raised from seedlings begin to bear 5-8 years after planting whereas vegetatively propagated trees fruit much earlier. In India, fruits that split on the tree exposing crimson coloured aril are harvested (Nair *et al.,* 1977). Peak harvesting periods vary in different places. In Konkan region, it is confined from June to October, where as in Kerala it is June-July.

Yield varies among trees and plantations. Average yield of a good tree in full bearing is 3000 fruits (Krishnamoorthy, 1987) between 15th and 30th year of growth. Average weight of a single fruit is 60 g of which the seed weighes 6-7 g, mace 3-4 g and the rest pericarp (Krishnamoorthy, 1987). Ratio of mace to nutmeg is 1:8. Immature fruit drop occurs to an extent of 74.4 per cent (Nazeem, 1979) due to degeneration of embryos resulting from inadequate pollination, water scarcity, root competition due to closer planting or interplanting with shade trees, nutritional imbalance, fungal attack and physiological reasons.

Fruits are collected from the tree by hand or with hooked sticks or allowed to fall naturally on the ground and are gathered everyday.

Fruit is composed of three parts: the pericarp or husk, the mace and the seed. The pericarp is removed and the mace which envelopes the shell is peeled off. The "blades of mace" as the peelings are called are flattened by hand or between boards and are spread out to dry in the sun. In good dry weather, drying operation is accomplished in two or three days. Exceptional care must be taken to prevent mace getting mouldy. A perfect sample of mace should consist of entire double blades, not broken, flattened and of large size, horny in texture and not too brittle and of a good, clear and bright colour. Nut is left in the shell and dried in the sun or in drying ovens. Drying is complete when the seed rattles. Nutmeg loses about 25 per cent of their weight by drying. Shell is then cracked with wooden hammers or mechanically in specially designed machines and discarded or used as fuel and the nutmeg removed and dried. The ratio of dried nutmeg to dried mace is approximately 20:3.

Manoj (1990) reported that the aromatic ethers which are the chief components that determine the flavour and drug action in nutmeg oil was the maximum one month prior to the fruit splitting stage. In mace oil, it was found to be high two months prior to the fruit splitting stage. If nutmeg and mace oils are intended for medicinal purpose, then it may be worthwhile to harvest fruits at the sixth month for extracting kernel oil and at the fifth month for extracting mace oil.

Processed products

Oil of mace and fixed mace oil are the mace products. Nutmegs products are volatile oil, nutmeg butter, volatile oil of bark and volatile oil from flowers. Various edible food products are made from the fruit rind of nutmeg.

Uses

Nutmeg is a stimulant, carminative, astringent, aphrodisiac and hallucinogen. Mace is used for flavouring cigarettes and chewing to mask foul breath. Oil of nutmeg or mace is employed for flavouring food products and liquors, soaps, tobacco, dental creams and perfumery products. Volatile oil from leaf has weedicidal properties. Nutmeg butter is used as a stimulant in ointments, hair lotions and is used in cases of rheumatism, paralysis and sprains. Fleshy pericarp of the fruit is used for making pickles and jelly. Pruthy and Krishnankutty (1984) obtained quality pectin from nutmeg waste (rind).

Cinnamon
(*Cinnamomum verum* Bercht & Presl., F. Lauraceae)

Cinnamon was one of the first spices prized and enjoyed by man since the early days of civilization. It was precious not only as a flavouring agent, but was estemmed as a medicine, as a perfume and as one of the aromatics burned as incense. Cinnamon (*Cinnamomum verum* Bercht & Presl.) is an important tree spice of the family Lauraceac. The dried bark of *C. verum* is the true cinnamon and cassia is obtained from various sources like *C. cassia, C. burmanii, C. laureirii* and *C. tamala.* Pure cinnamon is free from any admixture with cassia, which is considered inferior to the former in appearance, flavour and odour.

The genus, a native of South-Western tropical India and Sri Lanka, consists of more than 250 species distributed in South East Asia, China and Australia. Seychellus and Malagasy Republic are the major producing countries of cinnamon besides Srilanka. In India, cinnamon is mainly cultivated in Kannore (Kerala), South Kanara (Karnataka), Nilgiris, Lower Pulneys, Courtallam and Kanyakumari (Tamil Nadu) in India (Krishnamoorthy *et al.*, 1996). The domestic production is about 200 t of bark per year and about 60 t is imported annually.

Chemical composition and uses

Cinnamon bark

Cinnamon of commerce is the dried inner bark of *Cinnamon verum.* The bark, oil and oleoresin are the economically important products. Bark either as small piece or as powder is extensively used as spice or condiment. Increasing popularity of the preserved food products has ensured considerable growth of this spice. Cinnamon bark is used for domestic culinary flavouring and for industrially manufactured sauces, candy, pickles and some beverages. In ground form, it is used in curry powders and to flavour baked foods like bun and cakes and in the preparation of various preserves. Besides this, it is used in the manufacture of soaps, dental preparations and perfumes.

According to the indigenous medicinal preparations, cinnamon bark has carminative, astringent, stimulative and antiseptic properties. It is more powerful as a local than as a general stimulant. It is prescribed in powder and infusion but usually combined with other medicines. It checks vomiting and relieves flatulence.

Cinnamon oil

Cinnamon oil is obtained from bark (bark oil), leaf (leaf oil), root (root bark oil) and seed (seed oil). The yield of leaf oil is 1.6-3.4 per cent whereas that of bark oil is 0.95-3.55 per cent The main constituent of leaf oil is eugenol (75-80%) and that of bark oil is cinnamaldehyde (65 %). Root bark oil and seed oil are not products of commerce. However, a medicinal oil is obtained from unripe fruits.

Although cinnamon bark oil is sometimes used in perfumery and pharmaceutical preparations, its main value is as a flavouring agent and is substituted for the spice many a times. As the oil has a powerful odour and flavour, it must be used sparingly and most carefully. Some quantity of the oil is used in pharmaceutical preparations and to flavour dentifrice and mouth wash. As a powerful local stimulant, it is at times prescribed in flatulence and gastric disabilities. It has a high germicidal and fungicidal activity but the irritant nature prevents its use as such.

Cinnamon leaf oil almost equals clove oil in its eugenol content and in this respect cinnamon leaf competes with clove stem and leaf oil. Leaf oil of cinnamon is used for flavouring sweets and confectionery and is a common adulterant for the bark oil. Oil is also used as an embrocation in rheumatism.

Table 26. Physico-chemical properties of cylon cinnamon leaf and bark oil

Property	Leaf oil	Bark oil
Specific gravity	1.07	1.023-1.040
Optical rotation	1^0 96 ′ to 0^0 40′	Slightly laevo rotary
Refractive index	1.5288	1.581 to 1.591
Aldehyde content	4 %	65 to 76 %
Eugenol content	77.3 to 90.5 %	4 to 10 %
Solubility	1.5 volumes of 70 % alcohol	2-3 volumes of 70 % alcohol

Source: Gopalan, 1997

Oleoresin

Bark is distilled for the preparation of oleoresin and the recovery is about 10-12 per cent. It is a dark brown oil containing 50 per cent of volatile oil and has to be diluted before using it as a flavouring agent.

Other minor products of cinnamon are buds and wood. Cinnamon buds are also used for flavouring and spicing. Wood provides soft timber for low grade board wood.

Botany

Cinnamomum verum is a medium sized evergreen tree reaching a height of 15-20 m in a period of 25-30 years. But it is generally coppiced for the enhancement of side shoots.

Taxonomy

The wide variability in the morphological characters and the surprisingly uniform floral characters make this genus a taxonomically different one. The epidermal characters of the more important species occurring in Kerala have been studied by Ravindran *et al.* (1993) with the aim of using them as an aid in taxonomic characterisation. *C. verum* (wild and cultivated), *C. cassia, C. camphora, C. riparium, C. macrocarpum, C. nicolsonianum, C. perrottetii and C. malabatrum* were studied. All the species have anomocytic stomata except in which the stomata is paracytic. The stomatal characters showed absolute correlation with phyllotaxy which in turn were negatively correlated with stomatal frequency and epidermal thickness. These characters were helpful in delimiting *C. camphora* from all other species and *C. perrottetii* from *C. verum, C. malabatrum* and *C. cassia.* Trichome characters were useful in differentiating *C. perrottetii, C. riparium* and *C. macrocarpum* from other species.

In another study, cluster analysis was carried out in *Cinnamomum* species occurring in Kerala. Twenty eight collections representing eight species were used in the study. They were grouped using the centroid linkage technique and the clustering showed that *C. camphora, C. perrottetii, C. riparium* and *C. cassia* are very distinct types and hence formed independent clusters. *C. malabatrum* is more complex having considerable intraspecific variability and hence ended up in more than one cluster, pointing to the need for a re-examination of this species. These results are further supported by the flavonoid patterns also (Shylaja, 1984).

Seed grown population of cinnamon exhibit varied types of leaf size and colour. These can be grouped as small ovate light green, large ovate light green, medium narrow dark green, long narrow dark green etc. (Sahoo *et al.* 2000). Leaf is lanceolate or elliptic, opposite, triplinerved, strongly ascent, two of the basal nerves reach the tip, the juvenile leaves dark purple, red or pale green, oil and mucliage cells present in mesophyll layers.

Floral biology

Floral biology was reported by Mohankumar *et al.* (1984). Flowers are small, numerous in terminal and axillary panicles on current seasons growth. Peduncles are creamy white and five to seven cm long. Individual flowers are three mm in diameter with foetid smell. Each flower is subtended by a hairy bract. Perianth six; two seriate and pubscent. Stamens nine, in three whorls which stand at the base. Filaments hairy. Anthers are four celled, dehisce by four small vulvular lids. Staminodes three. Ovary superior, one celled with solitary anatropus ovule. Fruit is a berry surrounded by enlarged perianth.

Flowering is from November to March. On an average, 13.83 days are required for the flower bud development from the stage of its visible initiation. Peak anthesis is from 11 am to 12 noon. Stigma receptivity is maximum on the day of anthesis. Cinnamon flowers are cross pollinated. Ponnuswamy *et al.* (1982) reported variability in this crop which is indicative of cross pollination. Insects are the pollinating agents.

Crop improvement

Variability studies with 101 open pollinated accessions indicated that a wide range and high co-efficient of variability exist for number of peeler shoots per plant, plant height, plant girth and tree spread. (Ponnuswamy *et al.* 1982). It showed that there is scope for further improvement through selection.

Haldankar *et al.* (1994) identified four promising genotypes from among 300 cinnamon genotypes. Genotypes varied significantly for most of the characters of growth and yield. The bark yield varied from 33.35 to 84.5 g whereas the leaf yield (fresh) ranged from 1.47 to 3.69 kg. Oil percentage from bark varied between 1.88 and 3.2, cinnamaldehyde content from 66.77 to 70.23 per cent and eugenol from 5.71 to 8.61 per cent.

Paul and Sahoo (1993) isolated a promising selection RRL (B) C-6 from seedling progenies. It was the best genotype with stable high oil content and quality with 94 per cent eugenol in leaf oil and 83 per cent cinnamic aldehyde in bark oil. (Paul *et al.* 1998).

Krishnamoorthy *et al.* (1988) studied the relation between colour of young flushes and quality parameters. Plants with purple coloured flushses had 29 per cent more bark oil compared to green flush plants. The bark oleoresin and leaf oil content did not show any difference to such young leaf colours.

Varieties

Navashree & Nithyashree

Two high yielding and high quality cinnamon selections were released from IISR, Kozhikode namely "Navashree" and "Nithyashree" (Krishnamoorthy *et al.* 1996).

Table 27. Characteristics of elite cinnamon lines; Navashree and Nithyashree

S. No.	Character	Navashree (Sl. 63)	Nithyashree (Sl. 189)
1.	Regeneration capacity (shoots / plot of 4 plants)	25.45	18.90
2.	Fresh weight of bark (g)	488.95	511.15
3.	Dry weight of bark (g)	201.1	194.6
4.	Yield of bark/ha (kg)	55.56	54.16
5.	Recovery of bark (%)	40.6	30.7
6.	Bark oil (%)	2.70	2.70
7.	Yield of bark oil (l/ha)	1.50	1.46
8.	Bark oleoresin (%)	8.0	10.0
9.	Yield of oleoresin (kg/ha)	4.44	5.42
10.	Leaf oil (%)	2.80	3.00

YCD-1

It is a cinnamon selection from Horticultural Research Station, Yercaud. It comes to harvest from third year onwards and can be maintained economically for 20 years. It gives a bark yield of 359.75 kg quills and 3800 kg of dried leaves/ha with high bark recovery of 35.3 per cent. It has got 2.8 and 3 per cent volatile oil in quills and leaves, respectively. It gives a high regeneration capacity of 19.2 harvestable shoots. Quills are sweet and light pungent in taste (Kennedy and Balakrishnamoorthy, 2000b).

Konkan Tej

The variety "Konkan Tej" was developed by clonal selection at Konkan Krishi Vidyapeeth, Dapoli which has high oil (3.2 %), cinnamaldehyde (70.23 %) and eugenol (6.93 %) and yields 789.75 g fresh bark and 3.56 kg leaf per plant.

Sugandhini

It is a selection from Aromatic and Medicinal Plant Research Station, Odakkali, Kerala Agricultural University. It is recommended for cultivation in the midlands and highlands of Kerala both in open and as intercrop in coconut gardens for cinnamon leaf oil. Average leaf yield is 18 kg/tree/year and average bark yield is 1.2 kg/tree/year. Leaf oil yield is 295 ml/tree/year or 125 kg/ha and the eugenol content is 94 per cent in leaf oil (Joy *et al.*, 1996).

PPI (C)-1

It is a selection made at Horticultural Research Station, Pechiparai, Tamil Nadu. It gives a fresh bark yield of 980 kg/ha. Bark recovery is 34.22%, leaf oil recovery 3.3% and bark oil recovery 2.9%.

RRL(B)C-6

It is a selection made at Regional Research Laboratory, CSIR, Bhubaneswar. Bark oil yield is 250 kg/ha. Bark is of high quality with 83.0 % cinnamaldehyde in the oil. Leaf oil contains 94.0 % eugenol.

Crop management

Climate and soil

Cinnamon is a hardy plant which tolerates a wide range of climatic conditions. The crop performs well from 300 to 350 m above MSL and thrives up to 1000 m above MSL (Nair, 1970). It flourishes in places with an annual rainfall of 150-250 cm and an average temperature of 27 ^{0} C. A hot and moist climate is highly suited for cultivation of cinnamon (Radhakrishnan, 1992) and prolonged spells of dry weather are not conducive for its growth.

Propagation

Cinnamon is commonly propagated through seeds (Joseph, 1981). Seeds are extracted from ripe fruits from selected mother trees with desirable characters like smooth bark, erect stem, easy pealing of bark, vigorous growth, free from pests and diseases and having good qualities like sweetness, pungency and flavour. The seeds are sown immediately after collection, otherwise viability is reduced. Seeds are sown in nursery beds or in pots filled with a mixture of sand, cattle manure and soil in the ratio 2:2:1 (Nair, 1970). Seeds may be sown in rows of 12 cm apart in nursery beds and covered with thin layer of soil. Radhakrishnan (1992) observed July-August as the best time for sowing. Beds may be watered and shade provided during early stages.

Kannan and Balakrishnan (1967) obtained the highest germination of 94 per cent by sowing seeds on the third day after harvesting. After 40 days, there was complete loss of viability. Under normal conditions, seeds germinate within 20 days (Krishnamoorthy and Rema, 1988). From beds, seedlings are transplanted to polythene bags when they attain a height of 15 cm. Polythene bags of 30 cm x 15 cm size filled with soil, sand and farm yard manure (3:3:1) are used (Krishnamoorthy and Rema, 1988).

Cinnamon is also propagated by cuttings and layerings. Vadivel *et al.* (1981) reported 45 per cent rooting in hard wood cuttings treated with IBA 2500 ppm. Nageswari *et al.* (2000) obtained 50 per cent rooting when hard and semi hard wood cuttings were treated with IAA 100 ppm. Success of air layering was reported by Banerjee *et al.* (1982) and Hegde *et al.* (1990). Ranaware *et al.* (1995) reported that the most favourable season for air layering in cinnamon is the month of July, followed by June and August. Haldankar *et al.* (1996) reported 95 per cent success in air layering during September. The maximum period required for complete rooting was 270 days. Propagation by tissue culture is also successful.

Planting and aftercare

Ten to twelve months old seedlings are suitable for planting. (Nair, 1970 and Krishnamoorthy and Rema, 1988). A spacing of 3-5 m between plants is suggested by Radhakrishnan (1992). Shade and irrigation are essential immediately after planting. Weeding and mulching reduce weed growth (Nair, 1970 and Nazeem *et al.*, 1992).

Systematic manuring is rarely practiced in India for this crop. The Kerala Agricultural University (KAU, 2003) recommended a fertilizer schedule of 20 : 20 : 25 g of NPK along with 20 kg of compost per year during the first year of planting which is to be gradually increased to 200 : 180 : 200 g NPK and 50 kg of compost per year from 15th year onwards. Fertilizers may be applied in two splits during May-June and September-October.

Pests and diseases

Major pests affecting cinnamon include *Chilasa clytic* L, *Sorolopha archimedia* Meyr., *Acrocercops* sp., *Popillai coplanata* Newm., and *Singhale helleri* Ohs. The minor pests are *Sauris* sp., *Ceroplastes rubens* Mark., *Apoderes scitulus* Walker and *Oecophylla smaragdina* (Singh *et al.*, 1978).

Diseases reported include die back caused by *Colletotrichum gloeosporioides* (Karunakaran *et al.*, 1993), red leaf spot caused by *Colletotrichum capsici* (Prakashan, 1991) and Colletotrichum blight caused by *Glomerella cingulata* (Kumar, 1983).

Harvesting and yield

Cinnamon harvesting starts from the fourth or firth year after planting depending on the availability of peeler shoots. Nair (1970) and Rao (1991) recommend coppicing from 2nd or 3rd year onwards. Stems are cut during rains to facilitate peeling. Best time for peeling is when new flushes and leaves are hardened after a rainy season (Radhakrishnan, 1992). Plants are coppiced to a height of 12 cm from the ground level. Subsequent harvest is made between 12-18 months after previous harvest. Side shoots are also cut so that the plant assumes the shape of a low bush and a bunch of canes suitable for peeling are available subsequently. Regular peeling operations are commenced from fourth or fifth year depending on extent of peeler shoots available (Rao, 1991). Best quality is obtained from the thin bark of shoots in the centre from the middle portion of shoots. Fully developed cinnamon shoots of 1.5-3.0 cm diameter harvested during rainy season gives good quality bark and high yield (Pruthy *et al.*, 1978). The cut stems are collected, tied and bundled and carried to the peeling shed. Cutting is followed by scraping and peeling. After scraping the outer skin, the peeling of the bark is done carefully by piercing the bark with a sharp knife. The tube like barks thus obtained in lengths 20-28 cm are placed one inside the other to form a tube of 90 cm called quills. These peels are piled up within enclosures of sticks and wrapped up in mats and kept overnight in that condition. A little softening of the bark does occur, with the result that the peels become more pliable for the subsequent piping operation (Pruthy, 1992). These are then dried on a mat in the sun for about three days and thereafter in shade for another three days (Nair, 1970). During the process of drying, the quills should be pressed by hand to prevent them from opening up. Barks which cannot be takenout like tubes are called 'quillings' and the scraped pieces are called 'featherings'. About 70-100 kg of quills and 30 kg of quillings and featherings are obtained from 0.4 ha of cinnamon after four years (Nair, 1970). Yield from one hectare of cinnamon plantation after attaining maximum yield is 200-300 kg of bark and about 35 kg of leaf oil/ year.

Besides quill extraction, the leaves could be used for distillation for obtaining aromatic oil. Conditions for the utilization of cinnamon waste (leaves and twigs) with regard to recovery of essential oil were standardised by Pruthy *et al.* (1978). Yield of cinnamon leaf oil was 1.8-2.6 per cent in fresh leaves and 3.75 per cent in dried leaves. Recovery of oil from leaves dried for 2-3 days in shade is recommended to avoid loss of oil (Pruthy *et al.* 1978).

Cassia (Chinese cinnamon or Cassia Lignea) (*Cinnamomum aromaticum* C.G. Nees. Syn. *C. cassia* Bescht & Presl., F. Lauraceae)

The name "dalchini" meaning cinnamon is actually a derivation from Arabian term "Dar-al-Chini", meaning wood or bark of Chine; one of the oldest and largest producers of cassia, so well known in commerce as China. In pre-war years, trade in cassia used to be more important than that in cinnamon, but situation has now changed (Pruthy, 2001). Now India exports cassia bark and cassia bark powder. Cassia is generally cheaper and considered inferior to cinnamon, and therefore is often used as a substitute. Cassia and cinnamon, when whole can be very easily distinguished from each other, the former being with coarser skin of greater thickness and milder flavour. However, when in ground form, it is rather difficult to identify precisely admixture or purity of cinnamon. In international nomenclature, cassia is described as *C. inners* Reinw. Cassia is also obtained from *C. burmanii* (Indonesian cassia), *C. laureirii* (Vietnamese cassia) and *C. tamala* (Indian cassia). Cassia bark, bark oil, leaf oil and oleoresin are products of commerce which are important in international trade. Inner barks of these species have more intense aroma with high essential oil (1 to 4.5 % as against 0.5 to 3% in true cinnamon bark). Unlike cinnamon, various parts of *C. cassia* contains only one type of essential oil in which cinnamaldehyde is the main component.

Cassia is cultivated mainly in China. The tree is present in evergreen forests of Western Ghats. Cultivation and processing methods are same as that for cinnamon.

Tejpat (Indian Cassia Lignea) (*Cinnamomum tamala* Nees and Eberm., F. Lauraceae)

Leaves of *C. tamala* is the genuine source of "Tejpat" spice of commerce. It is an evergreen aromatic tree found in tropical and subtropical Himalayas up to an altitude of 2400 m. It is more common in North East India and Shillong of Meghalaya is the main market of Tejpat in India (Baruah and Nath, 2001). Yet another species of cinnamon, *C. impressinervium* Meissn. is also reported to give the best quality tejpat (Baruah *et al.*, 1998). *C. bejolghota* and *C. sulphuratum* are also used as tejpat in Meghalaya (Rema *et al.*, 2005).

Crop management

Commercial cultivation of *C. tamala* is very limited in India. Plantations occur in *Khasi* and *Jaintia* hills, *Garo* hills, *Mikir* hills of Manipur and Arunachal Pradesh and in limited areas in UP and HP. Normally tijpat is collected from wild and traded. Cultivation practices are similar to that for *C. zeylanicum*.

Harvesting

Leaves are collected every year from vigorous plants, dried under sun for 3-4 days and tied up into bundles for marketing. Harvesting can be done when trees have put on sufficient vegetative growth. Bark is also collected for adulteration with true cinnamon bark.

Uses

Leaves contain essential oil. Main constituent is linalool. Eugenol and cinnamaldehyde are present in trace amounts only (Nath *et al.*, 1994). Leaves of tejpat are widely used as spice for flavouring meats, sausages and sauces. Bark is also used as a spice for flavouring food.

Clove

(*Syzigium aromaticum* (Merrill et Perry)., F. Myrtaceae)

Clove is one of the most ancient and valuable spices of the orient, known as bark as 100 BC. Its source and place of origen were not known until the Portuguese discovered the Molucca island of Indonesia in the 16th century. The clove of commerce is the dried, unopened flower buds of the evergreen tree *Syzigium aromaticum* of the family Myrtaceae. It is native to Moluccas, the so called 'spice islands' in Indonesia and was first introduced in India around 1800 A.D. by the East India Company in their 'spice garden' in Courtallam, Tamil Nadu. Major clove producing countries are Indonesia, Zanzibar, and Madagascar, Zanziber being the biggest producer. Importing countries are India, USA, Germany, France and Singapore. The world production is estimated to be 63700 t and Indonesia alone accounts for 66 per cent of the world production.

Major clove growing regions in India are Kanyakumari (Keeriparai, Mahendragiri, Pechiparai), Nilgiris (Kallar, Burliar), Dindigal, Madurai, Salem (Shevroys) and Namakkal (Kolli hills) of Tamil Nadu, Calicut, Kottayam, Quilon and Trivandrum districts of Kerala and South Kanara of Karnataka (Balakrishnamoorthy and Kennedy, 1999).

Chemical composition and uses

Cloves are available as whole buds that range from ½ to ¾ inch in length and that have spike like shape. The flavour of clove is characteristic and is some times described as strong, pungent and aromatic. Clove flavour is also available in clove leaf oil, clove oil (prepared from flower buds) and clove stem oil. Clove oil is also called caryopil oil and amboyna (Sharma and Singh, 2000). A wide range of food products are being flavoured with cloves. Clove oil which is rich in eugenol is sometimes used in making synthetic vanillin, but it is a costly process. Except food applications, clove is used in pharmaceutical, flavour and some other house hold products.

The dried buds yield 14-21 per cent oil which contains 70-90 per cent eugenol and 5-12 per cent eugenol acetate.The yield and the physico-chemical properties of clove oil depend upon the quality of cloves and their condition prior to distillation and the type of distillation employed (Gopalan, 1997). Water distillation gives clove oil with 81 per cent eugenol, suitable for perfumery whereas dry distillation yields strong oil with 95 per cent eugenol. The physico-chemical properties and chemical composition of oil from clove bud, stem and leaves are given below :

Character	**Bud oil**	**Stem oil**	**Leaf oil**
Yield of oil	17.46 %	1.050 %	1.054 %
Specific gravity	1.051	$-0^{0}36'$	$-1^{0}20'$
Optical rotation	$-0^{0}32'$	1.5352	1.5379
Refractive Index	1.5318	1.5352	1.5379
Total phenol	91 %	91 %	88.5 %
Solubility	1 vol. of 70 % alcohol	1 vol. of 70 % alcohol	1 vol. of 70 % alcohol

Volatile oil obtained at lower maturity stage has more eugenol acetate and less eugenol. In fully mature clove, the eugenol acetate concentration decreases and eugenol increases. Clove oil from younger buds has mellow odour (Gopalakrishnan *et al.*, 1982).

Botany

Clove is a slender, evergreen tree reaching a height of 10-15 m. Leaves obovate, oblong to elliptic, opposite, simple, glabrous and possesses plenty of oil glands on the lower surface. The terminal trichotomous panicle is shortly pedunculated with 3-20 flowers per panicle. Flower bud is 1-2 cm long and has a cylindrical thick ovary consisting of four fleshy sepals. Fruit is a one seeded drupe and popularly known as mother clove. It contains one oblong seed with 1.5 cm length.

Floral biology

Clove flowers are bisexual. The male phase starts along with the opening of the flowers and the stamens shed after two days. The female phase starts before the male phase is over. In general, inbreeding is common in clove. Krishnamoorthy and Rema (1994) could not detect any appreciable variation for various seed and seedling characters except number of branches proving indirectly that self pollination may be predominant in clove. But controlled crossbreeding is also possible by adopting improved techniques. The maximum receptivity of the stigma occurs on the fifth day after opening of the flower. The best period for pollination is between 4th and 6th day after opening. The highest fruit set obtained was 66 per cent. A fertilized flower takes about three months for maturity.

Crop improvement

Variability in clove is very much limited and the present day population in India originated from a few trees with narrow genetic base originally introduced into India. Besides this, self pollination also limits the scope for variability. A survey was conducted by Horticultural Research Station (HRS), Yercaud during 1999 to the clove growing areas of Kanyakumari (Keeriparai and Maramalai). Morphological and flower bud characters were recorded and two distinct bud variants were identified. One is having bolder flower buds than the normal type (King clove) and the other one is smaller than the normal clove (Liliput clove/Mini clove). The above two types differed distinctly from normal cloves in morphological as well as yield characters (Table 28).

Table 28. Characteristics of two variants and common types of clove

Characters	King clove	Mini clove/ Liliput clove	Common clove
Leaf	Big	Small with curved tips	Medium
Number of buds / inflorescence	10-12	15-20	10-15
Flower bud size			
Length (cm)	1.0-1.2	0.75-0.80	1.6-1.8
Breadth (cm)	1.2-1.6	0.4-0.5	0.6-0.8
100 bud weight (g)	18-20	4-5	9-10
Weight of mother clove (g)	2.6	0.8-10	1.3-2.0
100 seed weight (g)	100	80	90-100
Yield (kg/tree)			
Green buds	15	25	15-20
Dry buds	5	8	5-7
Peduncle	5	8	6-7
Leaf (fallen)	15	10	8-10

The mother cloves of two king clove types (KC-1 and KC-2) and one Liliput clove (Lc-1) are collected and are being evaluated at Horticultural Research Station, Yercaud (Balakrishnamoorthy and Kennedy, 1999).

Studies on improvement of this crop are scanty. Being a perennial spice, having long juvenile period, development of superior type is a long term process.

Elite types in clove

Twelve high yielding types were identified from the survey based on five years performance from a private estate (Nagercoil). Mother cloves of this type were collected and are being evaluated at Horticultural Research Station, Yercaud. Range of variability among the elite types are presented below. (Balakrishnamoorthy and Kennedy, 1999).

S.No.	Characters	Range
1.	Tree height (m)	25-30
2.	Age (years)	61
3.	Green bud yield (kg/tree)	90-120
4.	Dry bud yield (kg/tree)	30-40
5.	Peduncle yield (kg/tree)	30-40
6.	Leaf yield (kg/tree)	9-12
7.	Weight of mother clove (g)	1.38-2.68
8.	100 seed weight (g)	85-120
9.	Seed length (cm)	2.2-2.62
10.	Seed breadth (cm)	2.8-3.4

Measurement of leaf area is one of the important methods of assessing the variability and productivity. A simple inexpensive and accurate method by linear measurements revealed that for clove, the constant K=0.649 can be used to measure the leaf area *in situ* without destroying the crop (Balakrishnan *et al.*, 1998). Experiments conducted at Horticultural Research Station, Yercaud revealed that for clove, girth of tree can be measured at 45 cm from ground level (Kennedy and Nageswari, 2000).

Crop management

Climate and soil

Clove is strictly a tropical plant and requires a warm humid climate (Pillai, 1972). It thrives well from sea level upto 600-1000 m above MSL with a rainfall of 1500-2000 mm and a mean temperature of 20-30^0 C. Deep red loam, sandy soil, black soil and deep gravelly subsoil are suitable but water logged conditions are undesirable. Clove can be conveniently grown mixed with commericial crops like pepper and areacanut.

Propagation

Clove is propagated through seed which is colled 'the mother of clove'. Seeds are collected from healthy and regular bearing trees (Pillai, 1972). Seed viability is short and hence to be sown immediately after collection. (Pillai,

1972; Amma, 1981; Krishnamoorthy and Rema, 1988; Nazeem *et al.* 1992 and Sable *et al.* 1992). Germination and vigour of seedlings are increased by keeping the seeds between two moist gunnies (Nair *et al.* 1977) and sowing during June, by using dehusked seeds (Kannan, 1972 and Amma, 1981). Medium to large sized seeds (>1.4 g) which were sown after dehusking gave upto 46 per cent success (Kumar *et al.* 1992). Germination studies conducted at Thadiankudisai revealed that higher percentage of germination (59.9 %) with taller seedlings (13.5 cm) were produced by seeds weighing 2-2.5 g. Hulled seeds treated with 0.1 per cent carbendazim resulted in maximum percentage of germination (24) (Chezhiyan and Ananthan, 1997). At Dapoli, clove seeds treated with IAA (20 ppm) gave 93.3 per cent germination within 14 days (Sable, 1991).

Gunasekharan and Krishnasamy (1999) reported that clove seeds sown in the horizontal position with micropylar end of the seed facing sideways registered increased germination compared to vertical and inverted sowing. Vigour of the seedlings was closely associated with larger sized seeds and seeds placed in horizontal orientation.

In a field trial, effects of different germination media on germination and seedling growth were studied. Sand and sand + FYM (1:1) media were the best suited for early, uniform and higher germination. Soil alone exhibited higher but delayed germination, whereas soil + FYM (2 : 1) proved to be the best medium for good vegetative growth of clove seedligs (Sable *et al.* 1995).

Age, colour of cotyledon and height of seedlings determine the time of transplanting. Nine to twelve months (Amma, 1981) to two year old seedlings (Krishnamoorthy and Rema 1988) are suited for main field planting. Bagade and Shinde (1993) reported that clove seedlings can be made ready for transplanting in one year by foliar application of GA (200 ppm). Nawale *et al.* (1995) observed that three months old clove seedlings sprayed with IBA (45 ppm) triggered the growth considerably followed by GA3 (300 ppm).

Planting and aftercare

The area selected for raising clove plantations is to be cleared of wild growth before monsoon and pits of 75 cm cube are dug at a spacing of 6-7 m (Krishnamoorthy, 1998). Pits are filled with compost and top soil and planting is done after the receipt of sufficient rains.

Shade is essential to protect the crop from adverse weather conditions. Banana, *Acacia* sp., *Albizia* sp. and Subabool (Krishnamoorthy, 1998), are the common shade plants to be established 6-12 months prior to clove planting (Pillai, 1972). Artificial shade with plaited coconut leaves (Pillai, 1972) is also useful. From third year onwards, shade is not needed and wind breaks like casuarina and japanese bamboo may be planted at the border of the plots.

Mulching the soil near the tree base during summer conserves moisture (Pillai, 1973a). Weeding is to be done when weed intensity is high. Cover crops and green manure crops like *Calapagonium* sp. (Krishnamoorthy, 1998), *Tephrosia candida, Mimosa* sp., *Stylosanthus* sp. and *Vigna hosei* (Balakrishnamoorthy and Kennedy, 1999) can be used to enrich and conserve the soil.

Regular irrigation is essential especially during summer and dry period. Drip irrigation studies conducted at Tamil Nadu Agricultural University indicated that dripping of eight litres of water/plant/day recorded the highest plant height and number of branches in the juvenile phase (TNAU, 1996).

Reddy *et al.* (2000) reported that clove can be successfully cultivated under coconut plantations aged over 20 years with assured irrigation facility during summer in the west coast of the country.

Manuring

Clove responds to applied fertilizer. Nutrient removal by adult clove tree showed the following trend : K > N > Ca > Mg > S > Mn > P > Fe > Cu > B > Zn (Nazeem, 1989). Fish and prawn refuses are used as manure (Pillai, 1972). The Kerala Agricultural University (KAU, 2003) recommended a fertilizer schedule of 20 : 18 : 50 g of NPK respectively along with 15 kg of compost per year during the first year of planting which is to be gradually increased to 300 : 250 : 750 g NPK and 50 kg of compost per plant per year from 15th year onwards. Kennedy and Balakrishnamoorthy (2000a) reported that addition of 50 g each of *Azospirillium* and *Phosphobacteria* along with 100 kg of FYM and 400 : 350 : 1200 g of NPK per tree per year is advantageous. Scattering of manures around the tree base to a radius roughly equal to canopy spread and to a depth of 10-15 cm, 15-20 cm away from the base of the tree and band application of fertilizers at a distance half the radius of canopy and quarter the radius away from the base of plants are suitable (Pillai, 1973b). Nutrient deficiency symptoms in clove were described by Nazeem (1989). Gnandas (1989) standardised the index leaves for assigning the nutritional status of clove in relation to soil fertility.

Plant protection

The insects attacking clove are mainly the black scale (*Saissetia nigra*), marked scale (*Mycetaspis personalta*), green scale (*Lecanium sp.*) and the mealy bug (*Pulvinaria psidii*) (Chandramani *et al.* 2001). Removal of affected leaves and destruction will prevent further spread of infection. Application of dimethoate 2 ml/l or monocrotophos 1.5 ml/l is essential in case of severe infection.

The most important disease of clove is sudden death disease caused by *Valsa eugeniae* (Rajappan *et al.* 2000). Cut off the branch cleanly below the stained infected area and paint the cut surface with a fungicide to control the disease.

Harvesting

Clove trees start flowering from fourth year after planting under good management conditions and full bearing stage is reached only after 15 years. Flowering season varies from September-October in plains to December-January at high altitudes. Flower buds are produced on young flush and it takes about 4-6 months for the buds to become ready for harvest. Flower buds are harvested when they turn pink or just before opening of bud. If they are allowed to flower they have no value as a spice clove (Pillai, 1973b). Clove oil obtained from flower buds harvested at lower maturity stages has more of eugenol acetate and less of eugenol (Gopalakrishnan *et al.* 1982). Harvesting is done using a step ladder and care must be taken not to break the branches. Three to four kilograms of dried buds can be harvested on an average from 15-20 year old clove trees (Krishnamoorthy, 1998). Alternate bearing habit is reported by Pillai (1973b). For harvesting seed, the optimum time has been identified as 78-91 days after fruit set (Nair *et al.* 1979).

The harvested flower buds are separated from the clusters by hand and spread in the drying yard for drying. It takes normally 4-5 days for drying. Fully dried buds develop characteristic tan colour and their stems. A dark brown colour and both are crisp. Is the product is uniformity good, 8-10 cloves could weigh just one gram only. Good quality cloves should be brownish-black with full and plump crown, somewhat rough to touch, without wrinkles and should not contain more than 12 % moisture.

Khoker cloves are inch cloves which have undergone furmentation due to improper drying. They are recognized by their pale-brown colour and whitish mealy appearance (Pruthy, 2001). Headless cloves and mother cloves also are considered as defectivr cloves. Well-dried cloves will be only about one third of weight of the original. A full grown tree gives 3-4 kg dried buds after 15 years.

Processed products

Clove bud oil (yield; 16 % constituents: eugenol, eugenol acetate and caryophyllene), clove stem oil (yield; 4 – 5 %), clove leaf oil (yield; 1- 2%), oil of mother cloves (yield; 6.5 %) and clove root oil (yield; 6 %) are the main products of clove.

Cambodge (Garcinia)
(*Garcinia gummi-gutta* L. N. Robson Desr., F. Clausiaceae)

Garcinia gummi-gutta L. N. Robson Desr. is one of the potential under exploited tree spices, currently gaining much agricultural, commercial, industrial and medicinal importance. The plant is indegenous to South Asia especially Thailand and India (Muthulakshmi and George, 1999) and is fairly common and abundant in the forests of Western Srilanka and Malaysia. It is found commonly in the evergreen forests of Western Ghats from Konkan southwards to Thiruvananthapuram. The tree often finds a place in the back yards of Kerala's homesteads, as one among the miscellaneous tree crops (George *et al.*, 1992).

Chemical composition and uses

The fruits are edible, but too acidic to be eaten raw. They are valued for their dried rind used in Kerala as an important ingredient for flavouring curries in place of tamarind or lime. The thick pericarp is cut and dried under sun and in smoke and can be preserved for future use. In Kerala, the dried rind is a principle ingredient in culinary preparations involving fish. It is also a valuable source of commercial 'gamboge' – a resin used extensively in paints and varnishes (George, 1988).

The fruit is rich in acids and possesses marked antiseptic properties. Garcinia rind is the only richest natural source of (-) –hydroxy citric acid (HCA). The derivatives of the acid are potent metabolic regulators of obesity. The unique acid lowers the blood lipids such as cholesterol and tri-glycerides by triggering fatty acid oxidation in the liver via thermogenesis (Muthulakshmi and George, 1998). It mobilises the body's fat stores and dissolves fat in the liver and also throughout the body paving way for weight management (Muthulakshmi and George, 1998).

Decoction of fruit rind can be given against rheumatism and bowel complaints. The rind is also used for polishing of gold, silver and as a substitute for acetic acid and formic acid in the coagulation of rubber latex. The translucent yellow resin of the tree has purgative properties.

Botany and flowering

Garcinia is a small to medium sized tree with oval shaped crown having orthotropic and plagiotropic branches. Sherly (1994) noticed that shoot growth in garcinia was seasonal with one main flushing period commencing from January and extending throughout the year upto May. However, scattered flushes occur throughout the year.

The species *Garcinia gummi-gutta* is androdioecious in nature. Two types of trees were observed by George *et al.* (1992). One type was functionally male bearing only staminate flowers while the other was functionally female with hermaphrodite flowers. Muthulakshmi (1998) observed that male trees had more number of flowers per cluster, higher petiole length, more stamens per flower, higher pollen viability and germination. Flowers in bisexual trees have higher flower weight, petal length and ovary size.

In both male and bisexual trees, the season of flowering was from the end of January upto early March. The flower buds are borne mostly on past season's growth and rarely on current season shoots. In both the tree types, flowers are seen at the shoot tips as well as in leaf axils. In a male tree, the flowers are borne in clusters of 10 to 30 in number occurring in leaf axils and shoot tips. In hermaphrodite trees where ever solitary flowers occurred, they were in the leaf axils whereas flowers in cluster of 3 to 5 were borne at shoot tips. Sequence of flower opening was similar to that of a cymose inflorescence with the central bud opening first followed by the ones at the two sides. Time of anthesis was at mid night. Both male and bisexual flowers have four sepals and petals each arranged imbricately. In male flowers, numerous two celled anthers are seen on short filaments. In the case of bisexual flower, 6 – 20 stamens often sterile, are present surrounding the ovary. Ovary is 6-10 celled with 6 – 10 lobbed stigma.

Pattern of flower development from bud emergence to anthesis was different in male and bisexual flowers. A Comparison of the male and bisexual flowers is given below :

Sex	Colour			Length (cm)				Average size (girth in cm)				
	1-10 days	10-16 days	Anthesis	4 days	10 days	16 days	O	4 days	10 days	16 days	O	TTA
Male	Green	Yellow	Yellow	0.3	0.9	1.4	1.9	0.5	1.0	1.4	1.9	26
Bisexual	Green	Yellow	Orange	0.2	0.8	1.1	1.6	0.5	1.7	2.3	2.9	19

TTA – Time taken for anthesis (days) O – Opening (days)

Sherly (1994) reported that in male flowers, peak anthesis was between 16.30 and 17.00 hrs and in bisexual flowers it was between 17.00 and 17.30 hrs. Anther dehiscence occurred 10 hrs prior to anthesis. Stigma became receptive 30 hrs before anthesis and retained receptivity upto 12 hrs after anthesis.

Garcinia pollen appeared as yellowish powder. Acetocarmine test recorded 56 per cent fertility in male and 28 per cent in bisexual flowers. On bagging the bisexual flowers 2-3 days prior to opening, it was observed that the percentage of set in the selfed bag was only 20 per cent. One of the reasons for the poor set in bagged bisexual flowers was the low rate of fertility of pollen of bisexual flower. That is why male trees are essential for better fruit set in garcinia (George *et al.* 1992).

Muthulakshmi (1998) observed wide variation in tree characters like shape of the tree and branching pattern. Early flowering, late flowering and sporadic flowering types were identified in this crop.

No significant differences between male and bisexual trees in terms of morphological characters were reported (Muthulakshmi, 1998). Pink and different shades of pink were more prevalent in the emerging flushes of bisexual trees as compared to male trees. Significant difference between male and bisexual trees, was noticed with respect to total phenol content of young and mature leaves. TLC profile of phenolic extract showed one more additional spot in male than in bisexual plants. Electrophoresis studies revealed similar banding pattern for peroxidase enzyme in both male and bisexual plants.

Pollination and fruit set

Garcinia is an often cross-pollinated crop (Shirly, 1994). Hand pollination with pollen from male flowers increased the percentage fruit set as compared to self pollination or open pollination indicating the importance of male trees in a population for improved fruit set. The fruit drop after set was 35.50 per cent and the major part of the drop occurred during the first 30 days. The fruit attained maturity in 130-140 days after fruit set. The developing fruits followed a sigmoid growth pattern. At ripening, the rind of the fruits had on an average 6.68 per cent acidity, 7.2 mg/100 g ascorbic acid, 8^0 Brix T. S. S. and 1.04 per cent reducing sugar. The mucilage developed around the seed towards ripening had 2.04 per cent reducing sugar and 3.3 per cent acidity.

Crop improvement

Kerala has tremendous diversity in the population of *Garcinia* particularly along the back water and river belts of the state (George, 1988). No released varieties are available in the crop. Studies conducted at the Kerala Agricultural University had helped to identify the following five groups in garcinia :

1. High yielding types in terms of bigger fruits, more number of fruits with thick rind
2. Quality types in which fruit are free from astringency and bitterness
3. Early flowering and early ripening types to avoid coincidence of peak harvest season with monsoon
4. Late ripening types in which the ripening coincides with clear weather at the end of monsoon
5. Types with higher (-) – hydroxy citric acid content

The National Bureau of Plant Genetic Resources Regional Station at Vellanikkara, Thrissur maintains 54 accessions of *Garcinia gummi-gutta,* seven accessions of *G. indica,* two of *G. xanthochymus* and two of *G. mangostana* (Abraham *et al.,* 1999).

Crop management

Climate and soil

Garcinia is adaptable to a wide range of climatic and soil conditions. It is a very hardy crop and can be grown in tropical and subtropical climates.

Sandy loam soils are ideally suited for its cultivation (Muthulakshmy and George, 1999). It tolerates fluctuating water tables and even prolonged inundation and hence is suitable for low lying reclaimed soils (George, 1988).

Propagation

The crop is both seed and vegetatively propagated. Plumpy and heavy seeds are ideal (80 % germination). Seeds have dormancy period of 8-9 months. Seeds with seed coat intact remain in the soil for 8-9 months and then start sprouting. On removing seed coat, germination occurs earlier (2-3 months after sowing). Nazeema (1992) reported 90.7 per cent germination by seed coat removal along with seed soaking in 500 ppm GA for 12 hours. About 10-20 per cent polyembryony is also noticed in the crop (Muthulakshmy and George, 1999). The seedlings segregate into productive bisexual and unproductive male trees. Sex is revealed only after flowering, for which the seedlings take 7-8 years. The unproductive male trees should be removed after they show their sexuality. To over come this problem, vegetative propagation can be adopted *Soft wood grafting*

Soft wood grafting has been standardised. One year old garcinia root stocks are used for grafting nine months old scion sticks, having 6-8 cm length (Nazeema, 1992). Soft wood grafting done during June gave maximum success. Besides, vegetatively propagated trees come to flower earlier (about 3 – 4 years after planting). It also results in dwarf bushy trees favouring high density planting. To get a normal growth habit, orthotropic shoots should be used as scion material . If plagiotropic shoots are used, the resulting grafts will be dwarf, bushy and drooping with reduced fruiting area. But orthotropic shoots are limited in number on mother plants. Therefore seedlings continue to be the major propagating material (Muthulakshmi and George, 1999).

Planting and aftercare

Planting can be done with the onset of first monsoon showers. One year old seedlings or soft wood grafts are used for planting (KAU, 2003 and Muthulakshmi and George, 1999). Seedlings are planted at a spacing of 10 m x 10 m and grafts at 5m x 5m in pits of 1m x 1m x 1m size. Refill the pits with mixture of top soil and compost or farm yard manure. Proper care should be given to avoid water stagnation in the pits. The tree does not require manures and fertilizers. However, addition of 50 kg farm yard manure / tree during early period of growth keeps the tree healthy.

Harvesting and processing

The fruit attains maturity in 130-140 days after fruit set (Muthulakshmi, 1998). For getting good quality non-astringent rind, fruits should be harvested at ripening stage. A well managed fully grown tree yields 400–450 fresh fruits or about 55-60 kg dried rind. The fruits vary in size weighing 50 to 180 g and the fruit rind is grooved into 7-10 segments (George, 1988). On picking, they exude a yellow resinous liquid which hardens slowly into a brown gummy mass. Fruits abscise on ripening and are collected from the ground and the seeds, enclosed in a pulpy translucent aril-which is often sweet and edible, but some times acrid, are separated. Rind is then dried. Due to the coincidence of south west monsoon rains with the period of fruit ripening, farmers resort to smoke drying. It can also be sun dried. In a week or two, the dry rinds attain brown colour and develop characteristic acrid taste. The blend of this with the acid components imparts an exceptional flavour to the dried rind.

The rind is then preserved by rubbing with salt and coconut oil which further improve the colour, texture and appearance (George, 1988).

Muthulakshmi *et al.* (2001) reported no significant differences in the quality of garcinia rind under different drying conditions namely oven, sun and smoke. However, the rind from smoke drying was soft and flexible, shiny black in colour with good shape and consumer appeal.

Prospects

In spite of its unique medicinal properties, no efforts are made in this crop for large scale cultivation. Exploiting the diversity available, with the ultimate aim of developing an ideal plant type with short stature, year-round fruiting and desirable fruit characters like edible pulp and seedlessness will bring about a radical change in garcinia cultivation (George, 1988).

Kokam
(*Garcinia indica* Choisy., F. Clausiaceae)

The 'kokam' of commerce comprises sun-dried rind of the fruit of kokam tree repeatedly soaked in its juice. Kokam (*Garcinia indica* Choisy) is one of the most important indigenous tree spices mainly grown in the south Konkan region of Maharashtra, Western Ghats, Coorg, Wynad, Goa, evergreen forests of Assam, Khasi and Jantia hills, West Bengal and Gujarat. Kokam plant grows widely in the tropical forests of Western Ghats of India, where it has originated. In the Konkan region, kokam trees are scattered over backyards, river and stream sides, waste lands and also in coconut and arecanut gardens.

No reliable statistical information are available as to how much kokam is available in India. However, in the konkan region alone about 4,000 t are produced (Sampathu and Krishnamurthy, 1982). Another survey report indicated that Western Ghat region contains about 15 lakh trees.

Chemical composition and uses

Kokam fruit has several uses. It serves as a flavouring substitute and is being used as a fish preservative. It contains substantial amount of malic acid and very small quantities of tartaric and citric acids. The ripend rind and juice of kokam fruit are commonly used in culinary for preparing 'Solkadi' which is a house hold preparation in Konkan region. Kokam syrup popularly called as 'Amrit Kokam' is commercially prepared from ripened fresh fruits and the syrup has potential demand in market. The dried and salted rind (Ansol) is being used as condiment in curries. Kokam seeds also have good commercial value. Kokam fat popularly known as 'kokam butter' is being prepared from seed kernels. Seed contains about 32 to 35 per cent fat. It is edible, nutritive, demulcent, astringent and emollient (Pruthy, 1976). Kokam fat is also used for local application to ulcerations, fissures of lips, hands etc. It is also used in pharmaceutical preparations and cosmetic products. Most of the kokam fat produced in the country is exported to Netherlands, Italy, Japan, Singapore, UK and Malaysia. Kokam rind is the richest source of natural red pigment (anthocyanin 2-3 %). Kokam fat is used in chocolate and confectionery preparations. It is also used in the manufacture of soap, candle and oilments.

A new fat soluble yellow pigment, garcinol has been isolated from the fruit rind (Sampathu and Krishnamurthy, 1982).

Composition of fresh kokam rind (values on moisture free basis)

Mositure (%)	80	Protein (NX 6.25) %	1.92
Crude fibre (%)	14.28	Total ash (%)	2.57
Tannins (%)	2.85	Pectin (%)	5.71
Starch (%)	1.0	Crude fat (%)(hexane extract)	10.0
Acid (as hydroxy citric acid)	22.80	Pigment (%)	2.4
Ascorbic acid (%)	0.06	Carbohydrates by difference (%)	35

The fatty acids present are myristic, palmitic, stearic, oleic and linoleic acids, stearic acid being the major acid (52 – 65.4 %).

Botany and flowering

Kokam is a slender, evergreen tree with drooping branches which attain a pyramidal shape on maturity. Kokam is dioecious, but seems to be highly variable in sex forms like papaya. Variability in sex and flower types in kokum were studied by Gunjate *et al.* (1984). Flowers are axillary or terminal, solitary or fascicled having 2 to 8 buds. Bud is 1.4 to 1.6 cm long and 0.7 to 0.8 cm broad. Flower is tetramerous and hypogynous. The calyx is sepaloid, consisting of four sepals arranged in decussate pairs, the inner pair being broader than the outer one. Corolla consists of four petals slightly larger than sepals and yellow to pink dorsally and dark pink ventrally. Male flower has generally long pedicel and numerous stamens forming short capitate column or collected in a ring surrounding the rudimentary pistil. Anthers are oblong, sessile on short thick filaments, adnate, four celled and very rarely two celled. Female flower has short pedicel, staminodes are arranged in four tufts and very rarely in two tufts around the pistil. Stigma is sessile, radiate, each ray with two lines of tubercles. Overy is two to eight celled, ovules solitary with axile placentation. Bisexual flower has long pedicel, four tufts of stamens surrounding the pistil. Stigma is sessile or sub-sessile.

Sex types

The trees could be designated into the following types on the basis of preponderance of particular type of flowers and the bearing tendency of individual tree.

Tree type - I - Staminate or male tree

The flowers have mostly long pedicels, mass of stamens crowded on receptacle and sometimes rudimentary pistle with pointed apex. They are incapable of producing any fruit and serve as pollinators only.

Tree type-II - Hermaphrodite or bisexual

Young fruits produced by the tree are generally irregular in shape containing 0 to 6 underdeveloped seeds. Yield per tree may vary from 1 to 3 kg of fruits.

Tree type III – Pistillate or female

Flower is identified by short pedicel, well developed pistil and two or four tufts of staminodes below. Fruits are round to globose, dark red when ripe and contain 1 to 7 well developed seeds. Adult tree bears heavy crop. In a population of 62 trees observed 37 per cent turned out to be male, 8 per cent bisexual and 55 per cent female.

Maximum anthesis occurs between 6 am to 8 am. Anthers dehisce just before anthesis. Stigma attains maturity one day before anthesis (12%) and the receptivity remains at peak (80%) on the same day of anthesis. Hand pollination resulted in 78 per cent and open pollination (natural cross) gave 68 per cent fruit set (Karnick, 1978). Fruit is spherical, as large as a small orange, purple throughout, not grooved, having 5 – 8 seeds compressed in an acid pulp (Raju and Reni, 2004).

Crop improvement

Among 36 kokam genotypes maintained and evaluated at Ratnagiri, five promising selections viz. S1, S3, S8, S11 and S14 were identified and growth, flowering, fruiting and quality aspects were studied. The kokam types varied considerably. Variation in height was 6.75 to 10.40 m whereas the crown volume ranged between 29.74 and 111.56 m^3. Flowering initiated during Oct-

Nov. Fruit retention ranged between 10 and 26 per cent. Harvesting period required was 78 to 91 days. Variation in fruit shape ranged from round to pear shape. Fruit parameters also varied considerably. Average weight of fruit ranged between 28.10 and 34.95 g. All the kokam types varied remarkably for yield. Average yield after seven years ranged from 29.00 to 138.28kg (Sawant *et al.* 1999).

Variety 'Konkan Amritha' was developed by clonal selection. This variety is considerably early having short harvesting period (78 days) with a few pluckings (3). The yield is high (138.28 kg) with medium sized fruits (34.15g) having rind of 17.55 g. Filled seeds were 3.55 per fruit. This variety is a pure female.

Crop management

Climate and soil

Kokam tree is very strong and sturdy in growth. It grows on wide range of soils ranging from marginal to deep alluvial soils which are well drained. Lateritic soils of the Konkan belt are very much suitable for its cultivation. Soils having good moisture retention capacity are the most favourable ones. Most favourable regions for kokum cultivation are warm amd moderate humid zones. Temperature range from 20 ^{0}C to 35^{0}C, 60 to 80 per cent humidity, well distributed annual rainfall ranging from 2500 to 4000 mm seem to be ideal for this crop. Extreme acidity is harmful to the crop (Nawale *et al.*, 1997).

Propagation

Propagation of kokam by seed is the prevailing practice (Nawale and Karmarkar, 1997). For raising of seedling, bigger sized fresh seeds from high yielding and regular bearing plants should be cleansed with water and then sown in polythene bags (12.5 x 20.0 cm) containing potting mixture. Two days prior to sowing, seeds should be soaked in water which give early and better germination (up to 90%). One year old seedlings should be used for planting in the field.

Vegetative propagation

Soft wood grafting technique is successful (Haldankar *et al.* 1991). October is the best season for soft wood grafting and grafts can be successfully maintained either in glass house or under open sun. The mature scion stick of 5-6 months age can be preferred for soft wood grafting. Pre-curing is not a pre-requisite for soft wood grafting in kokum. Age of root stock should be more than 22 weeks. Retention of leaves on root stock did not influence success of grafting (Haldankar *et al.* 1992). Length of scion stick from 2.5 cm to 15 cm did not exhibit any significant influence on survival of graft. Scion stick could be stored successfully in poly bag or sphagnum moss for seven days. To achieve maximum percentage of survival at the establishment stage, grafting operation should be performed *in situ* (Haldankar *et al.* 1993).

Cultivation

Before planting, all bushes, trees and stumps should be removed. Planting should be done at 6 m x 5 m spacing (333 plants/ha). Pits of size 60 cm x 60 cm x 60 cm should be dug out. The pits should be refilled with well decomposed farm yard manure and soil. Sevin @ 100 g/pit should be mixed in soil to protect the young plants from termites and also 1 to 1.5 kg single super phosphate per pit should be mixed in soil while refilling the pits. Planting should be carried out with the onset of monsoon. Strong and healthy seedlings or one year old grafts should be used for planting. Immediately after planting staking should be done.

After care

The young plants should be provided with partial shade in winter and summer months and should be irrigated at least once in a fortnight during winter and summer months, for better establishment in field. Orchard should be kept clean by removing weeds, grasses and bushes. Adequate supply of moisture in soil helps the bearing trees for producing higher yield.

Manures and fertilizers

Kokum trees are generally not manured by the farmers at present and hence production is very low. As the kokum tree is an evergreen plant and survives for many years and yields every year, it should be manured regularly for healthy and vigorous growth and for higher yield.

For one year old plant, 5 kg FYM or green manure and 50 g N, 25 g P_2O_5 and 25 g K_2O should be applied. This dose should be increased proportionately by 5 kg FYM, 50 g N, 25 g P_2O_5 and 25 g K_2O per tree every year. For ten year old onwards, 5 kg FYM or green manure, 500 g N, 250 g P_2O_5 and 250 g K_2O should be given per year. The fertilizers should be applied in August after heavy rains are over. A circular trench of size 30 cm depth and 30-45 cm width should be dug out below the tree canopy. Manures and fertilizers should be incorporated in the trench and refilled with soil.

Intercrops

Rainfed crops like cucurbitaceous vegetables, okra and sweet potato can be successfully grown in between two rows of kokam.

Removal of male plants

The kokum plant is dioecious in nature. If orchard is established by using kokum seedlings, it is very difficult to identify the male and female plants at early stage. Only after flowering, the male and female plants can be identified. It is advisable to remove excess number of male plants. For ensuring adequate pollination, better fruit set and higher yield, it is desirable to maintain 10 per cent male plants in the orchard. By following top working technique the extra number of male plants can be converted into good bearing female plants.

Pests and diseases

There are no serious diseases or insect pests damaging the plant. Among insect pests, leaf miner is important which causes damage to tender leaves. Spraying with 0.03 per cent phosphomidon or 0.03 per cent dimethoate are effective for its control.

Harvesting

Kokam 'seedling-tree' commences bearing after 7 to 8 years of planting whereas grafted tree starts bearing 5 to 6 years of planting. Flowering starts from October-November and continues upto February under agro-climatic conditions of the Konkan region. Harvesting of fruits commences from March and continues upto first fortnight of June. The peak period of harvesting is April-May. At fruit maturity, colour changes from green to red. The ripened fruits are harvested frequently by using harvester.

Yield potential in different kokum genotypes ranged from 30.60 kg to 173.00 kg per tree per year with the productivity of 9.27 t to 51.90 t per ha. It shows that there is tremendous yield potential for this multipurpose fruit crop. Further, the establishment of kokum orchard with vegetatively propogated trees (grafted) is ideal and the most beneficial from the viewpoint of higher yield and returns.

Processing and utilization of kokam

A number of important processed products are prepared from the fruit as detailed below:

Dried green kokam rind

This product is prepared from the green kokam fruit . The fruit is cut into four longitudinal pieces and dried under sun (Joshi *et al.* 1999).

Kokam syrup (Amrit kokam)

The product is prepared from the fresh ripe kokum rind by adding cane sugar at the rate of 1:2 proportion. The mixture strained thorough muslin cloth and the syrup is stored in the food grade polythene container and in glass bottles. If required, the preservative sodium benzoate is used at the rate of 610 mg/kg of syrup. This product is required to be diluted 5 to 6 times with water before consumption. A pinch each of common salt and cumin powder are added for taste (Joshi *et al.* 1999).

Dried ripe kokam rind

The ripe kokam rind is dried under sun or in a cabinet drier at 60^0 C. It is treated with brined kokam juice and again dried. This is done for five times. The final dried product is used in curry in place of tamarind (Joshiet al. 1999). The drying ratio of ripe kokam fruits in a cabinet drier was 6.24:1 and in sun drying it was 7.77:1. (Gosavi, 1988).

Kokam agal (Brined kokam juice)

This product is prepared either from ripe fruit rind and pulp or only from pulp. Fifteen to 16 per cent common salt is mixed with the fruit part, mixture is kept for 8 days by occasional stirring. Extracted brined kokam juice is stored in food grade polythene container or glass bottles. This product is used for making "Solkadi", a refreshing drink after lunch (Joshi et al. 1999).

Other products

Kokam rind contains 2-3 per cent anthocyanin pigment and is a promising source of natural colourant for foods. Processing conditions are standardized at CFTRI, Mysore for commercial scale extraction and purification of the pigment concentrate (Sampathu and Krishnamurthy, 1982). Fundamental studies showed that cyanidin- 30sambubioside and cyanidin-3-glucoside are the major pigments present in the ratio 4:1. Isolation of hydroxy citric acid present to the extent of 15 per cent in the pure form also was done at CFTRI, Mysore. The acid finds potential application in medicine.

Curry Leaf
(*Murraya koenigii* Spreng., F. Rutaceae)

Curry leaf (*Murraya koenigii* Ispreng) is a condiment of Rutaceae family used in South Indian culinary preparations, mainly to increase the flavour and taste of the food. Leaf is the economic part which is slightly pungent, bitter with feebly acidic taste and the leaves retain their flavour and other qualities even after drying. Curry leaf grows throughout India including the Andamon & Nicobar islands uo to an altitude of 1500 m. Both curry leaf and curry leaf oil are exported from India.

Chemical composition and uses

Leaves of the shrub are used extensively for seasoning and flavouring dishes. Leaves are a fair source of vitamin A. They are also a rich source of calcium, but due to presence of oxalates in high concentration (total oxalates 1.35 %, soluble oxalates 1.15 %), its nutritional availability is affected. Leaves on hydro-distillation gave 0.5 per cent essential oil on fresh weight basis, having dark yellow colour, spicy odour and pungent taste. Composition of the oil suggests that the oil may find application as fixative in heavy type of soap and detergent perfumes. Leaves and stalk in decoction or boiled in ghee is given as a remedy in rheumatism, leprosy, epilepsy, internal poison, piles, diarrhoea, fever and worms. It is a laxative, improves brain and cures eye diseases. Plant is credited with tonic stomachic and carminative properties.

Curry leaf contains rich amount of protein (6.1 %), carbohydrate (18.7 %), Vit. A (12600 I.U.) and Vit C. (4 mg/100 g) (Goudra *et al.* 1995). The peculiar aroma is due to the presence of sulphur containing essential oil (Raju, 1995).

Botany

Leaves are alternate, exstipulate, imparipinnate, 15-20 cm long, usually glabrous, sometimes slighlty pubescent when young and very strongly aromatic, leaflets number 9-12 or more, short stalked and alternate ovate to lanceolate, glands dotted and strongly aromatic. Leaves have a slightly pungent, bitter and feebly acidous taste. It is a diploid tree having the chromosome number 2n = 18. There are two related species.

Murraya paniculata - Chinese box wood (or) Andaman satin wood.
Murraya exotica - A popular hedge and well adapted for topiary.

Varieties

Senkambu

A local cultivar of Karamadai tract of Coimbatore District of Tamil Nadu. The petiole is purplish red in colour and the leaves are aromatic.

DWD-1

A clone obtained from root suckers having dark green, highly aromatic shining leaves. Leaves have an oil content of 5-22 per cent. It can be dehydrated without any loss in its quality.

DWD-2

An open pollinated seedling progeny with slightly pale green leaves having lesser aroma (5.09 %). Due to its winter intensitive nature, it can give an extra income to the planters.

Suwasini

It is a high yielding cultivar, gives 1065.74 g leaves/plant/year, and the oil yield is 29.36 kg/ha/clipping.

Cultivation

Soil and climate

Curry leaf is a hardy crop. It can tolerate higher temperatures but when the temperature falls down below 16 ^{0}C the vegetative buds become dormant arresting new growth of the plant. The crop is more dominant at an altitude range of 60-1200 m above MSL, though it can be cultivated upto a height of 1800 m above MSL. Curry leaf can establish very well in red sandy loam with good drainage facilities. Heavy clay with poor drainage is not suitable for cultivation of curry leaf.

Propagation

Curry leaf is mainly propagated by seeds. However, one year old root suckers are also used for propagation (Regeena and Ravi, 1995). Rudramurthy (1993) reported that soft wood cutting was better than semi hard wood cutting because of their higher total sugar, C/N ratio, total phenols, lower levels of carbohydrates and nitrogen. Biochemical analysis of cuttings revealed that the starch, reducing and non-reducing sugars, total sugars, CHO, C/N ratio and dry matter contents have significant effect in enhancing rooting of curry leaf cutting. Experiments conducted at Tamil Nadu Agricultural University indicated that semi hard wood cutting from past season shoot planted during May and July recorded 30 and 20 per cent rooting while semi hard wood cuttings from current season shoot recorded only 10 per cent in both seasons. Cuttings from the past season shoot in the month of May was found the best for rooting of cuttings in curry leaf (Mohanalakshmi *et al.* 2000).

For large scale propagation of curry leaf, micro propagation was also tried. High frequency of direct shoot proliferation was induced in intact seedlings of curry leaf on modified Murashige and Skoog medium supplemented with benzayladenine (5 mg/l) (Bhuyan, 1997).

Mathew *et al.* (1999) standardised micro-propagation of curry leaf using nodal segments from mature trees on MS medium supplemented with 6-benzylaminopurine, kinetin, NAA and IBA as growth regulators.

Use of growth regulators

Application of IBA with linolin paste at 600 ppm concentration recorded a maximum of 22.2 per cent rooting in soft wood cuttings of the variety Suwasini (Rudramurthy, 1993). GA (100 ppm) was effective in increasing length of new shoots, internodal length, number of new shoots and number of leaves in DWD-2 variety.

Nutritional requirement and manuring

Earlier studies on nutrition of curry leaf indicated that, curry leaf does not respond to applied phosphorous and potassium beyond 25 g per plant (Balaji, 1988). In case of nitrogen, a bush supplied with 150 g nitrogen produced a fresh biomass of 7.49 kg which is almost double that of unnourished bush. The method of application of nitrogen is found non-significant in curry leaf. Accordingly, it is suggested that a two year old bush could be fertilized with 150 g nitrogen. Interaction effect between nitrogen and cultivars has also been significant that a quantity of 150 g nitrogen per bush is specific for the variety DWD-1 (Goudra and Madalagiri, 1992). A complete mixture comprising of 150 g N, 25 g P_2O_5 and 50g K_2O per bush along with 20 kg farm yard manure is essential for curry leaf. The method of application does not influence the growth and productivity in any measure.

Diseases and pests

Leaf spot

Causes severe defoliation. Spray of carbendazim (Bavistin) 1g/l is effective.

Basal stem rot

It is caused by *Sclerotium rolfsi* and is reported from Orissa University of Agriculture and Technology, Bhubaneswar.

Citrus butterfly

Feeds on the leaves and causes sever defoliation. Application of malathion 1ml/l is effective.

Psyllid bug and scale insect

Causes defoliation. Spraying of dimethoate 1 ml/l controls the pest.

Harvesting

Plants should be pruned at the age of 8-10 months after planting at a height of 30 cm (Krishnarajan, 1991). At the end of first year, the yield will be 250-400 kg of leaves per ha. During second year, the average yield is 5400 kg/ha/year. From the fifth year onwards, 5000 kg/ha of leaves can be harvested once in three months which would work out to 20,000 kg/ha/yr. The leaves retain their flavour even after drying and hence these are marketed both in fresh and dried forms.

Dehydration

Oven temperature of 50^0 C was significantly superior to maintain acceptable quality viz., pleasant flavour and emerald green colour of dehydrated powder. Artificial oven drying was more advantageous than sun and shade drying methods (Mahadeva, 1993).

For oil extraction, solvent extraction method was superior to hydrodistillation method with regard to the time taken for extraction (90 min) and extractable oil yield (2.2%). However, a base temperature of 70^0 C was suggested at farm levels with some modification. Leaf maturity influenced oil composition, the youngest leaves being the best.

Tamarind
(*Tamarindus indica* L., F. Fabaceae)

Tamarind (*Tamarindus indica* L.) is one of the most important multipurpose trees in Indian subcontinent. Now it is recognised as an important tree spice. The name is derived from the Arabic word "Tamare-hindi" or "Tamarhindi" meaning Indian date, because its dark pulp resembles the preserved dates of Arabian countries. It is also called as 'Amlika' in Sanskrit meaning sour taste (Joshi, 1983). The monotypic genus *Tamarindus* has only one species, *indica*. It is considered as a native of Eastern Tropical Africa. India is one of the largest producers of tamarind in the world and it is the only country which produces a commercial crop tamarind. Tamil Nadu, Madhya Pradesh, Andhra Pradesh, Maharashtra and Karnataka are the major Tamarind growing states.

Chemical composition and uses

Lewis and Neelakantan (1964) reported that about half of the tamarind pod is fruit pulp. The fruit is composed of about 30 per cent pulp, 40 per cent seeds and 30 per cent shell. Pruthy (1976) and Chundawat (1990) reported that ripe fruits on an average comprises of 55 per cent pulp, 33 per cent seed and 12 per cent fibre. The total pulp content (including seed and fibre) of various tamarind clones varies from 56-87 per cent with a mean pulp content of 72 per cent and the shell content is 28 per cent (Kennedy and Thangaraj, 1996).

The approximate composition of ripe tamarind pulp and solubility percent is presented in Table29. The composition varies from sample to sample as reported by various authors. The reason might be the result of real differences in composition between samples or differences due to handling of the sample. It is the fruit with the lowest water content and probably as a consequence of this, tamarind has one of the highest levels of protein, carbohydrate and minerals of any fruit.

Table 29. Composition of tamarind pulp

Composition	Pruthy (1998)	Chundawat (1990)	Kennedy and Thangaraj (1996)
1. Moisture (%)	18.2	20.6	-
2. Solubility (water) (%)	-	-	72.06
3. Acidity (free) (%)	9.8	10.0	16.62
4. Combined acids (%)	6.7	-	-
5. Reducing sugars (%)	-	-	19.01
6. Total sugars (%)	38.2	-	27.87
7. Fat (%)	-	0.4	-
8. Protein (%)	2.8	3.1	-

9. Carbohydrates (%)	-	70.8	-
10. Pectin (%)	2.8	-	-
11. Fibre (%)	19.4	30.0	-
12. Mineral matter (%)	2.8	-	-
13. Calcium (%)	0.17	-	-
14. Phosphorous (%)	0.11	-	-
15. Iron (%)	0.01	-	-
16. Calorific value (cal/100 g)	283	-	-
17. Ascorbic acid (mg/100 g)	-	-	3.01
18. Vitamin-A (IU/100 g)	100	-	-
19. Niacin (IU/100 g)	0.2	-	-
20. Ash (%)	-	2.0	-

Botany

Tamarind is a large, handsome, long-lived evergreen tree, widely cultivated for its shade, ornamental value and fruit. It attains a height of 20 m or even more with a girth of more than 7 m when fully grown. Leaves are alternate, leaflets are almost stalkless and close together along a slender pale green axis.

Flowers appear in showy clusters of yellowish or pale yellow blossoms. Flowers are a few, in sub-terminal racemes. Normally the upper three flowers in a raceme develop into fruits.

Fruit is a pod, 10-20 cm long and 3-4 cm wide, 3-10 seeded and is strap shaped. At first, the pods are green when tender and the immature seeds are soft and whitish, but as they mature the skin develops into brown, brittle shell and the seeds are covered with dry, sticky pulp. The pulp and seeds are hard, flat and relatively square, measuring about 1.25 cm long and about 0.4 cm thick.

Floral biology

Tamarind trees flower from April to July in most areas of South India. Peak flowering occurs during May-June. New leaves appear in May and are closely followed by the flowers. Flower opening begins as early as 5.30 a.m. and continues upto 8.30 a.m. with peak anthesis occurring at 6.30 a.m. Mature anthers are reddish brown in colour, dehisce by longitudinal splitting of their lobes between 10.00 to 11.30 a.m. Pollens are round with average diameter of 15 m?. The percentage of fertile pollen is 96.5. Stigma is receptive from one day before anthesis to two days after anthesis. Tamarind is predominantly cross pollinated. Red ant plays an important role in pollination.

In tamarind, fruit set is very poor (1-2 %) resulting in large scale abscission of flowers as well as fruits during various stages of development. The main causes of poor fruit set might be the failure of fertilization, sterility or higher percentage of defective pollen grains or slow growth of pollen grains or its early degeneration (Lokesh *et al.* 1997).

Crop improvement

Genetic resources of under utilized crops like tamarind were discussed by Pareek and Sharma (1993). According to Singh *et al.* (1997), all the plantations in India are of seedling origin and hence there is immense scope for selection of desirable traits. Several authors documented data on various accessions/clones/varieties of tamarind (Paules, 1975 and Singh, 1992). A total of 85 clones at Horticultural College and Research Institute, Periyakulam (TN) and 120 accessions at College of Agriculture, Pune (Maharashtra) are being evaluated (Sharma and Bharadwaj, 1997 and Azhakiamanavalan and Vadivel, 1997). As reported by Karale *et al.* (1997), 5 elite types (T-9, T-10, T-11, T-12 and T-13) out of 195 plus trees were surveyed from Ahamadnagar and Satara districts of Maharashtra by Keskar *et al.* in 1989 and another set of seven types were selected out of 110 genotypes. Three hundred and fifty trees are under germplasm evaluation at Fruit Research Station, Aurangabad under Marathwada Agricultural University, Parbhani (Shinde *et al.*, 1997). The forest department of Karnataka identified 230 plus trees and planted forty trees in Karnataka for evaluation of agronomic characters in collaboration with University of Agricultural Sciences (UAS), Bangalore (Murthy, 1997). Institute of Forest Genetics and Tree breeding, Coimbatore has also identified 12 phenotypes based on various characters during their survey in 1994 in Kanyakumari district of Tamil Nadu (Bennet *et al.*, 1997). The Department of Forest, Govt. of AP, has surveyed tamarind trees in Chittur and Ananthapur districts and they have recorded data from 52 identified trees (Naidu *et al.*, 1997).

Tamarind is mostly self sown or sown with seeds of unknown parentage, which result in wide variation among seedling progenies (Paules, 1975). Thimmaraju *et al.* (1978) found it to be a protogynous, entomophillous and highly cross pollinated crop. Owing to its wide geographical distribution and adaptability to different agroclimatic zones, large genetic diversity is present in the seedling population. Lack of standard cultivars and long, juvenile period of the seedlings come in the way of development of systematic plantations. There is however, great scope for selection of superior types from the existing variability all over the country (Pareek, 1988).

Keskar *et al.* (1989) made some promising seedling selections viz. T9, T10, T11, T12 and T13. There is excellent potential for markedly improving the crop. The tamarind trees now grown throughout the tropics are not the products of any careful agronomic selection, they were derived from seeds picked up at random in Africa and India. In Thailand, a very sweet tamarind type is grown. Germplasm collections need to be made in a region that stretches from Senegal across Subbe sahelian Africa to Sudan and also from Thailand and other South East Asian countries where superior types are available (Irulappan and Anbu, 1994).

A wide variation exists among the seedling populations of tamarind in Tamil Nadu, Karnataka and Maharashtra. The survey of tamarind growing areas in and around Coimbatore brought out the existence of extensive variability for pod size, pulp colour, quality, yield and proportion of edible material. Besides, there is large variation for the white spongy tissue covering the pulp, which decides the keeping quality of fruits under ambient storage (Kennedy and Thangaraj, 1996).

There is wide range of variability among the types studied. Higher values of genotypic and phenotypic co-efficient of variations were observed for pod length, mean fruit weight, seeds per pod, pod girth, flesh content and ascorbic acid content. Heritability was higher for mean fruit weight, flesh content, solubility in water, pod girth and seeds per pod, while genetic advances were higher for mean fruit weight, flesh content and pod girth. These estimates indicated scope for tamarind improvement by resorting selection (Kennedy *et al.* 1997).

Further, the total experimental materials were divided into five groups based on their physical and chemical characteristics viz. (1) Long, bold fruited types (2) Medium fruited types (3) Small fruited types (4) Curved, twisted and irregularly shaped fruited types and (5) Sweet types. Solubility percentage of pulp in water varied from 79.48 to 61.07. Medium fruited types recorded the maximum solubility percentage and the sweet types the minimum. Acidity in terms of tartaric acid varied from 14.91 to 17.14 per cent. Among the fruit types studied, sweet types recorded the minimum values and the small fruited types recorded the maximum values. Ascorbic acid content varied from 2.72 to 3.34 mg/100 g. Twisted and irregularly shaped fruited types recorded the minimum quantity and the long fruited types recorded the maximum quantity. Reducing sugar content varied from 18.71 per cent in irregularly shaped fruits to 19.39 per cent in sweet types. Wide range of variation recorded among various types of tamarind in terms of physico-chemical characters provide good scope for selection (Kennedy *et al.*, 1998).

Cultivars

Tamarind cultivars are generally named based on their place of identification.

1. Periyakulam-1 (PKM–1)

It was released by Horticultural College and Research Institute, Periyakulam, Tamil Nadu Agricultural University, during 1992. It is a clonal selection collected from Endapuli village which is an early bearer. The grafts of this variety come to flower from third year as against fifth year in seedlings. It bears fruits in clusters of 3-7 and fruits are characteristically semicircle in shape. It has a high pulp recovery of 39 per cent (compared to 28 % in the local). Characteristic white colour will be observed in the inner side of pulp indicating its superior quality. This selection has high tartaric acid (17.1 %) and ascorbic acid (3.95 mg/100 g) contents. It is a very high yielder with 263 kg/tree as against 167 kg/tree in local type accounting for an increase of 59 percent over local types.

2. Urigam

It is a superior local genotype located near Urigam, a small village near Thenkanikottai in Dharmapuri district in Tamil Nadu. It is short in size and branches out like umbrella. Pods are lengthy, fleshy and tasty. Pods are flat and exhibit a typical inward "C" shaped curvature. Each fruit weighs about 150-200 g. Seed content varies from 10-12/pod. Flowering is slightly late, seen in March and fruits are ready for harvest by July. Seedlings start bearing from 6-8 years, while grafts yield from 4th year onwards. (Geetha, 1995).

3. Prathisthan

It is also a high yielder released from Fruit Research Station, Aurangabad, Maharashtra.

4. DTS-1

It is a selection made at University of Agricultural Sciences, Dharwad. A sweet red type with an average yield of 1.5 kg/plant (dry).

5. Yogeswari

A red pulped selection made by the Forest Department, Karnataka.

Other local selections are

i) *Cumbum selection:* Selection made at Cumbum in Madurai district of TN
ii) Rahuri selection from Maharashtra
iii) Karur urigam tamarind from Karur, TN
iv) Kangadevanapalli tamarind and
v) Dever Ullimangalam tamarind

Classification of tamarind

Tamarind is classified based on their colour, taste and place of origin (Kennedy and Thangaraj, 1996).

1) Based on colour (a) Red, black, brown etc. (b) Light and dark coloured types

2) Based on taste (a) Sour type : Prathisthan, PKM-1 (b) Sweet type : Thailand type

   Reddish pulp varieties are generally preferred as their pulp is almost acid free and hence sweet. However, pink and dark brown varieties are common but their pulp is sour.

3) Based on place of identification

   Urigam, Cumbum, Karur etc.

4) Based on place of origin

   (i) East Indian variety : Long pods with 6-12 seeds/pod.

   (ii) West Indian variety : Shorter pods with 1-4 seeds/pod.

Crop management

Climate and soil

Tamarind always prefers a warm climate. It is grown in semiarid, tropical, and sub-tropical regions and is suitable for growing in regions experiencing extended spells of dry weather. It is drought resistant but frost sensitive and requires early protection from winter. Ripening of fruits do not commence under cold climate (Irulappan and Anbu, 1994). In its natural habitat, the absolute maximum temperature varies from 36 to 47.5^0 C and the absolute minimum from 0 to 17.5^0 C. Optimum rainfall for the species, judged from its distribution in India is 750-1900 mm and it grows upto 100 m above MSL. It is a crop suited for the waste lands (Gupta, 1993 and Karuppaiah *et al.* 1997) for the black cotton soils (Devaraj *et al.*, 1997). However, hardly any survival was noticed as recorded by Patil *et al.* (1996) in their travels at Karnataka.

Tamarind is not exacting with regards to its soil requirement. It can be grown on a variety of soils, ranging from poor, degraded, eroded, gravelly, saline and alkaline soils. It is a popular tree found on marginal soils and waste lands. However, the productivity is higher when grown in red loamy, deep, well drained soils, which provide optimum conditions for the development of its long tap root (Gupta, 1993).

Propagation

Most common method of propagation is through seeds. It does not breed true to type due to heterozygosity since the flowers are cross pollinated and noticed to be out crossing. Prolonged juvenile phase has also been observed in plantations due to predominant seed propagation (Karale *et al.*, 1997). Seedlings aged between 1½ to 2 years are recommended for planting. In the recent years, vegetative propagation techniques like cuttings, root suckers, inarching, soft wood grafting and patch budding were standardized for large scale multiplication.

Seed propagation

Tamarind seedling trees start flowering and fruiting at the age of 10-15 years, and continue producing fruit for a very long period. Ripe pods are collected during March-April by shaking the branches and these are dried in the sun and seeds are extracted. The outer shell is removed by hand or by beating with a mallet and then the seeds are separated from the edible pulp by hand kneading and washing in water. Washed seeds are dried in the shade and stored in gunny bags in dry cool place. It retains viability for about six months. A healthy full grown tree produces about three quintals of fruit every year, which yields about 80 kg seeds. On an average 1800-2000 seeds weigh a kilogram. Germinative capacity is about 60-75 per cent. For direct sowing, about 20 kg of seeds are required per hectare of land. Germination starts in about a week and is completed in about a month. Sankaranarayanan *et al.* (1994) reported that the hard seed coat causes slow and poor germination. Soaking the seeds in 10 per cent cow urine or in cow dung solution (500g in 10 l of water) for 24 h increased the germination percentage from 37 per cent (untreated controls) to 72.6 and 82.6 per cent respectively. Fresh seeds are sown in nursery beds in March-April. In Tamil Nadu, two year old transplants are doing better than one year old ones (Joshi, 1983).

Vegetative propagation

Cuttings: Cuttings with heel of a small portion of the bark tissue intact at the basal portion roots better in October (39.9 %) followed by November and June. IBA (1000 ppm) produced better rooting (25.36 %) whereas control recorded only 8.36 per cent rooting (Navaneetha et al. 1990).

Grafting: Purushotham and Rao (1990) and Amin (1978) studied the prospects of veneer and soft wood grafting for the propagation of tamarind, which yielded 49 per cent and 68 per cent graft success respectively. Approach grafting was also studied and 74 per cent graft success was observed compared to 25 per cent using ordinary grafting (Swaminathan and Ravindran, 1989). Kulwal *et al.* (1997) reported soft wood grafting for multiplication of tamarind. Grafting on to 8 month old stock of seedlings having a height at the range of 22-32 cm and 0.3-0.4 cm diameter performed in March-April was highly successful. Similar results were observed by Sathisha *et al.* (1997). In another study, in green wood cleft grafting recorded maximum graft success during May (80 %) followed by April (76.7 %) and June (70 %) and the least success was observed in February (26.7 %) (Nachegouda, 1997).

Approach grafting appeared to be easy, very ideal and successful for propagation of tamarind in view of various advantages like cost effectiveness, easy and higher percentage of establishment and rapid multiplication of trees.

At Fruit Research Station, Himayatbagh, Aurangabad, Shinde *et al.* (1997) claimed the success of wedge grafting.

Air layering: Air layering or 'gootee' as it is alternatively termed was attempted in tamarind. Navaneetha et al. (1990) have reported 37 per cent of rooting, when the shoots were air layered with IBA (4000 ppm). Higher rooting percentage was also recorded when the concentration of hormones IBA, IAA and NAA were 7000 ppm (Pradhan, 1997).

Manuring

Tamarind is seldom grown in organized plantations, until recently it remains a crop of isolated patches, roadside avenues and barren and puramboke areas. Larger plantations are steadily coming up in the southern states. Little is done to the tree by way of improved cultural practices. Tamarind is seldom manured. However, beneficial effects are reported with the application of manures and fertilizers. A combination involving 50 g each of neem cake and groundnut cake along with 50 g of superphosphate promoted growth of six months old seedlings. Liquid manuring prepared by over night soaking of nine kg of groundout cake in 40 l of water when applied at 2 l per tree has a stimulatory role on growth (Sambandamurthy and Vadivel, 1996).

Organic manures like farm yard manure and neem cake when applied at 50 kg and 2 kg respectively per tree coinciding with rainy days was instrumental for getting higher yield of crops (Desikan, 1980). It has further been reported that a complete fertilizer mixture comprising 200 g of nitrogen, 150 g of phosphorus and 250 g of potassium along with 25 kg of farm yard manure applied annually to each tree is beneficial for a steady state of productivity (Irulappan and Anbu, 1994).

Use of biofertilizers

Biofertilizers influence the growth of tamarind seedlings at the initial stages. The tree responds favourably to soil inoculation of as many as thirteen VAM fungi collected from various parts of the world. The seedling response is manifested through greater plant height, more leaf number, increased stem girth and enhanced biomass production levels. showing an increased level of phosphorous and zinc in plants. There is also an elevated number of VAM spores in the soil, root colonisation and external hyphal network as increased by soil aggregation. Tamarind reacts positively to the soil inoculation of *Gigaspore margarita* followed by *Glomus fasciculatum* (Reena and Bagyaraj, 1990). VAM inoculation significantly improves the growth of eight months old container grown seedlings besides enhancing the uptake of phosphorous and potassium. The seedling growth is at its best when rock phosphate is added at 200 kg per 0.42 hectare.

Aftercultivation

The rootstock sprouts should be removed as and when they appear. The graft joint is to be examined after two months and if the wax cloth is intact without decomposition, it should be cut and removed. Shallow cultivations are essential to keep down weeds and to provide favourable conditions for plant growth (Singh, 1992).

Intercropping

Intercropping with leguminous crops, short duration vegetables, annual drumstick, sesamum and sorghum during rainy season is recommended.

Training and pruning

Early training and pruning is necessary to form a high head and uniform scaffold branches in all directions. Generally no pruning is necessary except to remove the dried, diseased or criss-cross branches (Irulappan and Anbu, 1994).

Harvesting and processing

Seedling trees start flowering in 8-12 years after planting and vegetatively propagated plants in 4-6 years and the economic returns start from 7-10 years. Fruits mature and ripen from February to April in India. Both immature and mature fruits are harvested depending upon demand in the market and their uses. At full maturity when ripening starts, the exocarp gets hardened and separated from the pulp (mesocarp). The pulp colour changes from green to brown and dark brown or black or red according to the types. It takes 10-11 months from flowering to harvest of the fruits.

A general decrease in fruit bearing during two or three consecutive years is reported as of common occurrence in tamarind (Challapilli *et al.* 1992). Hence only a three year average yield is reported to give the true value. Usha and Singh (1996) remarked that mid and late flowering types of tamarind should be selected for planting to obtain higher yields. This was based on their assumptions that in late and mid season flowering types increased cross pollination was promoted whereas early flowering types were mostly self pollinated.

The first harvest of fruits is by shaking the branches and collecting the fallen fruits. During the second harvest, the left over fruits are picked using fruit picker. The harvested fruits are semi-dried and do not spoil due to high acidity. The fruit crop is annually sold out to contractors in November-December, when fruits are already on the tree.

The ripe tamarind fruit can be separated into pulp and kernel, both having diversified uses in medicine and industry. Fruits contain about 30 per cent pulp, 40 per cent seeds and 30 per cent hull. Value added products from the pulp are edible pulp, sweetened pulp and tamarind concentrate. Seed gives the kernel powder, seed testa and seed oil which also possess excellent potential uses in pharmaceuticals, medicine and industry (Chempakam and Peter, 2000).

Processing methods differ in different places. Pulp is separated from the fibre and seed and mixed with about 10 per cent salt. This is beaten with mallets and packed in gunny bags matted with palm leaves. Sometimes they are rolled into balls and kept for a week in the sun and dew. During this period, the colour changes from light brown to dark brown.

Another way of preparing the edible pulp is to mix the fruit with minimum amount of water and pass through a pulper to remove the seeds, fibre and cellulose. This homogeneous pulp is then drum-dried and compressed into blocks like cheese.

Tamarind concentrate

For the preparation of concentrate, a method was developed by CFTRI, Mysore. The pulp is heated with boiling water and soluble solids are extracted to the extent of 20 per cent using counter current principle. It is sieved and concentrated under vacuum in a forced circulation evaporator and directly filled in cans or bottles, which sets like jam on cooling with a soluble solid content of about 68 per cent. Composition of the concentrate is tartaric acid (13.0 %), inert sugars (50 %), pectins (2 %), protein (3 %), cellulosic material (2 %) and moisture (30 %). Freshly harvested fruit pulp (not older than 5-6 months) gives a good quality product (Manjunath et al., 1991 and Nagaraja et al., 1975). The whole process is patented. Alternatively, the pulp can be extracted with Pectic Enzyme Concentrate (PEC) for giving an acceptable nonviscous product with about 70 per cent yield.

Tamarind kernel powder (TKP)

This is prepared by decorticating the seeds and pulverizing the creamy white kernels. Decortication is done either by soaking the seeds in water or roasting. The powder should be kept in a dry place to avoid contamination or stored with the addition of sodium metabisulphite to avoid enzymatic deterioration. Defatting is advised to get good storage quality and colour.

Seed testa

This is the residual byproduct in the preparation of kernel powder. It is used as dyestuff due to presence of leucoanthocyanins and also as plywood adhesive. Tannins from seed testa are black in colour and highly polymeric (Rao et al. 1997). Seed coat extract of tamarind possesses antioxidative property. It can be widely used as a low cost source of antioxidant in lipid containing foods.

Kernel oil

Seed kernel contains six to eight per cent oil, which resembles peanut oil. The oil can be used in paints and varnishes and also for lighting lamps.

Leaves, flowers and timber

Just like coconut palm, every part of tamarind can also be put to various uses. The leaves of young seedlings of tamarind are eaten as a vegetable. The flowers are also highly nutritious and are rich in vitamins and minerals. In addition, the tamarind timber has got very good potential in furniture industry.

Pimento (All Spice)
(*Pimenta dioica* (Linn.) Merr., F. Myrtaceae)

Pimento is the major spice of Jamaica contributing seventy per cent of the world's production. It is grown to a certain extent in Southern Mexico, Honduras, Gautemala, Cuba, Haiti and Costa Rica, besides a few West Indian islands. Among tree spices, it has a prominent position. However, its volume in the international trade is much less than clove and nutmeg.

Pimento of commerce is the dry unripe fruits of *Pimenta dioica* (L.), Merr. an evergreen tree which grows to a height of 8 to 12 m. The word pimento originated from "pimenta", the Spanish word for black pepper berry to which it resembles.

Dried immature berry is the main commercial spice product which combines the orders and flavours of four spices viz. nutmeg, clove, cinnamon and black pepper. In India, it is grown in Kerala, Tamil Nadu and Karnataka.

Chemical composition and uses

Dried berry contains aromatic steam volatile oil, fixed oil, resin, protein, pigments, minerals, vitamins etc. Eugenol, isoeugenol and b-caryophyllene are the major components present in all spice.

As a flavouring agent, the main outlets are in culinary preparations like soups, sauces, ketchup, pickles, meat, curry powder and in bakery items like cake. In liquor industry, it is used especially as a favourite ingredient for mulled vines (Rema and Krishnamoorthy, 1989)

Total yield of oil from berry varies from 3.3 to 4.5 per cent with a eugenol content of 65.89 per cent. There is good market for berry oil and it is being used widely in food industry, perfumery and pharmacy. Oil is also used as carminative and stimulant. Oil is extracted from the leaf and it is used as a cheap substitute for the expensive berry oil. Yield of leaf oil is about 0.7 to 2.9 per cent.

Allspice is aromatic, stimulant, digestive and carminative. It is an anodyne against rheumatism and neuralgia. Powdered fruits are used against flatulence, dyspepsia and diarrhoea. It is used in the preparation of tonics and purgatives.

Botany

All spice is a small, functionally dioecious evergreen tree, 7 - 10 m tall, slender trunk profusely branched at its extremities. Bark is smooth and shiny, pole silvery brown. Leaves are borne in clusters at all the ends of the branches, simple, opposite, entire, coriaceous, punctate with pellucid glands, aromatic when crushed. Inflorescence an axillary cyme. Flowers structurally hermaphrodite, but functionally male or female. Fruits is a sub – globose berry, green when unripe and deep purple when ripe. Allspice is polygamodioecious in nature, which makes it difficult for identification of functional male and female trees until it comes to bearing. Sreeja (1999) tried to characterise dioecy in allspice. The female plants were taller with lesser primary and secondary branches and recorded a higher leaf area than the males. Females had a higher content of total free amino acids, essential oil and oleoresin. On the other hand, phenols, nitrogen and carbohydrates were higher in males.

Genetic resources

Allspice is a recent introduction to India from Jamaica and hence the variability is very less. National repository of tree spices germplasm at IISR, Kozhikode maintains 137 cultivated accessions of allspice. A distinctly variant type of allspice seedling was identified by Mathew *et al.* (1999) in the field germplasm repository at Indian Institute of Spices Research, Kozhikode. This dwarf/semi dwarf plant type offers great potential in the crop improvement programmes.

Crop management

Climate and soil

Pimento grows from sea level to an altitude of 1000 m above MSL. An annual rainfall of 100 to 200 cm or more with a mean monthly temperature up to 27 ^{0} C are the best. Performance of allspice in plains is poor and fruiting is not observed (Krishnamoorthy and Rema, 1991).

Propagation and planting

Common method of propagation is by seeds. Ripe fruits are collected from high yielding and regular bearing trees. Seeds are extracted after soaking the fruits overnight in water and rubbing them in a sieve and washing with clear water. Drying of seeds is done in shade. Seeds are sown as soon as possible or else germination is reduced (Krishnamoorthy and Rema, 1988).

Seeds are sown either in nursery beds, boxes, pots or basins. Beds of 1.2 m width are prepared with light soil incorporated with organic matter or a mixture of sand and coir dust or coir dust alone. After sowing, nursery beds are mulched to hasten germination. Dried leaves, straw, paper and damp sacs are used as mulch. Watering is done as fine spray. Germination takes place 15 days after sowing (Krishnamoorthy and Rema, 1991). Devadas and Manomohandas (1988) observed differences in germination by storing seeds for different periods in polythene bags and 70 to 93 per cent germination was obtained when sown three weeks after collection at a temperature of 21.5 to 30.5 °C. The viability was completely lost after nine weeks.

Vegetative propagation has been tried in allspice. Haldankar *et al.* (1995) studied the feasibility of air layering and found 42 to 100 percent success during various months. January was the best season for air layering which recorded cent per cent success. However, it took about 18 to 28 months to separate all the rooted layers from mother plants. The air layers prepared during May recorded considerably low success and delayed rooting.

Sreeja (1999) reported 30 per cent rooting when leafy soft wood cuttings were used for propagation. Treatment with IBA (1000 to 1500 ppm) also enhanced rooting. Seedlings of 6 months (Shanmugavelu and Rao, 1977) or 9-10 months old having 25 to 30 cm height are ready for field planting. Seedlings are planted at a spacing of six m either way and in poor soils, a closer spacing is preferred. After planting, temporary shade, irrigation and weeding are essential for reducing casualities (Krishnamoorthy and Rema, 1991). Since the plant is dioecious, it is desirable to keep a male-female ratio of 1: 10 in the plantation (George, 1993).

Very little is known about the manorial requirements of pimento. A fertilizer dose of 20 g N, 180 g P_2O_5 and 50 g K2O/ tree in the first year after planting is progressively increased to 300:250:750 g/tree for a grown up tree of 15 years or more (Krishnamoorthy and Rema, 2004). Fertilizers are applied in two equal doses in shallow trenches dug around the plant about 1.0 – 1.5 m away from the tree.

Pests and diseases

A leaf rot disease caused by *Cylindrocladium quinquiseptatum* is reported in India (Anandaraj and Sharma, 1992). Larvae of red borer (*Zeuzera coffeae*) damage all spice by tunneling into the region(Abraham and Skaria, 1995)

Harvesting and processing

Clonally propagated plants start flowering in three year and seedlings in five to six years under well managed conditions. Male trees flower early when compared to females and the usual flowering time in India is during March to June (Krishnamoorthy and Rema, 1991). Paclobutrazol (1.25, 2.50 and 3.75 g) treatment induced flowering in allspice plants. Paclobutrazol treatments increased the number of panicles, flowers and setting of berries which in turn resulted in higher yield. From flowering, 3-4 months are taken for fruit maturity. Fully developed greenish berries are harvested to use as spice. (Shanmugavelu and Rao, 1977). A well grown tree yields 20-25 kg dry berries per year under good management. (Krishnamoorthy and Rema, 1991).

Harvested berries are taken to drying shed, heaped for five days to ferment. Berries are then dried in drying yards. It takes about 5 – 10 days for drying (12 – 14 % moisture content) depending up on the weather. About 55 – 65 dry berries is obtained from 100 kg green. Dried berries are cleaned and stored. Whole spice, ground spice, berry oil, leaf oil and oleoresin are the major products obtained from pimento.

Star-anise

(*Illicium verum* Hooker., F. Magnoliaceae)

Star-anise is the dried star shaped fruit derived from *Illicium verum*, an evergreen tree indigenous to tropical and subtropical East Asia. The crop is introduced to India from China. Commercial production of star anise is limited today to China and Vietnam (George, 2004). China is the major producer, supplier and exporter of star-anise. Commercial cultivation is not yet organized in India, a small quantity is produced at Arunachal Pradesh.

Botany and chemical composition

Star-anise is an evergreen tree attaining a height of 8-15 m and a diameter of about 25 cm. Leaves are entire, 10-15 cm long and 2.5-5.0 cm broad, elliptic to oblanceolate, flowers are solitary, white to red in colour. Fruits are star shaped, reddish brown, consisting of 6-8 carpels (follicles), arranged in a whorl, radiating from top of a central short central stem; each follicle is 12-17 mm long, boat-shaped, hard and wrinkled, containing a seed; seeds are brown, compressed-ovoid, smooth, shiny and brittle. The carpels have a pleasant anise like odour and an aromatic refreshing sweet, anise-like taste; the seeds have the same, but to a much lesser extent.

Fresh fruits yield an essential oil (3.0-3.5%) which has odour and aromatic taste of true anise oil (*Pimpinella anisum*). Anethole is the main constituent of oil. Leaves on steam distillation yield about 0.5 % essential oil and the decorticated seeds contain fatty oil (55%).

Crop management

No literature is available on the cultivation practices of star-anise. Pruthy (2001) gave a short account of its cultivation. Star-anise is propagated by seed which looses its viability soon and has to be planted immediately after stratification. The nursery period is very long (up to 3 years) after which seedlings are transplanted into main field. The tree comes to production in about 10 years. The tree produces flowers throughout year, but flowers develop into fruits only during certain seasons. Fruit matures all year round but in normal years, August-October harvest accounts for about 80 % of the total harvest.

Harvesting and processing

Star-anise fruit contains maximum essential oil before complete maturity and should be gathered at this stage to give higher yield of oil. Over ripe naturally dropping fruit can be used for drying. From 10th to 20th year, a tree yields 30 to 35 kg fruits (which yield about 600 g of oil). From 20th year onwards, a tree produces on an average, 40 to 45 kg of fruits. Fresh green fruits are sun dried to give dried brown star-anise of drug market. Fresh fruits are distilled to give the oil.

Uses

Star-anise fruit is used as a condiment for flavouring curries, confectionaries and spirits and for pickling. It is also used in perfumery. In medicine it is stomachic, carminative, mildly expectorant and diuretic and oil is used in pharmaceutical preparations and perfumery.

Caper
(*Capparis spinosa* Linn., F. Capparidaceae)

Capers of commerce are immature flower buds which have been pickled either in vinegar or preserved in granular salt. Semi-mature fruits and young shoots with small leaves may also be pickled for use as a condiment. The use of capers can be traced to pre-historic times. Citations of Capparis are available in early Rabbinic, Mesopotamian and Greco-Roman texts (Rivera *et al.*, 2002). *Capparis spinosa* is native to the Mediterranean basin. It is found growing wild on rocky and hilly localities in the Deccan Peninsula, Rajputana and North-Western India (Wealth of India, 1985).

Botany and chemical composition

Although *C. spinosa* is the most widely used species, there exist 23 species and subspecies which occupy large territories from the Atlantic coasts to the Pacific in the old world (Rivera *et al.*, 2003). *C. sicula, C. orientalis, C. aegyptica, C. cartilaginea* and *C. decidua* are the popular Mediterranean cultivars (Inocencio *et al.*, 2000).

Caper plants are small shrubs and may reach about 1 m in height. However, uncultivated caper plants are mare often seen hanging, draped and sprawling as they scramble over soil and rocks (Ravindran *et al.*, 2004). Leaf stipules are transformed into spines. Flowers are borne on first year branches. They are pink with long tassels of puple stamens.

Flower buds contain a glycoside, rutin. Caper seeds yield 34-36 % of pale yellow oil. Component acids of oil are palmitic and stearic, oleic and linoleic. Root bark contains a rutic acid and a volatile substance of garlic odour. Inocencio *et al.* (2000) reported that the average content of flavanoid glycosides in commercial capers was 5.18 mg/g by fresh weight.

Cultivation

Not much information is available on the cultivation aspects of caper in India. In the Mediterranean region, varieties have been developed for characters such as spinelessness, round, firm buds and flovour through selection. propagation is mainly carried out by seeds which generate high genetic flows resulting in low productivity and an increased variability of caper quality, in addition to low germination rate. Various factors such as unit fruit weight, fruit position on mother plant, maturity of the fruit etc. are reported to influence caper germination (Pascual *et al.*, 2003). Dried seeds should be initially immersed in warm water (40 °C) and then allow to soak for one day. Seeds are then wrapped in a moist cloth, placed in a sealed glass jar and kept in the refrigerator for 2 - 3 months. After refrigearation seeds are soaked again in warm water over night and then sown in soil. Further more, multiplication using lignified cuttings resulted in a very low rooting success (8%). The spineless caper varieties responded well (90% rooting success) to multiplication using herbaceous cuttings taken at the beginning of the vegetative cycle. About 20 % success was reported in herbaceous grafting when the spineless caper was used as the scion and the spiny one as the rootstock. Micropropagation was also successful. Plants are placed 2 - 2.5 m apart. Afull yield is expected in 2 - 3 years. Three year old plants will yield 1 - 3 kg of caper flower buds/plant. Caper plantings will live for 20 - 30 years. The unopened flower buds are hand picked regularly throughout the growing seasons. This labour expensive harvest makes this herb an expensive one. The capers are washed and allowed to wilt a day in the sun wilted buds used for pickling.

Uses

Capers have a sharp, piquant flavour and add pungency, a peculiar aroma and saltiness to sauces, pizza, fish, meat and salads. Caper reduce flatulence and has an antirheumatic effect. In Ayurvedic medicine capers are recommended as hepatic stimulants and protectors, improving liver function.

Sweet Bay
(*Laurus nobilis* L., F. Lauraceae)

The commodity, traded as sweet bay leaf is derived from the leaves of *Laurus noblis* L. It is a native of the Mediterranean and grows spontaneously in Europe and California. It is widely cultivated in Europe, America and in Arebian countries. It is sometimes grown in Indian garden but does not seem to thrive well. Dry berries of the tree, commonly called bay berries are imported to India for medicinal use. Bay leaves are among the oldest herbs of the world. Flavouring properties of bay leaves have been known since antiquity. Wreaths of laurel leaves were used by Greeks and Romans to honour their heroes.

Botany and chemical composition

Laurus nobilis is a heliophile and mesophile species native to the Mediterranean basin (Paolette and Rocca, 2003). It is a small evergreen tree attaining a height of 15 – 20 m. Laurel leaves or bay leaves are stiff, narrow and fragrant. It is dioecious, both leaves and berries are used for flavouring. Leaf is elliptical, green on the upper surface and pale green or yellowish on the under surface. Flowers are small, yellow in colour, unisexual, appear in clusters (Kumar *et al.*, 2004). Dried berries, commonly called bay berries are ovoid (about 1.5 cm long), black, coarsely wrinkled and contains a single seed. Laurel wood resembles walnut wood in grain and colour and is suitable for decorative cabinet work.

The leaves on steam distillation yield 1-3% essential oil, the principal constituent of which is cineol (25-50%). The berry (pericarp 30%, seed 70%) yields 20-34 % of an aromatic fat. Other constituents of the oil are eugenol, geraniol, a and b pinenes, linalool, a-phyllandrene and small amount of free fatty acids. Fat contains lauric acid 30-35%, palmitic acid : 10-11%; oleic acid: 33-40% and linoleic acid : 18-32%.

Cultivation

Laurel prefers rich and deep soils, especially ones that preserve a fair degree of moisture throughout summer. Sweet bay is propagated by cuttings or by seed, former being the best method. The plants lend themselves to decorous trimming to almost any shape.

Harvesting and processing

Laurel trees are pruned from September to March and the leaves are dried in shade to the desired moisture level. Laurel leaves are available whole or cracked, are not usually ground.

The culinary uses of bay leaves are for flavouring soups, stews, meat and fish and in confectionary also. Both leaves and fruits possessing aromatic, stimulant and narcotic properties, were formerly employed against hysteria, amenorrhoea and flatulent colic. Commercial oil and fat of laurel berry are mainly used in veterinary medicine.

Juniper

(*Juniperus communis* Linn., F. Pinaceae)

The common juniper, which is found in the Himalayas at an altitude of 1520 to 4270 m is an evergreen shrub, some times attaining the height of a small tree upto three metres. It grows wild in many parts of Europe, Asia, England and North America. In England, the juniper tree is threatened by extensive grazing, other competing tree species and lack of sites to colonise (Dearnley and Duckett, 1998). Italy is by far the largest producer of juniper berries. Italian berries are considered to be the best, followed by Hungarian berries. In European countries, large quantities of juniper berries are used domestically for fermentation and subsequent distillation of alcoholic beverages "Borovicka". Slovaks are particularly fond of this strongly flavoured beverage. The juniper berries of commerce are obtained by drying ripe fruits.

Botany and chemical composition

Juniper is an evergreen dioecious coniferous shrub or small tree with erect trunk and spreading branches, covered with a shreddy bark. The leaves are straight and rigid, oval shaped and are with sharp prickly points. Male flowers are short catkins and female flowers small cones. The fleshy berry - like fruit does not ripen until second year. The roundish fruit is sub-globose, bluish black, dark purple when ripe with three bony seeds embedded in the pulp. The plant flowers in March-April and the fruits ripen in August-September of the second year.

The ripe fruits yield 0.5 to 2.0 % volatile oil. Juniper oil is colourless or pale greenish yellow limpid liquid with a characteristic odour of the fruit and with a burning bitter flavour. The oil contains a-pinene as the major constituent, together with smaller amounts of camphene, cadinene, junene, turpineol and traces of esters. Besides the volatile oil, fruits also contain fermentable sugars (33%), resin (8%), Lannins, fixed oil, wax, pectins, organic acid and potassium salt. They are good sources of ascorbic acid also (35 mg/ 100 g).

Crop improvement

Five taxa of *Juniperus*, of which two with tree habit, are recorded from Western Himalaya (Kaushal, 1994) and were studied for their morphological variation and distribution. The genus is represented by two distinct tree species (*J. macropoda* and *J. excelsa*) and three shrubby forms (*J. communis, J. squamata* and *J. pseudosabina*). Several variants were observed in the polymorphic *J. squamata.* A putative hybrid, intermediate in characters between *J. squamata* and *J. pseudosabina* has also been reported.

Cultivation

Juniper thrives best on sunny heights and mountain slopes exposed to warm sun. The quality of berries varies with the geographical origin and local conditions like climate, soil, altitude and exposure to sun. Trees at elevated attitudes and in warm sunny places produce best berries.

Juniper is not commercially grown in India. The natural regeneration is through seeds. Propagation by rooted cuttings is reported by Houle and Babeua (1994). They also found that cuttings from female plants rooted more easily than those from male plants.

Ripe (blue) berries yield more oil than unripe green berries and hence only ripe berries are harvested. Harvesters carefully beat branches of the bush with sticks, until ripe (2 years old) berries drop into baskets, or upon sheets placed beneath the bush. Green berries remain on branches and may be collected the following year.

Processing

The freshly harvested berries are sifted, winnowed and spread out in a thin layer on the floor for drying in the sun or under shed. Care must be taken to prevent fermentation, thus to avoid inferior flavour and poor appearance. The dried berries are graded and packed.

Fruits are employed for flavouring gin and food products and for making alcoholic beverages. Juniper oil is largely used in compounded gin flavours, liquors and cordials. The fruits and volatile oil possess carminative, stimulant and diuretic properties.

ŽŽŽ

Nutmeg

Cinnamon

Cassia

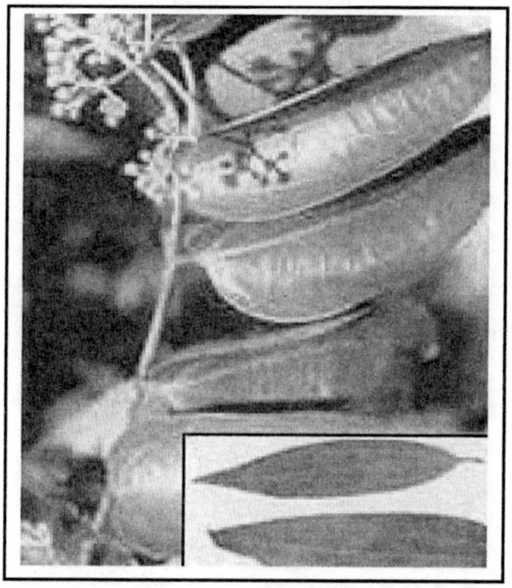

Tejpat

Clove

Cambodge

Kokam

Curry Leaf

Tamarind

Pimento

Star-anise

Caper

Sweet Bay

Juniper

Chapter 4

Seed Spices

Coriander
(*Coriandrum sativum* L., F. Apiaceae)

Coriander is used by man since long as coriander seeds were found in Egyptian tombs of 960 BC. Further, Persia grew coriander 3,000 years ago and it added fragrance to Hanging Gardens of Babylon. In 3rd century BC, Romans also found coriander seed as an excellent seasoning for popular foods.

Besides seeds, fresh leaves are being used for garnishing of food in many Latin American, Spanish, Russian, Chinese and Indian dishes since long. Fresh coriander leaves are also known as Chinese Parsely. In India, coriander seed is used almost daily in scores of curries for flavouring and leaves for garnishing. Coriander is a native of Mediterranean region.

India, Morocco, Former USSR, Hungary, Poland, Romania, Czechoslovakia, Bulgaria, France, Guatemala, Mexico, Argentina and USA are important countries where coriander is grown commercially. In India, coriander is grown in most of the states and constitutes an important subsidiary crop in black cotton soils of Deccan and South India under rainfed conditions and in rich silt loams of North India under irrigated conditions.

India also exports coriander seeds but the quantity is negligible compared to demand levels in major consuming countries like USA, Saudi Arabia and Germany (Raju, 1990). However, among seed spices, coriander export quantum is the highest, followed by cumin seed. But value wise, cumin seeds stand first, followed by coriander.

Botany and chemical composition

Coriander is a thin-stemmed, small bushy herb, 25-50 cm in height, with much branches, tops or umbels. Leaves are alternate, compound and petiole with a pair of stipules, which sheath the stem at base. Compound leaves become highly segmented and linear as they reach upper extremities. Both leaves and stems have a strong pleasant aroma. Inflorescence is a compound umbel and usually comprises about five smaller umbels. Corolla is made up of five petals. Flowers are both unisexual and bisexual, latter being mostly staminate. There are five stamens, ovary is inferior and bilocular. The crop is cross pollinated. Fruit is globular, round and small with longitudinal grooves and ribs on surface. When pressed, fruit easily breaks into two locules with one seed in each. Seeds are pale white to light brown in colour.

Aromatic odour and taste of coriander fruits (seed) is due to an essential oil. Amount of oil varies considerably according to source of fruit. Indian coriander is rather poor in oil content (0.2-1.0%). European coriander is fairly rich in oil (1.4-1.7%) and Russian coriander contains oil up to 2 per cent. Volatile oil is made up of hydrocarbons and oxygenated compounds. Hydrocarbons account for about 20 per cent of oil. Major oxygenated compounds present are d-linalool or coriandrol (45-70%). Main component of coriander seed oil from Brazil was reported as linalool (77.48%) by Figueredo *et al.* (2004). Besides the essential oil, seeds contain 19-21 per cent of fatty oil (Non-volatile oil). Coriander herb oil is obtained from stalks and leaves (0.1 - 0.95%). Herb oil has a pronounced odour of decylaldehyde or other higher fatty aldehyde. Coriander leaves also constitute one of the richest sources of vitamin C (125-250 mg/100 g) and vitamin A (5,200 IU/100 g).

Crop improvement

Most of the improvement work done in coriander are with respect to varietal development. Correlation studies carried out by Agrawal *et al.* (1990) revealed that essential oil content of seed was negatively correlated with seed yield. Another correlation study by Tripathi *et al.* (2000) identified plant weight, numbers of secondary branches, days to flowering, days to maturity and number of umbels/plant as major yield components. Multivariate analysis using 40 gemotypes of coriander using D^2 statistics showed that characters such as days to flowering, days to maturity and number of secondary branches were to be taken into account while selecting parents for hybridization (Srivastava *et al.*, 2000). In another experiment Srivastva *et al.* (2000) reported that day to flowering had highest direct effect on seed yield followed by days to maturity and number of umbels/plant. Kalra *et al.* (2003) reported negative correlation between seed yield and essential oil content and susceptibility to stem gall disease and a close relationship between days to flower and days to maturity, oil yield and late maturity.

Cultivated types of Indian coriander can be divided into two more or less distinct classes on basis of seed size - small and bold seeded. Small seeded types are highly aromatic and fetch better price. English, German and Russian varieties are of this type. Besides, they are hard to break, which reduces breakage during handling, transport and storage. Small seeded types are normally cultivated in irrigated areas. Bold seeded types are usually less aromatic and fetch comparatively low price. These types are cultivated in unirrigated areas. There is another classification based on part used viz. seed purpose varieties and dual purpose varieties.

Varieties for seed purpose

CS 2, MS 1, MS 3, CS 3, CS 4, CS 6, RCI, Karan (UD 41), UD 1, UD 2, UD 20, UD 21, GAU 1, CIMPOS 33, CIMAPS 20, S 15 and S 33.

Varieties for dual purpose

CS 5, Co 1, Co 2, IARI S 360, IARI Seethal 36-3, CIMPO S 28 and CIMPOS 52.

The CIMAP accessions CIMAP 2053 and CIMAP 2096 are recommended under late sown conditions in Indo-Gangetic plains for higher seed and essential oil yield (Kalra *et al.*, 2003). For irrigated tracts of Tamil Nadu, CS-12 and CS 203 with an yield of 5793 kg/ha and 561.0 kg/ha respectively are suitable (Selvarajan *et al.*, 2002).

Sengupta and Datta (2001) produced three highly productive dual purpose mutants namely Cone mutant type - I, Prostrate mutant type 1 and Lax branching [NP (D) 95], following gamma irradiations and EMS treatments of dry and filled seeds of coriander cultivars TNP (D) 92 and NP (D) 95.

Indian coriander is generally not preferred for distillation of essential oil because of poor quality and oil content. There is urgent need to evolve high yielding and disease resistant varieties rich in essential oil and adaptable to various agro-climatic regions of the Country.

Crop management

Coriander is a tropical crop; it requires a cool and comparatively dry frost free climate, particularly at the time of flowering and seed formation for good quality and high yields (Sharma and Sharma, 2004). Coriander is cultivated either as a pure crop or mixed crop under irrigated as well as rainfed conditions. As an irrigated crop, it comes up well in all types of soils, but as rainfed crop it has to be cultivated in black cotton soils or any other type of heavy soils which have better water retention capacity. In some areas, it is also grown as a *rabi* crop.

Seeds and sowing

Seed rate varies from 10-15 kg/ha for a pure crop and 5-6 kg/ha for mixed crop. Land is prepared to fine tilth. Before sowing, seeds are trampled or crushed into halves with feet or by rubbing with hands. Sowing is done either by broadcasting or by seed drills in 25 to 30 cm apart rows, dropping 3-4 seeds /hole, spaced 10-15 cm apart. Depth of sowing should not exceed 8.0

Table 30. Improved varieties of coriander recommended for cultivation in different states

Variety (year of release)	Parentage	Special characters	Crop duration	Average yield (kg/ha)	Institution/ University
Gujarat Coriander-1 (1974)	Selection from local type	High yield, more number of branches, seeds bolder and greenish colour	112 days (Medium)	1100	Jagudan (GAU)
Col 1 (1981)	Selection from germplasm	Tall plant, many umbels per plant, suitable for green leaves and grains, tolerant to grain mould	110 days (Medium)	500	Coimbatore (TNAU)
Co 2 (1982)	Reselection from Culture P 2 of Gujarat	High yield, dual purpose variety, both *rabi* and *kharif* crops, tolerant to drought, 0.3% oil	90-100 days (Medium)	600-700	Coimbatore (TNAU)
Gujarat Coriander-2 (1985)	Reelection from Co 2	High yield, more secondary branches, dense foliage, fistular, stem, umbels large sized, grain purpose variety, bold seeds, no lodging	110-115 days (Medium)	1500	Jagudan (GAU)
RCr 41 (1988)	Recurrent selection from UD 41 (Karan)	High yield, tall, erect plant type, round and small seeds, suitable for irrigated areas 0.25% oil, resistant to stem gall, tolerant to wilt	130-140 days (Late)	1200	Jobner (RAU)
Rajendra Swati (1988)	Mass selection from germplasm (RD 44)	High yield potential, medium height, suitable for intercropping, fine-seeded, rich in essential oil, resistant to stem gall disease, free from attack of fruit fly	100 days (Medium)	1200-1400	Dholi (RAU)
Swati (1989)	Mass selection from germplasm Nandyal (CS 6)	High yield, semi-erect plant type, suitable for delayed sowing, escapes powdery mildew, 0.3% oil	80-90 days (Early)	885	Guntur (APAU)
Sadhana (1989)	Mass selection from germplasm Alur (CS 4)	High yield, suitable for rainfed areas, semi-erect type resistant to aphids and mites, 0.2% oil	95-105 days (medium)	1025	Guntur (APAU)
Co 3 (1991)	Pureline selection from Acc. No.695, IARI	High yield, dual purpose, less susceptible to wilt and grain mould, medium-size grain, brownish-yellow colour, 0.4% oil	90 days (Medium)	640	Coimbatore (TNAU)
CS 287	Recurrent selection from CS 6 (Andhra Pradesh)	Short duration, small, bold grain, tolerant to wilt and grain mould, suited for rainfed tracts	78 days (Early)	510	Coimbatore (TNAU)
Lam Sel. CS 2	Mass selection (Lam Sel CS 2)	Medium tall, busy type, more number of branches, grain purpose variety, tolerant to pests and diseases, good quality grain, 0.4% oil	110 days (Medium)	1500	Guntur (APAU)
UD 20	Not available	Medium tall plant, bold grains	100-110 days (Medium)	1175	Jobner (RAU)

cm. In case of broadcasting, seed should be mixed with soil with help of rakes in such a way that all seeds are covered well with sand. To hasten germination process, seeds soaked in water for 12-24 hours can be sown after draining excess moisture by spreading over cloth or gunny bag under shade.

For seed purpose, sowing is done between 15 October and 15 November. Time of sowing depends much upon day temperature. For an early seed crop in southern areas, sowing is done in mid May and harvesting in June-July. For growing coriander for leaves, the crop can be sown any time during the year excepting the period when temperature is too high.

Manures and fertilizers

Application of 10-15 tonnes of well rotten FYM and NPK @ 60-90; 15-50 and 15-30 kg/ha respectively are recommended for Indian conditions (Singh and Gangwar, 1991). Application of S (12 ppm ha^{-1}) and Zn (205 ppm ha^{-1}) besides the recommended dose of NPK (80:30:30 kg ha^{-1}) resulted in higher seed and essential oil yield in coriander (Manure *et al.*, 2000). Channabasavanna (2002) obtained the highest benefit cost ratio with 60 kg N, 40 kg P_2O_5 and 40 kg K_20/ha under Karnataka conditions for an irrigated crop. In unirrigated crop, full quantity of manures and fertilizers are applied at time of field preparation and for irrigated crop ½ N, full P and K are applied as basal dose and remaining nitrogen in two equal splits at time of first irrigation and then at flowering.

Irrigation and interculture

For irrigated crop, 3-4 irrigations are required at 35, 60, 80, 100 days after sowing. Coriander cannot withstand excess moisture. For a good crop, weeding and hoeing are necessary.

Pests and diseases

Wilt caused by *Fusarium oxysporum* var. Corianderi, powdery mildew caused by *Erysiphae polygoni* and stem gall caused by *Protonyces macrosporum* are the three major diseases of coriander. Crop is also affected by aphids and leaf eating caterpillars or semiloopers.

Harvesting and threshing

Coriander crop ripens in 115 to 125 days depending upon the variety and agroclimate. Crop is harvested when 60 per cent of seeds in main umbels turn yellowish brown. Plants are uprooted or cut with sickles, tied in small bundles and stacked for drying in shade, keeping bundles upside down so that grains are less exposed to sun and their greenish colour is protected. When bundles dry, grains are separated by beating gently on threshing floor. Grains are cleaned by winnowing and stored in moisture free storage. In case of dual purpose varieties, the foliage is plucked when crop is 60-70 days old. Yield under irrigated cultivation is 1.0 to 1.5 tonnes/ha and under rainfed cultivation, 0.4-0.5 tonnes/ha and in a mixed crop it is 0.15-0.30 tonnes/ha.

Processed products and uses

Coriander is sold as whole coriander, ground coriander (powder), coriander seed soil, coriander herb oil and coriander oleoresin. Fresh green plant is used as a flavourant in chutneys and soups. Seeds are extensively employed as condiment with or without roasting. In medicine, seeds are carminative, diuretic, tonic, stomachic and aphrodisiac. Oil finds use in perfumery also.

Cumin
(*Cuminum cyminum* L., F. Apiaceae)

Cumin or "Safaid zeera" comprises the dried pale-yellowish seeds of a slender annual herb, believed to be a native of Egypt and Syria; Turkestan and Eastern Mediterranean region. It is one of the oldest spices known since bibilical times. Today the plant is grown extensively in Iran, India, Morocco, China, Southern Russia, Southern Europe and Turkey. According to recent statistics , India ranks first in the world in area and production of cumin. India is also the largest consumer of cumin and exports hardly four per cent of its total production. Major types of cumin in the international trade are Iranian cumin, Indian cumin, Egyptian cumin and Turkish cumin. In India, cumin is extensively grown in Rajasthan, Gujarat, Madhya Pradesh, Uttar Pradesh, Punjab and Maharastra.

Botany and chemical composition

Cuminum cyminum is a slender much branched annual herb about 25 cm in height, leaves 2-3 partite linear, deep green, flowers white or pink in colour produced in small compound umbels. The aromatic seed like fruit, commonly known as seed is elongated, oval, elliptical, deeply furrowed, approximately 6 mm long, light yellowish brown. The odour is peculiar, strong and heavy, while flavour is warm, slightly bitter and somewhat disagreeable.

Seeds yield a volatile oil (2.5 to 4.5%), the chief constituent of which is cuminaldehyde (20-40%) which is used in perfumery. In addition to this, seeds also contain a fixed oil (10%) with a strong aromatic flavour.

Crop improvement

There are four cultivars of cumin viz. (1) tall, (2) dwarf, (3) pink flowered and (4) white flowered. The pink flowered variety yields better than the white flowered variety.

Improved varieties

RS-1

Developed in Rajasthan. Slightly late maturing with bold hairy aromatic seeds.

MC 43

Medium duration variety (110-115 days) evolved at Gujarat Agricultural University, Jagudan. Tolerant to wilt & blight diseases. Seed yield 580 kg/ha.

Gujarat Cumin I

Short duration variety from GAU, Jagudan. Tolerant to wilt and blight diseases. Seed yield 735 kg/ha.

RZ 19

Medium duration variety evolved at Rajasthan Agricultural University, Jobner. Tolerant to wilt and blight. Seed yield 590 kg/ha.

S 404

Developed in Gujarat. Yield 350 kg/ha.

Gujarat Cumin 2

Seed yield 700 kg/ha.

Other promising varieties are Gujarat Cumin 3, Gujarat Cumin 4, Sel 7-3, CJS 182, Vc-19, Vc-198 etc.

Crop management

Soil and climate

Cumin thrives well in tropical and subtropical climate. Frost is the most dangerous and limiting factor for cumin cultivation. High humidity during flowering and fruiting induces development of diseases like powdery mildew and blight. Sandy loam or loam soil is supposed to be the best for its successful production.

Seeds and sowing

The land is ploughed to fine tilth and levelled. Field is divided to beds of convenient size. Seed rate is 90 kg/ha. In Gujarat and Rajasthan, sowing is done between 15 November and late December. Late sowing results in poor yield. Seeds are broadcast uniformly and covered with soil lightly. Soaking of seeds for 24-36 hours in running water improves the germination percentage. Light irrigation is given after sowing. Line sowing can also be done at a spacing of 30 x 15 cm. In order to avoid fungal diseases, seed treatment with fungicides is done. It takes about a week to germinate. Thinning is done when the seedlings attain a height of about 5 cm (usually one month) maintaining a distance of 10 cm (plant to plant).

Manuring and fertilizer application

FYM is applied at the rate of 15-20 t/ha basally. NPK at the rate of 70:60:40 kg/ha is applied basally and 30 kg/ha of nitrogen is top dressed 30-40 days after sowing (Srimathi and Malarkodi, 1998).

Irrigation and interculture

First irrigation is given just after sowing, second 7-8 days after sowing. After 2-3 irrigations, the crop is irrigated at an interval of 12 to 20 days depending upon the weather conditions and soil type. It is essential to provide irrigation at the flowering and seeding stage and to withhold at the time of seed maturity.

Interculture

First hoeing and weeding is done 30-40 days after sowing. Thinning is also done at this stage.

Harvesting and yield

The crop matures 100-110 days after sowing and leaves turn yellow. Plants are uprooted and stacked in small bundles for sun drying. The grains are separated by beating with light sticks and cleaned by winnowing. Yield - 550 kg/ha. The clean and dried seeds are stored in gunny bags.

Processed products

Processed products include cumin powder, cumin oleoresin, volatile oil and fixed oil.

Uses

Cumin seeds are largely used as a condiment and form an essential ingradient in all mixed spices and curry powders. In medicine, cumin seeds have long been considered a stimulant, carminative, stomachic and astringent. They are now widely used in veterinary medicine also. The volatile oil is used in perfumery and also for flavouring liquors and cordials. Fixed oil finds use in fat and soap industry.

Fennel
(*Foeniculum vulgare* Mill., F. Apiaceae)

Fennel is a native of the Mediterranean region where it is cultivated since early times. The world "marathon" comes from the Greek for fennel. The famed battle of marathon 490 BC was fought on the field of fennel. During the struggle, the Athenian athlete Pheidippides ran 150 miles to Sparta for aid and that is the root of our modern usage marathon (fennel). Fennel has become naturalized in many temperate countries and can be grown in the tropics. It is widely cultivated in India at altitudes up to 1825 m, Gujarat and Rajasthan being the major fennel growing states. India exports around 3000 t of fennel seeds annually. Major types of fennel seed in the world market are Indian fennel, Egyptian fennel and Chineese fennel. At present India is not only by far the largest producer of fennel seed, but its product also commands top price for its quality.

Botany and chemical composition

Fennel is a biennial or perennial aromatic, stout glabrous herb, 1.5-1.8 m high, with feathery much divided leaves. Small yellow flowers (June-September) with fine petals are born in flat topped umbrella like clusters. The flowers are followed by grey-green seeds (actually fruits) 6-8 mm long, straight or slightly curved. It possesses an agreeable aromatic and sweet aroma resembling aniseed. The fruits yield a volatile oil (0.7-6%), the chief constituent of which is anethole (50-70%). Fennel seeds also contain 9-13 % fixed oil.

Types/races or sub species

Varieties and races differing in size, odour and taste of the fruits exist among wild and cultivated fennels, but they are hardly distinguishable from one another. They are regarded as races, varieties or sub species of *F. vulgare*. The varieties which yield commercially important volatile oils are referred generally to the sub species *capillaceum* and placed under two distinct varieties. (1) var. *vulgare* - yielding bitter fennel oil and (2) var. *dulce* - yielding sweet fennel oil. var. *vulgare* is cultivated in India.

Chemotaxonomic classification of fennel varieties and sub species were attempted by many workers and distinct chemotypes have been identified based on the contents of anethole, fenchone, methyl chavicol, α-phyllandrene, limonene etc. (Bernath *et al.*, 1996; Rapisarda *et al.*, 1998 and Jimenez *et al.*, 2000b).

Crop improvement

The crop remained neglected in respect of genetic improvement. As a result, local types were being cultivated in many areas. Selection have been made from the local types and the improved varieties of fennel are available in India are the following.

UF 32

A selection from Rajasthan. Matures in 210-220 days.

PF 35

A selection from Gujarat Agricultural University. Crop duration is 216 days and yield 1,280 kg/ha with 1.9 % oil.

Gujarat Fennel I

A pure line selection made at GAU, Jagudan. Late variety (220-230 days) yielding 1,700 kg/ha.

Co 1

Re-selection from PF 35 made at TNAU. Crop duration 210-220 days and yield 570 kg/ha.

S 7-9

Dwarf selection from GAU Jagudan. Yield 1000 kg/ha.

Other released varieties are Gujarat Fennel 2, Hissar Swarup, Azad Sanuf 1, Pant Madhurika and Rajendra Sourabha (Johnykallupurackal and Ravindran, 2006).

Being an allogamous crop, it is easy to produce hybrid seeds. Dashora *et al.* (2000) reported high amount of heterosis for various traits including seed yield. But crossing is difficult due to small flower size and hence recurrent selection and development of composites is suggested to improve yield (Dashora *et al.*, 2000)

Crop management

Soil and climate

Fennel is a cold weather crop and comes up well in a fairly mild climate. It does not thrive in South India except at higher elevations. A dry and cold weather favours seed production. It thrives well on drained loamy clay soil, rich in plant nutrients and lime.

Seeds and sowing

Fennel is propagated readily by seeds (direct sown or transplanted), but can also be grown by root or crown division. The time of sowing is October-November in the plains of North India and March-April on the hills. Yadav and Khurana (2000) reported superiority of transplanting to direct sowing with respect to seed yield. In the case of transplanted fennel, nursery is raised during May-June and transplanting is done in August to September. For direct sown fennel , land is prepared to a fine-tilth and beds of convenient size are taken. Seeds are sown in rows at 45-60 cm apart. Seedlings are thinned to 25 to 30 cm spacing when they are 4-5 weeks old. Only shallow sowing should be done. Light irrigation is given after sowing. Seed rate is 9-12 kg/ha. Pre-soaking in water and seed treatment with fungicides improve the germination percentage and check the seed borne diseases respectively.

For preparing nursery, 3 to 4 kg seed is sown in 100 m^2 area of nursery to raise the seedlings for one hectare. Seeds are broadcast and mixed with soil. Beds are mulched and irrigated. The seeds germinate within 7-10 days. The mulch is removed after germination. Seedlings will be ready for transplanting after 6-7 weeks at a spacing of 80 x 60 cm.

Manuring and fertilizer application

Apply 10-15 tonnes of FYM/ha before preparatory tillage. Besides FYM, 60 kg N and 40 kg P_2O_5/ha are also recommended. Out of these, 30 kg N and full P_2O_5 are to be applied as basal dose. Remaining 30 kg N should be applied just before flowering.

Khan *et al.* (1992) recommended foliar sprays of N (2% urea) and P (0.2% sodium dihydrogen orthophosphate) for maximum oil and fenchone content in the oil. For getting maximum anethole content in the oil, soil application is recommended by them.

Interculture operations

Three to four weedings and hoeings before flowering are desirable to obtain a good crop. Earthing up of plants prevent lodging.

Irrigation

Fennel requires frequent irrigations and the number of irrigations depends on the soil type and prevailing weather conditions. Irrigate the crop at 20-25 days interval in winter and at 10-14 days interval in April-May at the time of seed formation.

Pests and diseases

Aphids attack fennel crop at flowering stage causing serious damage. Important diseases of fennel are blight and powdery mildew.

Harvesting and yield

The crop is harvested before the fruits are fully ripe. Since all the fruits do not mature at the same time harvesting of umbels is done 4-5 times at 10-15 days interval. Umbels are plucked when the fruits change their colour from deep green to light yellow. Umbels are dried under shade avoiding chances to ferment. Dried umbels are threshed and seeds are separated and cleaned by winnowing. On an average, seed yield of 500 to 900 kg/ha can be obtained. The harvesting and packaging of fennel absorb about 60 per cent of cost of production.

Uses

The plant is pleasantly aromatic and is used as a pot-herb. Leaves are used in fish sauce and for garnishing; leaf stalks are used in salad. The dried fruits are used as a masticatory or for chewing alone or in pans. They are also used as a flavouring agent. The fruits are aromatic, stimulant and carminative and are considered useful in diseases of the chest, spleen and kidney. Leaves are reported to have diuretic properties. The residue left after the distillation of essential oil from the fruit is used as cattle feed.

Fenugreek
(*Trigonella foenum-graecum* L., F. Fabaceae)

Originated in the Mediterranean region and Asia, fenugreek is among the oldest of medicinal herbs. Its seeds were a favourite curreal in ancient Egypt and India and later among the Greek and Romans. In religious rites, fenugreek was one among the components of Holy Smoke, an Egyptian incense, used in fumigation and embalming. It is understood that fenugreek was a part of Indian diet even 3,000 years ago. It is found wild in Kashmir, Punjab and the upper gangetic plains. Fenugreek is cultivated in India, Argentina, and Egypt and Mediterranean countries. The seed is produced as a spice and leaves and tender pods as leafy vegetable or as forage for cattle. India is one of the major produces and exporters of fenugreek. It is grown on an extensive scale in Rajasthan, Madhya Pradesh, Gujarat, Uttar Pradesh, Maharastra and Punjab. India exports fenugreek seed, seed powder, oil and oleoresin. This spice occupies the third place in area and fourth in production among all minor spices grown in India. More than 80 per cent of the area is in Rajasthan (Saini, 2005).

Botany and chemical composition

Fenugreek is an annual herb 30-60 cm tall with light green leaves with three finely toothed oval leaflets. Off-white flowers are produced singly or in pairs during April-May. It produces slender beaked or sickle shaped pods 10-15 cm long. Each pod contains 10-20 small hard yellowish-brown seeds, which are smooth and oblong about 3 mm long, each grooved across one corner, giving them a hooked appearance.

Fenugreek seeds contain many substances like protein, starch, sugar, mucilage, mineral matter, volatile oil, fixed oil, vitamins and enzymes. Seeds are rich in essential aminoacids and trigonelline for which fenugreek is so well known for medicinal uses. Seeds are also reported to contain the steroid "diosgenin" which is used in the preparation of contraceptives (Saini, 2005). Leaves and stems are also rich in calcium, iron, carotene and ascorbic acid. Leaves contain 18.6-20.9 % protein on dry weight basis. Seeds contain 7 % fixed oil and 0.02 % volatile oil.

Crop improvement

Genetic variability and correlation studies were carried out in fenugreek. Ahmed *et al.* (1989) reported that number of pods per plant has got the greatest direct effect on seed yield. Seed yield was also influenced by plant height, number of leaves per plant and weight of pods per plant. Sastry *et al.* (2000) reported that seed yield could be improved if selection is done for days to maturity, number of branches/plant, number of pods/plant, protein content and biological yield/plant. Estimates of phenotypic and genotypic coefficients of variation were high for straw yield, biological yield and grain yield per plant and harvest index; and low for the remaining characters (Saha and Kole, 2001). The genotypic and phenotypic correlations of days to flowering, plant height, pods per plant, pod length, grains per pod, straw yield, biological yield and harvest index with grain yield per plant were significant and positive indicating the importance of these characters in seed yield improvement.

Divergence analysis in fenugreek carried out by Kole and Mishra (2002) revealed that seeds per pod, 100 seed weight, grain yield, straw yield and branches per plant were the major characters contributing towards divergence.

Cultivars

Main cultivars in fenugreek are

1. Non-scented or Deshi or Common methi

Quick growing and upright plants with white flowers. Pods are straight, 8-10 cm long containing 10-15 seeds.

2. Scented

Slow growing types, cultivation is confined to North India. Kasuri methi, Marwari methi and Champa methi come under this group.

Improved cultivars

Saini (2005) gave a list of improved cultivars of fenugreek and their characters.

1. Lam Selection I

A high yielding variety recommended for Andhra Pradesh. Matures in 90 days with an average yield of 740 kg/ha. Plants are tolerant to powdery mildew, root rot, caterpillars and aphids.

2. Pusa Early Bunching

An early cultivar evolved at IARI. It is a common methi. Takes about 125 days for seed maturity.

1. Rajendra Kranti

It matures in 120 days with an average yield of 1.25 kg/ha. Moderately resistant to powdery mildew, aphids and caterpillars.

2. Hissar Sonali

It matures in 140-150 days with an average yield of 1.9 kg/ha. Moderately resistant to leaf spot and root-rot complex diseases.

3. H.M. 103

It takes about 140-150 days to mature with an average yield of 2 t/ha. Moderately resistant to leaf spot diseases.

4. C.O. I

A dual purpose variety evolved at TNAU. Matures in 95 days with an average yield of 680 kg/ha. Tolerant to root rot.

5. Kasuri Selection

It is a late cultivar of Kasuri methi evolved at IARI. It takes about 156 days to mature.

6. R.M.T. I

Matures in 140-150 days with an average yield of 1.47 kg/ha. Moderately resistant to root rot and tolerant to powdery mildew.

7. R.M.T. 143

Seeds mature 140-150 days after sowing. Average yield 1.6 t/ha. Moderately resistant to powdery mildew. It is specially recommended for heavier soils of Chittoor, Bhilwara, Jhalawar and Jodhpur.

Other cultivars recommended for different states are

Prabhat (NCM) is recommended for Gujarat and Rajasthan; Methi No.14 and Methi No.47 for Maharastra; UM 50 for Himachal Pradesh, ML 150 for Punjab and Prabha for Rajasthan. Rajendra Abha for Bihar, Hissar Suvarna, Hissar Madhavi and Hissar Mukta for Haryana, Pant Ragini for Utter Pradesh and Gujarat Mettul for Gujarat (Johnykallupurackal and Ravindran, 2006).

Crop management

Soil and climate

Fenugreek requires moderately cool climate for proper growth, yield and good quality. Heavy rainfall areas are not suitable for its cultivation. It can be grown in all types of soils which are rich in organic matter with good drainage.

Seeds and sowing

Fenugreek is propagated by seeds. It is either grown pure or mixed with other crops. Last week of October to first week of November is ideal sowing time for fenugreek in North India. In Southern part, it is grown in *kharif* and *rabi* seasons. Sowing time in the hills is from March to April and October.

Seed rate for common methi is 25 kg/ha and for Kasuri methi it is 20 kg/ ha (Saini, 2005). For irrigated crop, the seed rate is higher. Thorough soil preparation is necessary before sowing. Beds of convenient sizes are prepared and seed is generally broadcast and surface is raked to cover it. Sowing should be 30 cm apart in rows with a plant to plant spacing of 10 cm. Halesh *et al.* (2000) obtained higher seed yield under doses spacing (15 x 15 cm), while wider spacing (30 x 30 cm) recorded maximum growth and yield parameters except seed yield /ha.

Irrigation and interculture

Frequent irrigation is necessary to obtain quick growth of the crop. About 5-7 irrigations are required. The crop should be irrigated at IQ/CPE ratio of 1.0. Special care should be taken to avoid water stress at pod and seed development stages.

Two hoeings and weedings are enough to keep the crop well aerated and weed free. First hoeing and weeding should be done at the time of thinning (25-30 days after sowing) and 50-60 days after sowing.

Manuring and fertilizer application

It is commonly grown on the residual fertility of previous crops. To get good yields, add FYM @ 15 t/ha at the time of field preparation also. A dose of 40 kg each of N and P_2O_5 is also given at the time of sowing. Nitrogen may also be given as top dressing. Halesh *et al.* (2000) recommended a fertilizer dose of 60:90:50 kg NPK ha^{-1} under Bangalore conditions to obtain maximum growth and yield.

Diseases and pests

Diseases affecting fenugreek are leaf spot caused by Cercospora traverisiana, downy mildew caused by Perenospora trigonella, powdery mildew caused by *Erysiphae, Oidium* and *Oidiopsis* sp, rust caused by Uromyces anthyllidis and damping off of young seedlings. The plant is infested by mites and aphids.

Harvesting and yield

When common methi is used as leafy vegetable, the young shoots are nipped off in about three weeks after sowing. A number of cuttings may be taken from Kasuri types. First cutting is possible in about 25-30 days after sowing. Leaves are nipped two centimeter above ground level leaving the stump which produces new shoots. Subsequent cuttings may be taken at an interval of 12-15 days. If cuttings are late, leaves develop a bitter taste. After 2-3 cuttings, plants are left for seeding. Seed yield is high in case of crops grown only for seed. Harvesting for seeds is done when the lower leaves start shedding and pods become yellowish. Plants are cut with sickles. Plants are tied in bundles and allowed to dry for 4-6 days. Threshing should be done on clean cemented floor. Grains are separated by beating followed by winnowing or threshing using mechanical thresher. Yield varies with method of harvesting. Green leaf yield of Kasuri methi is 9 to 10 t/ha and that of common methi is 7 to 8 t/ha. Average seed yield is 1.5 to2.0 t/ha for a pure seed crop.

Uses

Fenugreek is used both as food and food additive, as well as in medicine. It is one of the principal odorous constituents of curry powder in India. Fenugreek stimulates digestive process as well as metabolism in general. It is also useful in hypertension and diabetics. Seeds contain substantial quantities of diosgenin which is used as a starting material in the synthesis of sex hormones and oral contraceptives. It is used in cattle feeds and in veterinary medicines. It is also used as a hair tonic and as a cure for baldness. Fixed oil finds use in perfumery also.

Anise
(*Pimpenella anisum* L., F. Apiaceae)

Anise is an annual herb native to Mediterranean region. It is extensively cultivated in Central, South and South-East European countries and also in Egypt, Syria, Iran and Mexico. Anise is used since ancient times as a medicinal plant. The ancient Romans were so fond of anise seed, with its spicy aroma and liquorice taste. It is one of the oldest known herbs mentioned in record long before birth of Christ. By 16th century, Europeans discovered that anise was an irresistible bait for mice. The seed oil is used to poison pigeons. Anise seed and essential oil obtained from seed and foliage are widely used in food industry (Kolak *et al.*, 2002).

Botany and chemical composition

Anise ($2n = 18$) is an annual pubescent aromatic herb which attains a height of 60-80 cm under cultivation in India. Basal leaves are long and dissected with long petioles whereas, leaves in stem are pinnately decompound, emitting characteristic anise aroma when crushed. Flowers appear in umbels after 120 days of sowing. They are small, white in colour, borne centripetally in umbels. Fruit is small, ovoid and laterally compressed.

Crop is grown for its seeds which are rich in volatile oil containing about 84-87 per cent anethole and 12-15 per cent methyl chavicol. Other compounds include p-methoxy acetone, terpenes and sulphur containing compounds.

Cultivation

Production technology of anise is given by Bhuvaneswari *et al.* (2001). A fairly warm weather during sowing with a temperature of 20-25°C is desirable.

Low temperature & higher atmospheric humidity are conducive for vegetative growth, while the crop requires warm sunny weather during seed formation and development stages. Crop is grown on a wide range of soils, ranging from sandy loam to clay loam, having a pH of 6.0-8.5 and of moderate fertility.

Seeds and sowing

Anise is grown from seed as a cold weather crop in India. Seeds are small and need a meticulous land preparation for obtaining good germination and high seedling stand. Seeds are sown at spacing of 45 x 20 cm at one cm depth in late October or early November. To obtain uniform germination, seeds are soaked in water overnight and then sown, followed by light irrigation. Seed rate is 7-10 kg/ha. Germination will be over in 10 days.

Irrigation

For better growth and good yield, irrigation is given at an interval of 3-4 days. At flowering and grain filling stages, crop needs copious application of water. El-Din (2003) reported that water deficiency resulted in increased essential oil content of seed.

Manures and fertilizers

Crop is moderate in its requirement of manures and fertilizers. Farm Yard Manure @ 10 tonnes/ha is applied at time of land preparation. Application of 80 kg N, 60 kg each of P_2O_5 and K_2O/hectare is followed. Out of total quantity, 50 per cent N and full P_2O_5 and K_2O are applied when seedlings are 20 days old. Remaining 50 per cent N is applied at full vegetative stage. Weeding, preferably hand weeding is carried out at monthly intervals.

Pests and diseases

Semilooper (*Thyranopluria orichalsia*) and cigarette beetle (*Hasioderma serricome*) are minor pests reported.

Harvesting and yield

Umbels come to maturity from March to mid April. Terminal umbel matures first and lateral ones 15-20 days later. When 80-90 per cent fruits begin to turn grayish green in colour, tops are cut along with branches and are tied in bundles. Bundles are stacked keeping fruiting heads downwards. Later they are threshed, seeds cleaned and stored. A good crop yields 0.7-0.8 tonnes/seeds/hectare.

Processing

Aniseed should be crushed and distilled immediately to obtain higher oil yield. Fresh seed crop generally yields 2.1-2.4 per cent oil.

Uses

Fresh leaves of plant are used as a garnish and for flavouring salads. Essential oil is used in perfumery, soaps and other toilet articles, in perfuming sachets, in dental preparations and in mouth washes. Seeds are used for flavouring food, confectionaries, beverages and in processed food and liquor industries. Oil is used externally as an insecticide against small insects and also has fungicidal properties. Medicinally, oil is used as an aromatic carminative, to relieve flatulence and as an ingredient of cough syrups in combination with liquorice. It also finds use in preparation of gripe water.

Celery
(*Apium graveolens* Linn., F. Apiaceae)

Celery is an umbelliferous aromatic herbaceous plant grown for its leaves, seeds, oleoresin and essential oil. Celery seed is known for over 3,000 years. The native habitat of celery extends from Sweden to Egypt, Algeria, and Ethiopia and in Asia, to India, Caucasus and Baluchistan. It is also grown in France and China. The French celery seed is of a fine quality but still Indian celery holds the fort in the world trade in quality.

In India, celery is cultivated both for salad and seed-raising in the North-West Himalayas, Punjab, Uttar Pradesh and Haryana. India dominates the world exports of celery seed. Certain quantity of celery seed oil and oleoresin are also exported from India.

Botany and chemical composition

Celery is an umbelliferous herb 30-60 cm high, erect with conspicuously joined stems, bearing well developed leaves on long expanded petioles. Its shallow root system has a tapering thick main root penetrating upto 15 cm below the soil. Flowers are produced in compound umbels. They are hermaphrodite, the aromatic seeds are light brown in colour and bitter in taste. The epicarp is intercepted with oil ducts. It is widely used as a spice. In colder climate and on hills, celery is a biennial plant and produces seeds only in the second year, but in plains it becomes an annual and produces seeds about 4-5 months after transplantation.

Celery seeds on distillation yields 2-3 % of a pale yellow volatile oil with a persistent odour characteristic of the plant. Principal constituents are d-limonene (60%), d-selinene (10%), sedanonic acid anhydride (0.5%), sedanolide (2.5-3.0%), the last two being responsible for the aroma of the oil. Celery herb oil is also distilled from the leaves, but it is quite different from the seed oil.

Crop improvement

The IARI, New Delhi has recommended the following varieties of celery for salad purpose.

1. Standard Bearer: Early, medium tall plant, medium pink stems with white longitudinal streaks, stalks solid and of good size and flavour.
2. Wright Grove Giant: Medium, late, tall growing plants, produces large white stalks of fine quality, immense cropper.

There is only one variety recommended for the production of seed crop of celery (Saini, 2004). Selection RRL 85-1: It is developed by the Regional Research Laboratory (CSIR). It yields 2-3 % yellow volatile oil.

Cultivation

Celery thrives the best when the weather is relatively cool having a moderate and well distributed rainfall during its growing period. Places having low humidity are ideal. Higher temperature causes bitterness in the leaves. The crop bolts when temperature falls below 15°C. Celery thrives well on loamy soils of 6.6 pH (Saini, 2004).

Celery is somewhat exacting crop and requires a carefully pulverized and heavily manured seed bed. Seeds are sown during September end, in rows in well prepared flat nursery beds. One kilogram of seed is required for planting one hectare. Seeds germinate in 10-21 days. The seedlings 10 cm tall (60-70 days after sowing) are ideal for transplanting during November-December. Delay in planting causes early bolting, reducing seed yield. Transplanting may be done in moist soil at a spacing of 40 x 15 cm.

Celery requires heavy manuring. Farm Yard Manure (10-20 t/ha) and half the dose of N, full P and K of the fertilizer dose (60:60:40 kg/ha) are added basally. Remaining Nitrogen is applied in two splits 30 and 60 days after planting.

Calcium deficiency expressed as "black heart" (showing tip burn on young leaves is prevented by foliar application of 0.1 M solution of $CaNO_3$ or $CaCl_2$. Boron deficiency expressed as "cracked stem" is controlled by borax application @ 12 kg/ha. In the event of magnesium deficiency expressed as "chlorosis" 10 kg of magnesium sulphate is added per hectare.

The crop requires frequent irrigation with good drainage. Soil should be hoed lightly but frequently. Only shallow cultivation should be practiced.

Diseases and pests

Damping off caused by *Pythium spp.*, early blight caused by *Cercospora apii* and late blight caused by *Septoria apii* are the common diseases.

Leaf minor, green celery worm, thrips, celery looper, celery caterpillar, carrot rust fly and aphid are some of the pests which attack celery.

Harvesting

Crop is harvested when 80 per cent umbels begin to turn light brown in colour. Plants are cut at the ground level. Harvesting is done in the early morning to avoid shattering of seed. Umbels are dried over threshing floor before threshing and winnowing. Seed yield is 1.4 t/ha. Seeds are graded and stored in gunny bags in a cool dry place.

Processed products and uses

Celery seeds on steam distillation yield 2-3 % of a pale yellow volatile oil known as celery seed oil and is much valued both as a fixative and as an ingredient of novel perfumes, pharmaceutical preparations and medicines. Celery herb (leaf) oil and celery chaff oil are other volatile oils obtained from the plant.

Celery seed oleoresin is prepared by extraction of crushed dried celery seeds with suitable food grade solvents.

'Celery salt' and 'celery pepper' are other processed products of celery seed. Celery seed and its products are used as food flavourants, in medicine and also as a bird feed. Seeds are stimulant and tonic and used in asthma and for liver diseases. It is also used as a domestic remedy for rheumatism, nervine, sedative and tonic. Commercial use of celery seed oil is as a fixative and as an important ingredient of novel perfumes.

Dill
(*Anethum graveolens* L., F. Apiaceae)

Dill was known as a garden plant in ancient literature and the use of fruits as condiment was known since biblical times. Dill is of value for both its leaves and seeds. Leaves are also known as 'Dill weed', fresh or dried. When fresh, the leaves have a pleasant aromatic odour and warm taste. Dill seed with its pleasantly aromatic and warming flavour, laced by woody and menthol notes, is excellent for flavouring and seasoning.

The western world, Russia and parts of Middle East grow the dill classified as *Anethum graveolens*, the so called true dill. This is a native of the Mediterranean and Southern Russia. India and certain areas in Asia grow a type called *Anethum sowa*, or the Indian dill which is native to Northern India. The principle difference is that the sowa seeds are especially bold (large and good looking). They also tend to be longer and narrower in shape and their oil has a somewhat different profile (Pruthi, 2001).

European dill was introduced to India and it acclimatized well under Indian conditions and now India exports dill seeds. Small quantities of the seed oil are also exported. However, the performance of European dill was not comparable with the Indian dill crop with respect to flowering, fruit set and yield (Wealth of India, 1985).

Botany and chemical composition

Dill is a herbaceous annual, glabrous aromatic herb upto 1.2 m height, cultivated throughout India, chiefly in Punjab, Uttar Pradesh, Gujarat, Maharashtra, Assam and West Bengal. Leaves are compound, flowers pale yellow in compound umbels and fruits sub-elliptical and dorsally compressed, glabrous and winged. The ripe, light brown seeds emit an aromatic odour faintly resembling that of caraway and possess warm aromatic and slightly sharp taste almost akin to caraway seeds.

The following are the anatomical parts and essential oils of dill.

Dill herb - fresh weed (stalks, leaves and seeds)

Dried dill herb - dried aerial parts of the plant

Dill seed - dry fruit

Dill herb/weed oil - obtained by steam distillation from fresh whole plant or dried leaves and seeds.

Indian dill yields 1.2 to 4.0% volatile oil while European dill grown under Indian conditions yields 2.5 to 4.0% oil. The major constituent of seed oil is carvone (19.5 to 69.7%). Carvone content is the highest in European dill. The sowa herb yields 0.06% of essential oil, which has a higher proportion of terpenes (a-phellandrene), but no carvone. European dill herb oil contains both carvone and a-phellandrene. Badoc and Lamarti (1991) identified three broad chemotypes of European dill by the presence of carvone, and the presence or absence of myristicin and dillapiole.

Crop improvement

A number of high yielding selections were identified in Indian dill. Of the selections, N.P. 127 and N.P. 179 were recommended for growing in the Indo-Gangetic plains, N.P. 115 for Gujarat and N.P. 15 for the hills. The average yield of these selections is around 1 to 14 t/ha.

Intra specific variability was reported in Indian dill. Two prominent land races from Gujarat called 'Ghoda Sowa' and 'Variyali Sowa' having differences both in morphological and chemical characteristics were identified. Another race 'Vizag Sowa' from Andhra Pradesh is also reported.

Cultivation

The sowa crop is raised as a cold weather crop all over India. It grows well on sandy loam soil. Crop is raised from seeds. Seed rate is 8-10 kg/ha. Seeds are broadcasted in rows 40 cm apart and the plants in a row are maintained at 12 cm by thinning. The seeds germinate eight days after sowing with 32 per cent germination. In most parts of India, sowing is done during September to October. Sowing is done in well prepared land incorporated with Farm Yard Manure @ 20-30 t/ha. Application of 60 kg N and 45 kg P_2O_5 /ha produced maximum seed yield in potash rich soils (Gupta, 1982). Crop requires one or two weedings in the early stages of growth. The soil should be kept moderately moist and even light irrigations (3-4) may be given, heavy watering should be avoided.

Diseases and pests

Powdery mildew caused by *Erysephe anethi* causes heavy damage to the crop between flowering and early seeding stage. The crop is susceptible to root rot caused by *Fusarium sp.* throughout the growing period.

Harvesting

Harvesting is done when the earliest fruit ripens to obtain the best quality and maximum quantity of oil from the seeds. At this stage umbels turn brownish in colour. Harvested fruits (seeds) are dried in the shade and threshed by hand or with a small stick. Seeds after drying and cleaning are packed in gunny bags and kept at a cool and dry place. Seed yield ranges from 700 – 800 kg/ha (Gupta, 2004).

Processed products and uses

There are six types of dill products in commerce today. Dill seed, dill seed products, dill leaves (dill weed), dill-seed oil, dill-weed oil and oleoresin . In India, only seed powder and oil are traded.

Sowa is in great demand as a condiment for flavouring pickles, soups and salad, for confectionery and for making tea. Seed is used in medicine mainly as an aromatic carminative, antipyretic and anthelmintic. It is used in the preparation of gripe water.

The oil of Indian dill seed is employed on a large scale in pharmacy as a substitute for the European dill oil. Dill waste (cake) left after the distillation of seed oil is used as a cattle feed. Sowa seed oil is also used as a synergist for insecticides.

Mustard

(*Brassica juncea* (Linn.) Czeranjajev., F. Brassicaceae)

Mustard can be considered as one of our most ancient spices. It is also an important oil seed in India where it is known as 'rai'. *Brassica juncea,* popularly known as Indian mustard or brown mustard, is possibly of African origin, but was taken early to China. It is extensively cultivated from Eastern Europe to China and in Africa. Other species of importance are *B. nigra* which provides black or true mustard and *B. hirta* (Syn. *B. alba, Sinapis alba*) which provides white or yellow mustard. Yellow mustard is believed to have originated in Europe. Indian mustard is a common field crop of Bihar, Uttar Pradesh and West Bengal.

Botany and chemical composition

Genus *Brassica* consists of 150 species of annual or biennial herbs, several of which are cultivated as oil seed crops or as vegetable or fodder crops. Seeds of only three species have condiment value.

B. hirta (Syn. *B. alba, Sinapis alba*)

White mustard or yellow mustard is a self sterile species. It has got hairy stem, irregularly pinnate leaves, large yellow flowers and spreading a few seeded and hairy pods, which have long empty, knife like beaks. Seeds are large, white and lightly pitted and give considerable amount of mucilage with cold water.

B. nigra

Black mustard or true mustard is a highly self-sterile species and it is quite distinct from other *Brassica* species. Fruits at maturity are closely appressed to inflorescence axis. It is a recent introduction to India and is grown to a limited extent only. It does not contribute to supply of mustard oil.

B. juncea

Indian mustard or brown mustard is an erect, much-branched annual up to 1 m tall, basal leaves stalked, upto 20 cm long, with very long ovate terminal segment. Flower pale yellow in colour, fruits 3-5 cm long. It is a self fertile species and lot of variability exist in India.

Indian mustard contains 30-38% fixed oil and upto 0.68% volatile oil. Main constituent of volatile oil is allyl isothiocyanate and related compounds. Seeds also contain glucoside 'sinigrin' and 'sinalbin' and enzyme 'myrosin' which hydrolyzes these glucosides in presence of water to release flavour and pungency.

Crop improvement

Lot of work were done in genetic improvement of mustard in India. Information on genotypic coefficients of variation, heritability and genotypic and phenotypic correlation coefficients are available. Larik and Rajput (2000) found that branches/plant and siliqua/plant are the most important yield components based on which selection could be done for improving yield. Short duration cultivars maturing in about 100 days were developed in India mustard (Singh, 2003). Some progress was also made in quality improvement.

Low erucic acid and low glucosinolate lines are now available. However, there appears to be a yield penalty for these lines. Good progress was made in molecular mapping of genes and transgenics development. Identification of a stable cytoplasmic male sterility and its restorer system brightened scope for hybrid development (Malik *et al.*, 1999; Singh, 2003). Isozyme banding pattern in parents and hybrids of mustard were compared using esterase isoenzyme analysis (Joshi and Patil, 2002). Variation in occurrence of bands between hybrids and their respective parents was observed, indicating differential gene expression in hybrids. Cross Varuna x TM-17 is a potential one for useful segregations in future generations.

Varieties

Through pure line breeding superior strains were developed in mustard. Varuna, PCR 7, Pusa Bold, Pusa Jaikisan, RH 30, RH 819, RH 8113, RL 1359, Rohini, Alankar, Pusa Bahar, Vaibhav, Kranti and Laxmi a few some of high yielding cultivars of Indian mustard. Yield and yield components of these high yielding strains were studied and established under rainfed and irrigated conditions (Singh *et al.*, 2002).

Crop management

Mustard in India is mostly grown mixed with wheat in North and with ragi in South. In North India, it is grown as a rabi crop and in the South from July-November. Loam and clay loam soils give better results. Land is prepared well. Farm yard manure is incorporated into soil and land is leveled. Seed rate is 5-7 kg/ha. Seeds are broadcast and covered with soil by light harrowing. Sowing behind country plough followed by rolling is also common practice. Spacing is about 30 cm between rows and 10 cm between plants. Water requirement of mustard was studied in detail. Water use efficiency was the highest for the 0.4 ID:CPE (ratio of irrigation depth to cumulative pan evaporation) irrigation treatment under Haryana conditions during winter (Yadav *et al.*, 1994). Rathore *et al.* (1998) observed that zero tillage along with mulching with paddy straw increased mustard yields by moisture conservation. Bharti *et al.* (2001) obtained the best response and the highest returns at 0.4 and 0.8 IW/CPE levels under Bihar conditions. Mustard responds well to fertilizers. Application of 60 kg N/ha gave higher seed yields in Haryana (Yadav *et al.*, 1994) whereas maximum returns were obtained at 80 kg N/ha under UP conditions (Singh *et al.*, 1998). Hatmode *et al.* (2001) obtained the highest seed yield at N:P ratio 60:50 kg/ha and the highest seed oil yield at 40:30 kg N:P/ha. Mustard also responds well to added Zinc and Sulphur. A combination of 40 mg S and 20 mg Zn/1 kg of soil gave the highest yield (Singh, 2003). Application of $ZnSO_4$ (1%) also gave higher yields in Mustard (Mankar *et al.*, 2003). Abraham (2001) obtained the best results by split application (3 splits) of 20 kg S/ha. Khan *et al.* (2002) studied effect of various growth regulators on mustard and found that GA3 application at 10.5 M increased seed, and oil yields by 33.3% and 31.2% respectively, compared to control.

Harvesting and processing

When crop is grown for dual purpose (green leaves and oil seed) leaves are picked about one month after sowing retaining some leaves on plant. In seed crop also 30% defoliation is found to improve seed yield (Khan and Ahsan, 2000). Seed crop will be ready for harvest in about 4 ½ months, when pods loose their green shade. Entire plant is pulled out, dried and mustard seeds are beaten out or threshed. Seed yield is 1-2 t/ha. Seeds are cleaned by winnowing and packed in gunny bags.

Processed products and uses

Major processed products from condiment mustards are mustard flour or powder, compounded mustard, mayonnaise (a creamy emulsion of vegetable oil in vinegar with mustard and egg as the emulsifying agents), ground mustard or mustard meal, and mustard cake. Mustard seed is an important condiment used for flavouring various food products, in making of mayonnaise and as a preservative.

Poppy
(*Papaver somniferum* L., F. Papaveraceae)

Poppy seed is derived from *Papaver somniferum*. Native to Mediterranean, poppies were taken to India and the East, by the Islam missionaries more than a thousand years ago. It was exploited first by Portuguese, then by Dutch and English, but later, attempts were made to regulate production and trade. It is now cultivated in Netherlands, France, Germany, Hungary, Rumania, Czechoslo-vakia, Turkey, Iran, Argentina, India, Australia and United States. Poppy is cultivated either for manufacture of opium, used as a narcotic agent or for seeds which is used as a spice. There is absolutely no narcotic content in opium seed. In India, var. *album* of poppy with white seeds is cultivated for many years for production of seeds under license. The Dutch variety is reported to be the best, characterized by uniform slate-blue colored seeds (Verghese, 2003b). Poppy seed is exported from India.

Botany and chemical composition

Poppy is an erect annual glaucos herb, with latex in all parts, 30-100 cm tall, leaves undulate ovate-oblong, often shallowly pinnately lobed, coarsely toothed, lower leaves shortly stalked, upper leaves sessile, clasping stem. Flower hermaphrodite, actinomorphic, with whorled parts, upto 18 cm in diameter, sepals - 2, falling before flower opens. Flowers 2 + 2, stamens numerous with bluish anthers. Ovary superior, one celled with numerous ovules, stigma disc with deep marginal lobes, fruit, a capsule when dehiscent opening by valves or pores; seeds small with minute embryo in oily endosperm. Depending upon variety, seeds differ in colour - white poppy with white seeds and black poppy with state-blue colored seed (Verghese, 2003b).

Poppy seed has a high protein content (22.3-24.4%), a major component being a globulin. It also contains fat, carbohydrate, fibre, calcium, iron, magnesium, phosphorus, potassium, sodium and zinc. Seeds also contain about 60 per cent fixed oil. White seeds are reported to yield the finest oil. Seed oil contains vitamin F for use in skin treatment. Notably, poppy seed is distinguished by freedom from essential oil. Seeds also contain lecithin (2-80%).

Crop improvement

Selection, hybridization, mutation, polypoidy breeding and production of synthetics have lead to evolve improved varieties in poppy, mainly for medicinal purpose. These are BROP I, Kirtiman, Chetak, Trishna and Jawahar Aphim-16. Sujatha is a variety with low morphine content and hence has got high value as a spice.

Crop management

Fairly rich, well drained loamy soil with adequate moisture and sunshine are ideal for successful cultivation of poppy. Light black cotton soil is the best for growth of poppy. Land is ploughed well and field is divided into small beds. Sowing is done in November and seed rate is 4.5-5 kg/ha. Broadcasting or line sowing is done 30 cm apart. One light irrigation is given after sowing. Crop requires well-planned judicious irrigation during growth period. Crop requires nearly 10-12 irrigations during its life cycle. Interculture, weeding and thinning are important operations. Thinning is of much importance. Healthy plants are set at a distance of 20-25 cm. No standard practice is followed in manuring this crop. Crop flowers after 1 ½ months of sowing. Fruits are fully ripe by February-March. For seed purpose, it is better not to scarify capsules for opium as they give higher seed oil yield than those from scarified.

Processed products and uses

Poppy seed oil is extracted by either cold or hot expression. Poppy seed cake is the cake or meal left after extraction of oil.

Poppy seeds are utilized as food and as a source of fatty oil. They are considered nutritive and are used in breads, cakes, cookies and pastries. They are also used in production of lecithin. Poppy seed oil is edible without refining. It is used in production of artists paints, soft soaps, ointments and emulsious and compositions of skin care. It is used against diarrhoea, dysentry and scalds.

Caraway
(*Carum carvi* L., F. Apiaceae)

Caraway is a native of Europe and Western Asia and is now cultivated in temperate regions extending to Sudan and Northern India. Seeds are used as spice. It is found wild in North Himalaya regions. Caraway sold in Indian markets is mostly imported, only a small quantity being home grown. It is cultivated in plains as a cold season crop and in hills of Kashmir, Kumaon, Garhwal and Chamba, at altitudes ranging from 2,500 to 4,000 m as a summer crop. Spice is known in market by name of country of origin. Dutch caraway is the best for distilling the oil.

Botany and chemical composition

Plant is a biennial herb, 30-80 cm tall, with thick tuberous roots, leaves compound, 4-5 times pinnately divided with linear cylindrical segments. Flowers small, yellow, in compound terminal umbels. Seeds (mericarps) slightly curved and tapering and brown in colour.

Black caraway or 'Kala Zira' (*Carum bulbocastanum* W. Koch) is grown wild in hilly areas of Himachal Pradesh.

Dried fruit or seed has pleasant odour, aromatic flavour, is warm with some what sharp taste (because of presence of carvone). Seeds on steam distillation yield an aromatic essential oil (4-6%). Carvone content varies from 45-60 %. Choudhury (2000) reperted that 81.5 % carvone content in caraway oil from Uttar Pradesh. Decarvonised oil consists of limonene with traces of carvone and is sold in market as light oil of caraway.

Crop management

Plant requires a dry temperate climate and thrives well in well tilled soils rich in humus. There are annual and perennial caraway varieties (Heine, 1998; Pank *et al.*, 2001). Studies on crop improvement are numerous in main growing countries and several improved varieties are evolved in caraway which are widely grown in temperate areas. In India, commercial cultivation has not been picked up. Crop is raised from seeds (fruits). Seeds are sown either broadcast or in rows 30 cm apart. About 20-25 kg of seed is required for one hectare. Biennial varieties are intercropped with annuals like dwarf peas, mustard or field beans. Manuring, weeding and irrigation improve yield of caraway. Crop is harvested before ripening of fruits. Crop is prone to shattering and so, correct time of harvesting is important to avoid losses. Well ripened fruits are harvested in early mornings when plants are bathed in dew. Non shattering varieties are now available in Germany and they could be harvested both mechanically or manually without any loss. Average yield of caraway is 1 t/ha. Plants are dried, and fruits threshed out, cleaned and stored in bags.

Processed products and uses

Caraway is mostly sold whole. It has good storage life like pepper. Volatile oil, decarvonised oil and carawy chaff oil (distilled from husks and stalks) are processed products of caraway.

Caraway is widely used as a spice for culinary purposes and for flavouring bread, biscuits, cheese etc. It is a mild stomachic and carminative. Carvone, isolated from caraway oil is used as anthelmintic against hookworm. Seed oil is also used in medicine and for scenting soaps.

Ajowan or Bishop's Weed
(*Trachyspermum ammi* (Linn.) Sprague., F. Apiaceae)

Ajowan, also known as bishop's weed, carum seed or carum ajowan is a native of Egypt. Although cultivated in Mediterranean region and in South West Asian countries such as Iraq, Iran, Afganisthan and Pakistan, ajowan is chiefly produced in India. Seed is the economic part used. It is grown throughout country, mainly in plains, but flourishes equally well at higher attitudes in plateaus and hills. It is grown on a commercial scale in Madhya Pradesh, Andhra Pradesh, Gujarat, Maharastra and Uttar Pradesh and to considerable extent in Rajasthan, Bihar and West Bengal (Sharma *et al.*, 2000). Large quantities of ajowan are consumed in India and entire demand is met by indigenous production. Considerable quantities are exported also.

Botany and chemical composition

Ajowan is an erect, glabrous or minutely pubescent branched annual, up to 90 cm tall. Stem striate, leaves rather distant, 2-3 pinnately divided, segments linear, ultimate, segment 1.0-2.5 cm long, flowers in terminal or seemingly-lateral, pedinculate, compound umbels, white, small, fruit ovoid, muricate, aromatic cremocarps, 2-3 mm, long, greyish brown, mericarps compressed, with distinct ridges and tubercular surface, one seeded.Ajowan is a cross pollinated crop (2n = 18). Flowers are self fertile, but cross pollination occurs through insects (Malhotra and Vijay, 2004).

Ajowan owes its characteristic odour and taste to presence of an essential oil (2-4%). Other constituents in fruits include sugars, tannins and glycosides. Ajowan oil was principal source of thymol for quite a long time. Since introduction of synthetic thymol, production of oil, and thymol from it, has considerably decreased. Oil is concentrated in husk. Principal constituents of oil are phenols, mainly thymol (35-60%) and some carvacrol. Thymol easily crystallizes from oil on cooling and is known in trade as "Ajowan ka phool" (Flowers-of-Ajowan) or 'Sat ajowan'. Remaining oil is called "thymine", because of its close similarity with corresponding portion of thyme. Semi-dried herb on distillation yields 1.52 per cent essential oil.

Crop improvement

Improved varieties of ajowan released in India are Gujarat Ajowan 1, Pant Ruchika, Ajmer Ajowan 1, Ajmer Ajowan 2, Lam Sel. 1, Lam Sel. 2 and Rajendra Mani (Johnykallupurackal and Ravindran, 2006).

Crop management

Ajowan is cultivated extensively as a cold season crop in the plains and as a summer crop in the hills. Farmers grow mostly local cultivars. NRC on Seed Spices have developed AA 19 and AA 61 high yielding cultivars suitable for both irrigated and rainfed production systems (Malhotra and Vijay, 2003). Crop is grown on all kinds of soil, but does well on loams and clayey loams. For dry crop, black-cotton soils which store moisture of earlier heavy rains are very suitable.

Cultivation practices of ajowan differ to a certain extent in various parts of India according to soil and climatic conditions. Sowing period extends from September to November in different parts of India. For irrigated garden crop, field is ploughed to a fine tilth and seeds are sown broadcast at rate of 2.5-3.5 kg/ha. Germination takes 5-15 days, depending upon climatic conditions. Field has to be irrigated carefully. Crop stand is made uniform by thinning. Optimum spacing is 45 x 30 cm. An organic manures of 10 - 15 tonnes, 80 - 100 kg N, 30 - 50 kg P_2O_5, 30 - 50 kg K_2O and 50 kg S/ha is recommended. Half dose of N, full P, K and S should be applied as basel dose whereas remaining dose of N should applied in two equal parts at an interval of 30 and 60 days after sowing (Malhotra, 2004) Two weedings are necessary. Crop flowers in about two months time. Harvesting period extends from February to March. Fruits become ready for harvesting when flower heads turn brown. Plants are pulled out and dried on mats on threshing floor. When completely dried, fruits are separated by carefully rubbing by hands or feet. A good crop yields 330-350 kg fruits/ha. Herb is not affected by any diseases and pests. However, collar end-rot caused by *Slerotium rolfsii* Sacc is recorded.

Processing and uses

Several commercial types of ajowan are known, of which 'Desi ajowan' (large), 'Desi ajowan' (small) and 'Nadiad ajowan' are the most common. The best quality small seeded variety is produced in Madhya Pradesh.

Major processed products are oil of ajowan prepared by steam distillation of fruits; thymol produced by treatment of ajowan oil with aqueous alkaline solution and regenerating thymol from it by extracting with ether or steam distillation; dethymolised oil or thymene and fatty oil from seed.

Ajowan, with its characteristic aromatic smell and pungent taste is widely used as a spice in curries. It is employed either alone or in mixture with other spices and condiments. Another use of ajowan is as an ingredient of pan mixtures. More important use of ajowan is medicinal and its is a household remedy for indigestion. It is much valued for its antispasmodic, stimulant, tonic and carminative properties. Ajowan oil is also used in medicine as an antiseptic and aromatic carminative. Dethymolised oil or thymene is used for industrial purposes.

Pomegranate

(*Punica granatum* L., F : Punicaceae)

Pomegranate is an ancient and popular fruit symbolic of plentyness and prosperity. It is known to have cultivated in middle east more than 5000 years ago. Wild or semi-wild pomegranate still exists in Northern Syria and Mount Carmel. It is a native of Iran, Afganisthan and Baluchistan. It is the seeds with pulp on, which when sundried or dehydrated, constitute the condiment, 'anardana'. It grows wild in warm valley and outer hills of Himalayas between 900 and 1,800 m lattitude and is also cultivated at selected sites almost throughout India. A major source of 'anardana' is wild pomegranate trees growing in hill slopes of Jammu, Himachal Pradesh and Punjab. Cultivation is mainly concentrated in Maharastra, Karnataka, Andhra Pradesh, Uttar Pradesh, Tamil Nadu, Gujarat, Rajasthan and Himachal Pradesh. India exports 'anardana' to other countries.

Botany and chemical composition

Pomegranate is a shrub or small tree 5-10 m high. Bark smooth, dark grey, branchlets some times spiniscent, leaves 2.0-8.0 cm long oblong or obovate, shining above. Flowers usually scarlet red, at times yellow, 3.7-5.0 cm long and much across, mostly solitary or 2-4 together. Fruits globose, crowned by persistent calyx with a coriaceous woody rind and an interior septae with membraneous walls, containing numerous seeds. Seeds angular with a fleshy testa which is pinkish or whitish.

Analysis of fresh edible portion (68%) of pomegranate (seeds with flesh) shows: moisture 78%; protein 1.6%; fat 0.1%; fibre 5.1%; other carbohydrates 14.5%, mineral water 0.7% and vitamin C 14 mg/100 g. In cultivated types, sugar content varies from 12 to 16% and acidity from 1.5 to 2.5%. Oxalic acid is major acid and glucose and fructose are the major sugars in juice.

Crop improvement

In India, there are many types of pomegranate, including double flowering, ornamental type and a white flowering type. White flowering type has a very poor fruit quality. Most of commercially cultivated varieties in India fall under two groups - the white fleshed and pink fleshed. Sour pomegranates generally grow wild in forests, they are not used fresh, but used for making 'anardana' to use as a condiment. Commercially important varieties are the following.

Alandi

Medium sized fruits with blood-red or deep pink arils slightly acidic pulp. Seeds hard.

Dholka

Fruits are large, rind greenish white, arils white or pinkish white, seed soft.

Ganesh

Most important commercial variety of pomegranate. Fruits large, yellowish to pinkish skin, Grains are soft with sweet pulp. Pulp colour during winter is pink and white during summer.

Mridula/Arakta

Developed at Mahathma Phule Krishi Viswa Vidhyalaya, Rahuri, fruits round in shape and medium in size with an attractive dark red skin. Grains are bold with soft seed, blood red colour in cool season and pink colour in hotter months. Arils are juicy and sweet. It has got good keeping quality and is also tolerant to anthracnose disease.

Jyothi

Developed at GKVK, Bangalore. The fruits are medium to large in size, have attractive yellowish colour with deep-pink arils. Seeds are soft. Popular in Karnataka.

Ruby

A hybrid variety developed at IIHR, Bangalore. Fruits resemble Ganesh having reddish brown skin colour with yellowish stripes here and there. Thin skin, large grains with soft pulp. Juice sweet and attractive.

Other Varieties recommended by the All India Co-ordinated Research Project on Arid Zone fruits are Ganesh, G-137, P-23, P-26 and Muscat for Maharastra, Bassin Seedless and Jyothi for Karnataka, Dholka for Gujarat, Jalore Seedless, Jodhpur Red and Jodhpur white for Rajasthan, Kabul Red, Vellodu, Yercaud-I and Co-variety for Tamil Nadu.

Crop management

Pomegranate grows the best in tropics, below an elevation of 1,000 m, with long, hot dry summers and cool winters or in those areas which are continuously warm and dry, provided irrigation is available. High temperature should accompany ripening season. In humid weather, quality of fruit is not good. It tolerates alkaline and wet soils.

Propagation

Many pomegranate plantations in India were raised from seedling populations and the variability is so high. For production of quality pomegranate plants, vegetative propagation is the best method. Cuttings are obtained from one year old shoots (20-30 long). Leaves are removed. Basal portions of cuttings are treated with rooting hormones and planted in soils with six nodes buried in soil. Root initiation occurs within a fortnight and cuttings will be ready for potting in nine months. One year old rooted cuttings are used for planting. Pomegranate can also be propagated by ground layering, air layering or root suckers.

Field planting is done in pits (60-75 cm^3) taken at a spacing of 5 m x 5 m. Eventhough pomegranate grows well in soils of low fertility, productivity can be increased greatly by adequate application of manures and fertilizers. Following doses of fertilizers have been recommended for Rahuri region of Maharashtra state.

Age (year)	Fertilizer dose (g / tree / year)		
	N	P_2O_5	K_2O
1½ - 2	250	125	125
2½ - 3	500	125	125
3½ - 5	500	125	125
6 and above	625	250	250

Along with above fertilizers 10 kg of FYM is to be applied in first year and quantity may be increased to about 40 kg from fifth year onwards. In drought prone areas, minor nutrient deficiencies appear frequently and these have to be corrected by application of commercial formulation of nutrients concerned.

For attaining good maturity and proper color development of fruits, potassium requirement is more during maturity stage. Foliar application of potassium dihydrogen orthophosphate @ 2 g per litre twice at 15 days interval is recommended to meet this requirement.

Pruning

In initial years, tree is trained to a desired shape. Mostly it is trained to a single stemmed or multi stemmed system. Fruits are borne terminally on short spurs arising from mature shoots. Shoots bear fruits for 3-4 years. Only a limited pruning of bearing tree is therefore required. Gradual growth of new shoots should be encouraged by restricted cutting back of bearing shoots. Suckers are removed regularly.

Flowering seasons (Bahars)

In Central, Western and South India, pomegranate flowers almost throughout year. There are however three distinct flowering and fruiting seasons. January-February ('Ambe Bahar'), June-July ('Mrig Bahar') and October ('Hasta Bahar'). One has to decide the season depending upon market demand, irrigation facilities available, soil type, climatological factors and pest and disease problems. After deciding crop season, flowering can be induced in pomegranate by 'Bahar treatment'.

For induction of flowering, methods adopted are root exposure, root pruning and withholding of irrigation until leaf fall. Among these methods withholding of water for 1-1½ months until leaf fall, is mainly adopted in several orchards. After this period, if favourable conditions are created like manuring and watering, tree flowers abundantly at once and produces a uniform crop.

Flower and fruit thinning

Flowering and fruit set in pomegranate are enormous. For getting large sized and quality fruits, crop load has to be regulated. Depending on age one can retain 70-100 fruits on a fully mature tree and rest are to be thinned out. In general, fruits born on strong short spurs are to be retained. Fruit cracking is also a problem in pomegranate, especially in 'Mrig Bahar' crop when grown as rainfed. In such situations, fruit cracking can be minimized by giving irrigations during rainless periods. Cracking due to deficiency of micronutrients can be checked by supplementing these nutrients.

Irrigation

Though pomegranate withstands drought, it responds well to irrigation. Irrigation schedule depends largely on bahars which farmer is likely to take.

Pests and diseases

Pomegranate butterfly (*Deudorix isocrates*), Shot hole borer (*Xyleborus similes*), Stem boring beetles (*Coclosterna spinator*), Thrips (*Rhipiphorothrips cruentatus*) Pomegranate white fly (*Siphoninus* sp.) and Mealy bugs (*Icerya purchase* & *Planococcus* sp.) are the pest complex reported in pomegranate. Leaf spots caused by *Colletotrichum, Phomopsis* and *Pseudocercospora*; fruit spots caused by *Colletotricum* and scab caused by *Sphaleloma punicae* are diseases which attack pomegranate. Post harvest fruit rots are also reported.

Harvesting and processing

Fruits are ready for harvest in 4½ - 5 months after appearance of blossom. Fruits are harvested when skin turns slight yellow and fruit gives a metallic sound when tapped or pressed. Tree starts yielding from fourth year onwards giving 20-25 fruits. Crop increases progressively and in 11th year, a tree gives 100-150 fruits/year and economic yield is up to age of 20-25 years.

Wild fruits are hand picked when they are ripe and brownish red in colour. Fruits are cracked and seeds along with their flesh are taken out by hand and sun-dried on roof-tops or platforms or on mats till desired moisture level is reached (10-15 days). Reddish brown dried seeds (anardana) are packed in gunny bags or in friction top tins stored at a cool dry place. A wild type of pomegranate called 'Daru' grows abundantly in lower Himalayas. It gives good quality 'anardana'.

Uses

Pomegranate is a favourite table fruit in tropical countries. It is very much liked for its cool refreshing juice and also valued for its medicinal properties. Anardana is mostly used as a condiment for acidification of chutneys and certain curries just like tamarind. Seeds are stomachic while pulp is both cardiac and stomachic. Bark and rind of fruit are commonly used against diarrhoea and dysentry. Rind is also used as a dyeing material for cloth. A kind of wine is prepared from pomegranate juice which is considered superior to grape wine.

Coriander

Cumin

Fennel

Fenugreek

Anise

Celery

Dill

Mustard

Poppy

Caraway

Ajowan or Bishop's Weed

Pomegranate

Chapter 5

Herbal Spices

Marjoram
(*Marjorana hortensis* Moench., F. Lamiaceae)

Marjoram is native of Southern Europe. It grows in West Asia, South and North America, France, Germany, Hungary, Spain, Portugal, England and North Africa. It is an important kitchen herb and its dried leaves with or without flowering tops in small proportion constitute the spice of commerce. In India, the crop finds its cultivation only in small scale in Karnataka, Andhra Pradesh and Tamil Nadu. France and Egypt are the main exporters of marjoram oil in the world.

Botany and chemical composition

Marjorana hortensis is an aromatic branched perennial, 30-60 cm high. Though a perennial, it is treated as an annual under cultivation. Main stem is woody with many softer branches. Leaves are round, soft, mat green and with a marvelous sweet smell. Flowers small, whitish or purplish in terminal clusters, seeds minute, oval and dark brown in colour.

Steam distillation of leaves and flowering heads yield a volatile oil known as the oil of sweet marjoram in trade. Average oil recovery is 0.3-0.4 per cent on fresh weight basis. Compositional pattern of the oil is governed by the origin of the herb, climatic and drying conditions, production procedure of the oil and many other factors (Verghese, 2000b). A sample of Indian oil consisted of carvacrol (36.4%), eugenol (6.7%), chavicol (4.6%), d-linalool (30.6%), methyl chavicol (3.2%), d-α-terpineol (4.8%) and caryophyllene (7.6%). Oils of European origin consist of 40 per cent terpenes, but are free from phenols. d-α-terpineol and terpineol are also present.

Cultivation

Marjoram prefers well-drained and nutrient rich soil for its successful growth. It has a strong flavour when grown in a rich soil (Vadivel and Kuruvilla, 2003). It is one of the a few crops which can be grown even on saline or alkaline soil as well. The plants resist drought, but prolonged periods of drought are harmful. It requires ample sunshine for best growth.

Marjoram is commercially propagated through cuttings. Cuttings of 15 cm length are made from healthy plants. They are first kept in polybags for rooting and rooted cuttings are used for planting in the main field. Marjoram can also be propagated by seeds. The tiny seeds are mixed with fine sand in the ratio of 1:10 and broadcasted in rows of nicely prepared nursery beds. Germination is slow. Seedlings are transplanted in rows when they are 50-60 days old at a spacing of 20 x 20 cm.

Application of nitrogen (240 kg/ha), phosphorus (40 kg/ha) and potash (80 kg/ha) in three equal split doses will increase the herbage and oil yield.

Pests and diseases

Incidence of pests and diseases are not severe in this herb. However, reports are available on the occurrence of leaf blight caused by *Colletotrichum gloesporioides* and *Phomopsis.*

Harvesting

First harvest of the leaves and tender tops of the herb is done as flowering commences. While harvesting, about 10 cm of the shoot above the ground level is left for further growth. Subsequent harvests are carried out after 45 days of first harvest. On an average, 4-5 harvests are taken annually. The herb has a tendency to become woody if pruning is not attended in the advanced stages of growth. To avoid this, it is advisable to prune the old wood at the fag end of winter before the new growth appears. On an average 16-18 t/ha of fresh herbage is obtained which in turn may yield about 35-40 kg marjoram oil.

Cut plants of modest quantities may be tied in bunches and dried in the open air or spread on wire trays in ventilated rooms and dried by regulated circulation of warm air. Leaves are largely freed of stems/stalks by rubbing on hand sieves; of chaff by fanning and finally of extraneous sand, earth and dust by shaking in wide-meshed sacks. Final moisture content should be 10-12 %.

Uses

Sweet marjoram is a versatile herb. Dried leaves and floral tops are superb for seasoning meat, poultry, seafood, fish, egg, soups, stews, fruit salads, in flavouring vinegar and in the formulation of liquors, vermouths etc. Medically it is used as external application for sprains, bruises, stiff and paralytic limbs and toothache.It is considered to be a carminative, expectorant and tonic, leaves and seeds astringent. It is reported to be useful in asthma, hysteria and paralysis. Seeds of marjoram find use in meat products and confectionaries. Marjoram oil is used in high grade flavour preparations and perfumes and in soap and liquor industries.

Savory
(*Satureja hortensis* L., F. Lamiaceae)

Summer savory is blessed with thyme like odour and flavour. It grows wild or may be cultivated. It is cultivated in France, Spain, Germany and other parts of Europe, England, Canada and United States. It also finds birth in Kashmir. Yugoslavia provides the premier grade of the spice. Savory of commerce consists of dried leaves and flowering tops, the best class comprising only leaves.

Botany and chemical composition

Satureja L. is a large genus of hardy annual or perennial aromatic herbs and subshrubs distributed in the warmer regions of both the hemispheres. There are two species viz. *S. hortensis* L. - summer savory and *S. montana* L. - winter savory (Varghese, 2003a). Summer savory is a small erect herbaceous annual, 25-30 cm high. It has got pinkish branches, dark green leaves which are narrow, elliptical, dotted with glands; flowers are lavender, pink or white in colour.

Leaves yield 0.5-1% essential oil, light yellow or dark brown with fresh herbaceous, spicy, slightly sharp, phenolic notes reminiscent of oregano and thyme. Phenol content (carvacrol) varies from 23.8 to 55.0 % (Guenther, 1974). Savory seed contains on a dry basis 45 % of fixed oil and 24 % protein.

Cultivation

Savory grows wild and is also propagated by division, cuttings or seeds, the best route is by seeds. Temperate climate, full sun and light rich soil are preferred. Seeds are sown in rows taken 30 cm apart. The spice is harvested at the full flowering stage which is usually done 75-120 days after sowing. Cut herbage is carefully dried in shade or at 35°C and stored in closed containers. Moisture content of dried herb should not be more than 11 %.

Uses

Savory is only a herb of minor importance. Traditionally, the fresh, tender young leaves and tips are eaten with meat loaf and hamburgers, egg dishes, fish, soups etc. Essential oil and oleoresin of savory find application in ppm level in baked foods, meat products, condiment and soft candy. The spice oil is used in the manufacture of liquors, bitters and vermouths.

The herb is antibacterial, antispasmodic, astringent, carminative, digestive, diuretic, expectorant, laxative, stomachic, sedative and antifungal, vermifuge and is a stimulating tonic after fevers.

Tarragon (Estragon)
(*Artemisia dracunculus* L., F. Asteraceae)

Tarragon is a vigorous herbaceous perennial shrub of Asteraceae family. There are two types of tarragon: Piquant, true French tarragon which is priced for its superior flavour quality and the less flavorous, false Russian tarragon. French tarragon thrives in Southern France and Netherlands and parts of USA, while the Russian counterpart, in Germany and in several eastern European countries. More than half of the crop of this spice is contributed by France. Flowering tops and leaves are the parts exploited. It is so well known to cornoisseurs the world-over for its unusual intriguing flavour.

Botany and chemical composition

French tarragon is *A. dracunculus* and Russian tarragon is *A. dracunculoides* (Werker *et al.*, 1994). The French tarragon is a herbaceous perennial. It grows upto 75 cm height, its roots are coiled and serpent like and densely branched with stocks carrying flowering tops. The fragrant leaves with incurled edges are tender, alternate about 4 cm long and 3-3 ½ cm wide. Compared to Russian tarragon leaves, that of the French variety are smoother, glossier, darker in colour, more pungent and aromatic. Flowers are greenish white, but are generally sterile.

Oil of tarragon is obtained by cautious steam distillation of the tender leaves and tops of *A. dracunculus*. The oil yield is 0.3 per cent. Oil is light yellow to pale yellow in colour with its sweetly camphoraceous and strong spicy odour. Methyl chavicol is the chief constituent of tarragon oil (60-70%). Apart from this ?-pinene, ocimene, linalool, myscene, phellandrene, eugenol, camphene, sabinene, methyl eugenol and anethole are also present. Werker *et al.* (1994) reported the main component of the leaf essential oil of *A. dracunculus* as methyl chavicol and that of *A. dracunculoides* as elemicin. Another form of tarragon, Polish tarragon is cultivated in Poland. This also contains elemicin as the main constituent in leaf oil (Olszewska and Jaroszewska, 1999).

Cultivation

The plant is propagated through division of roots. Root segments are planted in the field during February-March. Crop may be harvested twice or thrice an year during the growing period by cutting the stem, a little above the ground level. Plants should be cut at the beginning of the flowering stage; otherwise it loses some of its aroma and essential oil content. Freshly cut crops are dried indoors or shade dried to preserve the anise like flavour and aroma and leaf colour. Oil of tarragon is obtained by cautious steam distillation of the tender leaves and tops and its takes 1-1½ hour to complete the operation. Oleoresin tarragon is derived exclusively from French tarragon. Encapsulated oleoresins are also available.

Uses

Tarragon may be used in moderation in hot and cold soups, salads, soft candy, leafy green salads, fats and oils, mayonnaise, sauces, pickles, fish and meat dishes. Tarragon vinegar is a celebrated item. Tarragon oil is used in fragrance industry. The aromatic leaves are credited with stomachic, stimulant and febrifuge properties.

Thyme
(*Thymus vulgaris* L., F. Lamiaceae)

Thyme is one of the best known European culinary herbs. Thyme (*Thymus vulgaris* L.) commonly called "Common Thyme" or "Garden Thyme" yields "Red Thyme oil" of commerce. Dried leaves and floral tops constitute the thyme of commerce known as "Thymi herba".

Thyme grows wild in almost all the countries bordering the Mediterranean and also over much of Asia and in parts of Central Asia and parts of Central Europe. In India, it is found in the western temperate Himalayas, from Kashmir to Kumaon. There are mainly three major types of thyme viz., French Thyme, Spanish Thyme and American/California Thyme. Bulk of the world demand for red thyme oil is met by Spain and Turkey.

Botany and composition

Besides *Thymus vulgaris*, the other Lamiaceae members which yield thyme oil are *T. serphyllum* L., *T. zygus* L. and *T. satureioides* Cross and Bal. (Farooqi *et al.*, 2000). *Thymus vulgaris* is a low evergreen perennial under shrub reaching a height of 20-30 cm, whereas the wild thyme (*T. serphyllum* L.) creeps on the soil surface and has broad leaves with a weaker odour. The roots are fairly robust and the stems are branched. The former will have oblong lanceolate sessile leaves of 10 mm x 3 mm size with orange-brown glandular dots. The young leaves are slightly wooly. Flowers are small, purplish or bluish to almost white, united in spikes at the tip of the branches and have a bilabiate, tube - like calyx and a bilabiate tubular corolla with a three lobed lower lip. Fruit is a nutlet, brown, smooth and is found in the remains of the calyx. The entire plant is aromatic.

Cultivated thyme yields about 2.5 % volatile oil whereas wild thyme yields 0.5 % oil. Composition and quality of thyme oil vary in different geographical areas. Main constituent of the oil is thymol which is high in Spanish oil (60%) and low (20-35%) in French and Moroccan oils. Carvacrol, amylalcohol, ß-pinene, camphene, linalool, 1-borneol, geraniol and caryophyllene are the other components present in the oil.

Cultivation

Thyme prefers a light but fertile and calcareous soil for good growth and oil content. On heavy, wet soils, the aroma of the leaves will be less. A warm climate is the best suited for this crop. It can be grown both in the hills and the plains. Plants are susceptible to frost.

The crop is propagated through seeds. The seeds may be sown in nursery during April. Seeds are sown directly in rows or in well prepared nursery beds and then transplanted. In the case of direct sowing, rows are taken 90 cm apart and later, when the seeds germinate the seedlings may be thinned out to 30-45 cm within the row.

Thyme can also be propagated by cuttings or by layering of side shoots in March-April. Under Bangalore conditions where thyme does not flower and set seeds, Prasad *et al.* (2000) obtained early and better rootimg in semi hard wood cuttings soaked in IBA at 150 ppm. While planting rooted cuttings or layers , they are planted 30-45 cm apart in rows of 60 cm apart. A light irrigation is provided after planting.

Application of farmyard manure and nitrogen promotes the formation of numerous leafy shoots. Interculture operations and weeding are done at regular intervals to encourage good growth of plants. In the hills, inorder to avoid frost injury in winter, mulching is done.

Pests and diseases

The plant is not damaged by any pests of serious nature, but wilt disease is a major problem in this crop.

Harvesting and processing

Leaves and flowers which are used for culinary and medicinal purposes are harvested five months after sowing/planting. Either they are plucked from the plants or shoots of about 15 cm length are cut off from the plants and dried in shade or in dryer immediately after harvest and stored in air tight containers to prevent the loss of flavour. The dried leaves are curled of brownish-green in colour, usually not longer than 6-7 mm. The dried shoots may also be powdered and packed. Average yield is 1100-2200 kg/ha. The yield is comparatively low during the first year. Plant becomes woody and replanting becomes necessary after 3-4 years.

Processed products

Both essential oil and oleoresin extractives are prepared from the flowering tops of the thyme plant. As with other aromatic herbs, the volatile oil is the most important extractive. The Spanish thyme oil is marketed as red thyme oil. There is also a white thyme oil which is a redistillation product of the red oil. It tends to be smoother in flavour than red oil and pleasingly aromatic.

Thyme and its extractives should be stored under cool, dark, dry conditions to protect them against fragrance and flavour losses. Under good storage conditions, the thyme leaf will retain its flavour qualities and strength long enough to meet all normal food manufacturing requirements.

Thymol

The chief constituent of the oil is thymol (about 45%). In the pure state, thymol consists of colourless translucent crystals possessing an aromatic thyme like odour and a pungent taste. Thymol is also produced synthetically.

Uses

The herb is used as a food flavourant. Thyme oil is mainly used in perfumery industry. The oil of thyme is well recognized for its medicinal properties. It is used in the treatment of bronchitis and whooping cough.

Origanum
(*Origanum vulgare* Linn., F. Lamiaceae)

Origanum (Pizza herb) consists of dried leaves of an aromatic perennial herb found in temperate regions. It is a native of Southern Europe. Colour of dried herb is light green. It has got a strong comphoraceous aroma resembling marjoram. Oreganum is cultivated in European countries and is one of the most popular herbs in Mediterranean cooking. It grows abundantly in Mexico and is also known as Mexican sage. The herb is traded both as 'whole' dried leaves and in ground form. In India, the plant is found in temperate Himalayas from Kashmir to Sikkim at altitudes of 1500-3500 m.

Botany and chemical composition

Origanum vulgare is a branched perennial herb growing 0.3 to 1m tall. Leaves are broadly ovate, entire or rarely toothed and light green when dried. Purplish red flowers occur in heads at end of stems and branchlets.

Origanum contains both volatile oil and fixed oil. Oil of origanum appears to be sold under this name from a number of related species in Lamiaceae family. Main components of oil are gamma-terpinene (0.6-3.6%), P-cymene (17.3-51.3%), thymol (0.4-42.8%) and carvacrol (1.7-69.6%) (Kokkini *et al.*,

1997). Esential oil content varied from 0.25-1.51 per cent in inflorescence and from 0.09-0.98 per cent in leaves (Radusienei *et al.*, 2002).

Mockute *et al.* (2003) reported sabirene (14.2-18.8%), beta-ocimene (6.9-17%), beta-caryophyllene (14.2-18.8%) and germacrene D (9.6-15.5%) as main constitutents of origanum oil. Ozcan and Chalchat (2002) reported carvacrol as the dominant constituent of origanum oil.

Crop management

Temperate to sub tropical climate is ideal for the plant. It is propagated by seeds, cuttings, layers or root divisions. Because of very small size of seeds and perennial habit, crop is propagated by stem cuttings planted directly in field (Putievsky *et al.*, 1997). If the plant is well nourished and looked after, it will multiply to grand proportions. Leaves and tops are cut at blooming stage, dried in shade and are used to flavour foods in same way as sweet marjoram. Stalks may be picked as they bloom and hung up in small bunches in shade for drying. Dried leaf is about 1.5 cm long and light green in colour. Generally it is available as rubbed (crushed) or ground form.

Processed products and uses

Oregano is the spice that made 'pizza' famous and hence called 'Pizza spice'. It is used for flavouring food. It is also used as a vegetable. Oil possesses carminative, stomachic, diuretic, diaphoretic and emmonagogue properties. It is used in homoeopathy and in veterinary medicine and also in cosmetic and soap industry.

Sage
(*Salvia officinalis* Linn., F. Lamiaceae)

Sage is a native of Mediterranean area .It is cultivated in European countries and in North America. Sage is the dried leaf of "*Salvia officinalis*". Name sage originates from a latin world "Salvane" meaning 'to cure'. Dalmation sage from Yugoslavia is the best quality sage in trade followed by Albanian sage. Cultivation of sage was successfully tried in Jammu & Kashmir in India.

Botany and chemical composition

There are about 300 species of sage and all species which are traded throughout world as sage are not the classic *Salvia officinalis*. *S. trilobula* and *S. tomatoes* are also widely traded and these are also accepted as sage. *Salvia officinalis* is an aromatic evergreen perennial, grows up to 0.8m high with oblong woolly, grey green leaves that grow in opposite pairs on square stem. Violet-blue flowers are arranged in whorls at end of stems and branches in simple racemes.

On steam distillation, dry sage leaves yield an essential oil (1.3-6%). Chief constituents are α-pinene (1.8%), cineole, linalyl acetate, thujone, borneol, bornyl acetate and camphor. Quality of sage oil is determined on basis of its thujone content; higher the thujone content, better is the oil quality. In addition to essential oil, leaves contain tannin (3%) fumaric, malic and ursolic acids.

Crop management

Not much information is available on cultivation aspects of sage in India. Cultivation was tried at Jammu and also in NBPGR station at Bhowali in Uttaranchal, India (Negi *et al.*, 2004).

Crop can be grown as an annual or perennial. It prefers a cooler climate, hot dry climate produces an inferior crop. Better quality crop is produced on rich clayey and loamy soils.

Sage is commonly raised from seeds. Seed rate of 1 kg/ha is sufficient. Seeds are drilled directly in field or seedlings are raised in nursery or transplantation later. In case of transplanting 20 x 25 cm spacing is suitable (Atakisi *et al.*, 2001). For getting uniform crop stand, sage plantations are preferably raised from rooted stem cuttings which are planted 30-35 cm apart in rows 90 cm apart. Plants which have not come to flower are harvested, as they yield the highest amount of quality oil. Foliage is cut and dried in shade under air circulation. The highest yields of leaves and essential oil are obtained in second and third years by taking two cuts in an year. Production starts declining from fourth year and replanting may be done after five years (Pinzaru and Toma, 1994).

Dried bunches of sage can be sold without further treatment, or leaves may be cut, rubbed or pulverized for packeted herb trade. Sage should be stored under cool dry conditions.

Uses

Sage is the most popular herb for use in culinary preparation in west, it is particularly good with fatty meat and fish. In medicine, it is used as a mild tonic, astringent and carminative. It has a reputation for strengthening brain and nervous system. It is also used in mouth washes and tooth pastes. Essential oil of sage is used extensively for flavouring sausages, canned meat, seasoning and liquors. Sage extracts are also well known efficient antioxidants (Varadarasan, 2003). Antioxidative effect of sage is related to presence of carnosic acid, carnosol and rosmarinic acid.

Basil
(*Ocimum basilicum* L., F. Lamiaceae)

Basil, otherwise known as French basil or sweet basil is a native of old world tropics and has now been widely distributed. It is spicy when fresh and sweet when dried. Even though it is native to India, it is grown in other countries also. Pl ant is a rich source of essential oil. In India, there is wide spread belief that if planted around homes and temples, it ensures happiness. It is considered sacred by Hindus. The places of large scale cultivation are France, UK, USA, Reunion and to a certain extent, India. Next to French basil, other important Ocimum species cultivated on large scale are the camphor basil (*O. kilinandscharicum*), followed by *O. sanctum, O. americanum, O. canum* and *O. gratissimum*.

Botany and chemical composition

The genus *Ocimum* has been broadly classified into two viz. *Basilicum* and *Sanctum* on the basis of comprehensive studies on morphology, cytogenetics, breeding and chemical characteristics.

Sweet basil belongs to *Basilicum* group. It is a herbaceous much branched plant. Stem and twigs are quadrangular, greenish, purplish or brownish in colour. Exstipulate leaves are opposite and decussate. Leaves are simple, petiolate, ovate or subovate with serrate or entire margin. They possess glandular hairs which secrete essential oil. Flowers are small, conspicuous, arranged as a spicate or recemose inflorescence. They are hermaphrodite, seeds are black, ellipsoid and become mucilaginous when wetted. Major constitutents of essential oil (0.5% on fresh weight basis) are methyl chavicol (35%), methyl cinnamate, eugenol and linalool (40%).

Crop improvement

RRL, Jammu conducts intensive study on crop improvement in *Ocimum*. Floral biology of genus *Ocimum* was studied in detail by many workers. Flowers are mostly cross pollinated. Intervarietal and interspecific hybrids are not rare in Ocimum. But interspecific hybrids are either completely sterile or with partial seed setting. Inheritance pattern of chemical constituents in ocimum oil has been studied and three chemical races viz. rich in camphor, rich in eugenol and rich in methyl chavicol were isolated from *O. basilicum* var. *glabratum* (Pushpangadan and Bradu, 1995). RRL-011 is a strain developed by hybridization of selected genotypes from two different varieties of *O. basilicum*. Oil content of green herb is 0.45 per cent and main constituents are linalool and methyl chavicol.

Crop management

Ocimum species thrive well in a variety of soils and climatic conditions. It flourishes well under high rainfall and humid conditions. Long days and high temperature are favourablefor plant growth and higher oil production.

Seeds and sowing

Ocimum is generally propagated through seeds. Selected elite lines can be multiplied vegetatively also by planting tender shoot tips. Direct sowing is more beneficial compared to transplanting. Seed rate is 75-250 g/ha. Seeds are mixed with fine sand to ensure even distribution. Land is prepared well, levelled and sowing is done in long narrow furrows at a spacing of 50-60 cm and seeds are lightly covered by soil and then irrigated. Seeds germinate in two weeks time. After about one month, thinning/gap filling is done. In case of transplanting, nursery is raised and at 4-5 leaved seedlings are transplanted into the main field. Spacing is 60 x 30 cm.

Manuring and fertilizer application

Generally no manure is applied for *O. basilicum*, but application of 20-25 kg of N and 10-15 kg of P_2O_5/ha ensures good vegetative growth, herb and oil yield.

Harvesting and processing

Time of harvest plays an important role in quantitative and qualitative production of oil. *O. basilicum* is better raised as annual. First harvest can be taken 90-95 days of planting. Harvesting is done on bright sunny days. Crop is cut 15-20 cm from ground level. Herbage is distilled in fresh form. However, oil quality and its yield do not diminish up to 6-8 h after harvest.

Pepper Mint

(*Mentha piperita* L., F. Lamiaceae)

Mints are a group of plants belonging to family Lamiaceae which yield essential oil on distillation. They are distributed mostly in Northern hemisphere (Europe, America, Japan, China and Brazil). Commercial cultivation of mints in India is of recent origin. World demand for mint oil and menthol is met from three species of mint viz. *Mentha arvensis* (Japanese mint), *M. piperita* (Pepper mint) and *M. spicata* (spear mint). Pepper mint oil obtained from *M. piperita* is exported from India and its demand is increasing day by day. The crop was introduced to Nilgiris as early as 1818 and since then it has been raised in Karnataka, Jammu and Kashmir and Uttarpradesh. Uttarpradesh is contributing almost entire production of *M. piperita.*

Botany and chemical composition

Pepper mint is a glabrous perennial, strongly scented herb, occurring spontaneously or cultivated in temperate regions and introduced in parts of tropics. Plant is erect, 30-90 cm high, purplish or green, leaves ovate or oblong, coarsely serrate, smooth and dark green above, pale or sparingly hairy below; flowers are purplish in thick terminal spikes. Dried herb is commonly used as spice.

Pepper mint is source of true pepper mint oil extensively used for flavouring and in pharmacy. Oil yield varies from 0.3 to 1 per cent. Oil contains menthol (50-55%), methyl acetate, menthone and small amount of several other compounds. On chilling oil, menthol separates out as crystals.

Crop management

Pepper mint prefers a temperate climate. It is sensitive to drought (Thomas, 2003). Medium deep soil, rich in humus is the best suited. Soil should have pH range of 6-7.5 with good water holding capacity. Water logging is detrimental. Adequate and regular rainfall during growing period and good sunshine during harvesting are ideal for its cultivation.

Varieties

EC 41911: Exotic cultivar introduced in India and found promising under Indian conditions.

Pranjal: A high oil yielding selection with higher content of menthol. Tolerant to Bihar hairy caterpillar infestation (Dwivedi *et al.*, 2001).

Propagation

Mint is propagated mostly by rooted suckers. Stolons too are used for propagation. Land is ploughed deep as mints are deep rooted crops. Then it should be levelled and laid out into beds of 10 m x 10 m size. Method of planting is similar to that of sugarcane. Rows are opened up at a distance of 45 cm. Suckers are cut into pieces of 10-12 cm length and planted 30 cm apart. About 500 kg of sucker is sufficient for planting one hectare of land. January-February is the best planting season. Depending upon edaphic factors, FYM @ 15-25 t/ha and 40-50 kg N, 60-75 kg P_2O_5 and 60-80 kg K_2O/ha should be well incorporated into soil during final land preparation. Apart from basal application, 70-80 kg N should be applied in 2-3 equal splits, first when plants are about 15 cm high and remaining after each harvest (Datta *et al.*, 2001). On an average, 8-12 irrigations are required for commercial cultivation. For obtaining a good yield of crop, field should be weeded thoroughly, the most critical period of weed control is between 30-90 days after planting. Mulching during winter also improves herbage and oil yield.

Crop is harvested when it comes to flower in temperate climate and in plains of North India, where mentha does not flower, crop is harvested at the time when lower leaves start falling. Two to three harvests are possible from a crop. The crop should preferably be harvested on sunny days. Yield of fresh herbage is 14 tonnes/ha on an average. Harvested materials are staked, and kept as such for one day and then distilled. It helps to achieve more oil yield rather than fresh herbage distillation. Oil yield varies from 80-100 kg/ha.

Processed products and uses

Commercial pepper mint oils are derived from cured, partially dried plants, while official oils are obtained from fresh material; oil is rectified if necessary. Pepper mint oil is adulterated with dementholized Japanese mint oil, synthetic menthol, eucalyptus oil and spearmint oil.

Dried herb is commonly used as spice. Fresh leaf tops are used in beverages, sauces, meat preparations and salads. Herb is considered aromatic, stimulant, stomachic and carminative. Pepper mint oil is one of the most popular and widely used essential oils employed for flavouring pharmaceuticals, dental preparations, mouthwashes, cough drops, candies, confectionary and alcoholic liquors. Oil has mild antiseptic and local anaesthetic properties.

Parsley

(*Petroselinum crispum* (P. Miller) Nymann A.W. Hill.,F. Apiaceae)

Parsely is one of the most familiar and widely cultivated ogarden herbs. It is a native of Southern Europe and is now naturalized in most of the temperate countries. Greeks and Romans used it for flavouring. Leaves and seeds are used as spice. Dried herb is available as whole, rubbed or in ground form. Parsely grows wild or is cultivated almost all over world. USA is the largest commercial producer of parsely.

Botany and chemical composition

Parsely is an erect glabrous biennial or short-lived perennial, 30-70 cm tall, with a stout taproot, rosette leaves in a dense tuft, dark green, shiny, three pinnate leaflets 1-2 cm long cuneate, lobed, often much crisped, umbels 2-5 cm in-diameter, flat topped, flowers yellowish, 2 mm in diameter, fruits ovoid, laterally compressed, 2-3 mm long, carpels with five slender ridges.

Parsely is a rich source of vitamin C (281 mg/100 g) and vitamin A (8230 iu/100 g). Plant yields a volatile oil of which active principle is apiol. Fruits yield about 20 per cent of a greenish fatty oil which contains 76 per cent petroselinic acid.

Crop management

There are two main types of horticultural parsleys: one cultivated for leaves and other grown for turnip like roots. Only the former type is cultivated in India. Within leaf varieties, parsely has three types of foliage: plain leaf, double curled leaf and moss curled or triple curled leaf. Parsely prefers cold weather, rich moist soil and higher altitudes. It is propagated by seeds. Sowing is done in March-May on hills and in August-September in plains. Seeds are slow to germinate. They are sown in nursery beds and then transplanted when they have six leaves at a spacing of 15-20 cm in rows, 30-40 cm apart. When grown as a biennial, both foliage and seeds can be harvested. In Uttar Khand region (UP) leaves are ready for harvesting after 65 days of transplanting and seeds, after 185 days of transplanting (Khan and Ahmad, 2000). Yield is 1.1 tonnes/ha of foliage and 0.89 tonnes/ha of seeds.

Processed products and uses

Dehydrated parsely is the main product. It is usually done in dehydration plants. Moisture content is brought down to about 4 per cent in about two hours. Dehydration ratio is 8:1. Oil of Parsely is recovered by steam distillation. Oil from flowering top is of the finest quality, but yield is low (0.06%) for commercial production. Commercial parsley oil is distilled either from aerial parts (0.25%) or from mature fruits (upto 7%). Herb oil and seed oil differ in their properties.

Fresh leaves and oil are commonly used for garnishing and seasoning of foods. Leaves are also employed to make a sort of tea which is considered to possess antiscorbutic properties due to presence of vitamin C. Herb is reported to possess diuretic, carminative, emmenagogue and antipyretic properties and has long been in use for uterine troubles.

Rosemary
(*Rosmarinus officinalis* L., F. Lamiaceae)

Rosemary is a ritual herbal spice, indigenous to the Mediterranean region. Ancient Greeks believed that rosemary had the reputation for strengthening the brain and memory. It is used as an emblem of love, remembrance and friendship and generally used in weddings, christenings and funerals. The plant gives pungent, resinous taste with a delightful perfume. The British introduced rosemary in India during the 19th century.

Botany and chemical composition

The plant is botanically named *Rosmarinus officinalis*. It is a perennial evergreen shrub, highly aromatic, attaining upto two m high. The plant has a woody trunk. Leaves are linear and revolute, slightly in curved, green on the dorsal side and woolly greyish on the ventral side. On the ventral sides, the oil glands are situated. The blue, lavender or white flowers are born on old growth in late winter. The calyx contains lot of oil (Ravikumar, 2002).

When grown outside in warm climate, most rosemary cultivars grow upto 2 m height. *R. prostrata* is the prostrata rosemary (15-20 cm) with shiny green leaves and this is the smallest plant in rosemary genera.

Rosemary of commerce comprises dried leaves of *R. officinalis*. Leaves contain about 1-2.5 per cent essential oil which is of pale yellow colour, and with a camphoraceous taste. Chief components of the oil are 1, 8-cineol (30%), camphor (15-25%), borneol (16-20%), bornyl acetate and alpha-pinene (Suresh, 2002).

Cultivation

Rosemary is cultivated in the Nilgiris and Bangalore. It grows well in warm climates and well drained soils at attitudes above 750 m MSL. Plant is propagated by seeds, cuttings, division of roots and layering. Propagation by cuttings is easier. Seeds are difficult to germinate. From the healthy bushes, 15 cm long twigs just above the woody position is selected. Three tiers of leaves are trimmed from the bottom and the cuttings are planted 1 cm apart in the nursery bed containing sand. After 21 days, the rooted cuttings are pulled out and bound in moss for better growth of the cutting. Planting materials with the moss ball are stacked in a shady place and irrigated lightly with a rose can. After 20 days, they are lifted for planting in the main field. Treatment of cuttings with growth hormones such as IBA, IAA or Saponin enhance rooting (Shah *et al.*, 1996). An improved variety, Ooty RM 1 was released from Horticultural Research Station (TNAU), Ooty. It gives 24.40 tonnes green leaves/ ha/ year. Leaves contain 0.90 % essential oil (Johnykallupurackal and Ravindran, 2006).

In the Nilgiris, cultivation of rosemary is done organically. Land is prepared well. Pits of 30 x 30 x 30 cm are dug with a spacing of 45 cm between rows and plants. Rooted cuttings are planted as such without removing the moss ball. About 100 g of vermicompost is added in the pit. Organic manures are applied at the rate of 0.5 kg per plant three times an year. The plant should be protected from cold winds and frost. During summer the field is mulched heavily. A fertilizer dose of 40:40:20 kg NPK/ha is rrecommended for rosemary in India (Farooqui and Ramulu, 2001). Beneficial effect of biofertilizers viz. Azospirillum, Azotobacter and VAM applied in combination with inorganics has been reported by Anuradha *et al.* (2002). Rosemary is not affected by any pest or disease.

Harvesting

Rosemary can be harvested six months after planting. It is harvested when the plant produces flower buds. Shoots are cut just above the hardwood and taken immediately for drying. From a well grown plant three harvests could be made in an year. Yield of fresh leaves is 2 t/ha. Materials are rinsed in water and dehydrated in a well ventilated drying chamber. Drying temperature should not exceed 50^0 C. Care should be taken to retain the green colour. Dehydrated leaf needles are separated by gentle rubbing and are stored in air tight containers. The stalks rich in oil are used for distillation.

Processed products and uses

The only product is rosemary oil obtained by steam distillation of the leaves, flowering tops and twigs. Fresh tender tops are used for garnishing and for flavouring cold drinks, pickles, soups and other foods. It is a potent and strong herb; hence it should be used with restraint for obtaining the best results in the culinary art. Oil of rosemary is used mainly in cheap perfumery. Superior grades of oil are used for flavouring meant and other food products. Oil also finds use in medicinal preparations. It is a carminative, diaphoretic, circulatory and cardiac stimulant and is used in treating headache and irregular menstruation. The plant itself is an astringent and an excellent stomachic.

Antioxidant activity of rosemary extracts was shown to inhibit carcinogenesis and it has got anti-tumor and anti-inflammatory activity too. Carnosol and carnosic acid are the antioxidants present in rosemary oil (Chen *et al.*, 1992; Bauman *et al.*, 1999 and Varadarasan, 2003). Rosemary is also widely used for making medicinal tea and in aromatherapy.

Lovage
(*Levisticum officinale* (Koch.), F: Apiaceae)

Lovage is one of the old English herbs that was formerly very generally cultivated, and is still occasionally cultivated as a sweet herb, and for the use in herbal medicine of its root and to a lesser degree, the leaves and seeds. It is a native of the Mediterranean region. Centres of lovage cultivation are located principally in central Europe.

Botany and chemical composition

Lovage is a stout plant with a thick and fleshy root shaped like a carrot, of a grayish-brown colour on the outside and whitish within. It has a strong aromatic smell and taste. The thick, erect, hollow and channeled stem grows 9 to 11 m height. Leaves are dark green and divided into wedge like segments, shining, with an aromatic odour. Stems divide towards the top to form opposite whorled branches which bear yellow flowered umbels. Fruits are extremely aromatic, ridged and golden brown in colour (Clevely and Richmond, 1999). Odour of the whole plant is very strong.

Chemical constituents of lovage oil are mainly phthalides and terpenoids, the principle component being angelic acid. A colouring principle, ligulin has been isolated from lovage. Ibrahim (1999) reported the major component of leaf as b-phellandrene, a-terpenyl acetate and myrcene which accumulated for approximately 73 – 92 % of the total oil. Lovage root oil was characterized by high amount of Z-lingustilide, Z-butylidene phthalide, a-terpinyl acetate and benzene pentyl. Najda *et al.* (2003) studied the composition of various compounds in various plant parts of lovage. The phenolic acids in various plant parts were as follows: roots 0.12 - 0.16 %, herb 0.88 - 1.03 %, stem 0.30 - 0.39 %, leaf 1.11 - 1.23 % and fruits 1.22 - 1.41 %. Quality of tannins in various plant parts was: roots 6.6%, herb 5.3%, stem 7.4%, leaf 2.7 % and fruits 1.8 %.

Crop management

Lovage is not found in India. In the growing countries it is propagated by division of roots or by seeds. Mheen and Mheen (1996) recommended direct sowing by precision drilling. Mature plants require wider spacing as they are large and bulky. Plants are pruned during summer to get continuous supply of tender leaves. Rich, moist but well drained soil and a sunny weather is required for cultivation. Under proper management the plants are retained for several years as a perennial. Harvesting is done in the secand or third year of the crop. Young leaves, hollow main stems before flowering, sliced dry roots of 2 – 3 year old plants and ripe seeds are the useful parts. Fresh roots are generally first harvested from 2-3 year old plants subsequent harvests take place every third year. Fresh roots are washed, sliced and shade dried (Ravindran and Pillai, 2004).

Uses

Young stems and leaves are used for flavouring and confectionary whereas root, leaves and seeds are used for medicinal purpose, powdered root is some times used as a spice. A herbal tea is made of the leaves, when previously dried, the decoction having a very agreeable odour. Roots and stem are aromatic and stimulant and have diuretic and carminative action.

Marjoram

Savory

Tarragon (Estragon)

Thyme

Origanum

Sage

Basil

Pepper Mint

Parsley

Rosemary

Lovage

Chapter 6

Other Spices

Garlic
(*Allium sativum* L., F. Liliaceae)

Garlic is recognized all over the world as a valuable condiment for food and medicine. It is a native of Western Asia and Mediterranean area where it is in cultivation for centuries. China, Korea, India, USA, Spain, Argentina and Egypt are the major garlic growing countries, China being the largest producer (Pandey, 2004). Europeans, especially the Italians and the Spanish used this spice regularly for 2000 years and more. Garlic is one of the important spicy bulbs grown commercially in India. The plant is cultivated in temperate and tropical climates. Important states producing garlic on commercial scale are Gujarat, Maharashtra, Tamil Nadu, Karnataka, Andhra Pradesh, Madhya Pradesh, Orissa, Uttar Pradesh and Haryana. Demand of garlic is increasing day by day in domestic and export markets. Garlic and garlic products, especially dehydrated garlic and garlic powder are gaining considerable economic importance recent times.

Botany and chemical composition

Garlic, belonging to onion family is a hardy, bulbous rooted perennial plant with narrow flat leaves and bears small white flowers and bulbils. The compound bulb consists of 6-34 smaller bulblets called cloves, surrounded by a common thin white or pinkish papery sheath, skin or membrane.

Garlic has a strong characteristic odour and an acrid taste. Active principle in garlic is an antibiotic allium, which is an enzymatic cleavage product from its precursor 'allin' naturally present in garlic.

Oil of garlic is obtained by steam distillation of crushed garlic (0.1%). Garlic oleoresin is also available. Diallyl disulphide and diallyl trisulphide were isolated from garlic oil which were identified as the principles which give larvicidal properties to the oil. According to an American patent, the residue of garlic, obtained by alcoholic extraction and distillation contains a bacteriostatic and bactericidal substance, identified as allyl disulphide oxide (Pruthy, 2001).

Crop improvement

Clonal selection is the most important method for garlic improvement. A good number of both indigenous and exotic germplasm collections are maintained at National Horticulture Research and Development Foundation (NHRDF) Nashik, National Research Centre for Onion and Garlic Pune and various State Agricultural Universities. Varieties with large white bulbs, uniform in size and shape, bold and compact cloves, high yield and resistance to diseases and pests are preferred in cultivation and trade. Garlic bulbs 40-60 mm diameter with 10-15 cloves in each bulb are preferred for export. Promising selections are made from these collections for cultivation (Table 31).

Crop management

Soil and climate

Garlic favours a rich well drained soil. Moderate clay loam is the best suited for its cultivation. It requires a cool and moist period during growth and a relatively dry period during crop maturation. Extremely short or long dry periods are not favourable for bulb formation. The crop comes up well at slightly higher elevation (900-1200 m). In South India, it is grown in rotation with finger millet, chillies, maize, potatoes and beans.

Propagation

Garlic is propagated by cloves which are carefully detached from composite bulbs without any injury to get higher sprouting in field. Size of mother clove varies in different parts of country. Generally, cloves of medium to big size are recommended for bulb production for consumption, whereas smaller cloves could be used for further propagation (Sankar *et al.*, 2005).

Studies carried out in Maharashtra revealed that larger cloves produced higher marketable yields in garlic (Singh *et al.*, 1996;; Ramnivas *et al.*, 1998). Brar and Gill (2000) obtained a bulb yield of 20.92 tonnes/ha with sowing of mother cloves of size 3.30 g. According to Bhonde *et al.* (2002) and Verma *et al.* (2004) mother cloves of 3-4 mm size were optimum for garlic planting. Healthy medium to big sized uniform bulbs procured from high yielding variety after proper storage (to undergo sufficient dormant period) are used for planting. Depending upon variety and availability of planting material, seed rate should be 500 to 950 kg/ha.

Table 31. Improved selections of garlic made in India

Variety/ Year of release	Pedigree/ Parentage	Institution/ University	Av. yield t/ha	Salient features
Agrifound White (G-41) 1989	Mass selection from Bihar collection	National Hort. Research & Devt. Foundation, Kanada Batata Bhavan, 2954E, New Mumbai, Agra Road, Nashik-422011, Maharashtra	130	Bulb compact, silvery white with creamy flesh, bigger elongated cloves, 20-25 in number, duration 150-160 days, susceptible to purple blotch and *Stemphyllum* blight. TSS 41%, dry matter 42.78% and good storer.
Yamuna Safed-1 (G-1) 1991	Mass selection from local collection obtained from Delhi	-do-	150-175	Compact bulbs, silvery white skin with creamy flesh, sickle shaped cloves, 25-30 in numbers, 150-160 days duration. Tolerant to insect pests and diseases like purple blotch, *Stemphyllum* blight, onion thrips. TSS-38%, dry matter 39.5% and good storer.
Yamuna Safed-2 (G-50) 1996	Mass selection from local collection in Haryana	-do-	150-200	Bulbs compact, attractive white creamy flesh, 35-40 cloves. Duration 165-170 days. Tolerant to *Stemphyllum* and purple blotch disease as well as thrips. TSS-38-40%, dry matter 40.41%.
Yamuna Safed-3 (G-282) 1999	Mass selection from germplasm collection from Tamil Nadu	-do-	175-200	Bulbs are creamy white and bigger sized (5-6 cm diameter), compact 15-18 cloves. Duration 140-150 days. Tolerant to thrips. TSS 38-42%, dry matter 39-43% and medium storer suitable for export.
Agrifound Parvati (G-313) 1992	Selection from exotic collection from Hong Kong	-do-	175-225	Bigger sized bulbs of 5-6.5 cm, creamy white colour with pinkish tinge, 10-16 bigger cloves. Duration 250-270 days. Tolerant to common diseases, medium storer. This variety is suitable for export, developed for long day condition and very popular in hills of Uttar Pradesh and Himachal Pradesh.
G-323 1990	Mass selection from local collection obtained from UP	-do-	175-220	Bulbs silvery white and bigger sized, 30-35 cloves. Plants vigorous with wider green leaves, TSS 40-42%, dry matter 41-42%.

Contd...

Godavari (selection-2) 1987	Selection	Mahatma Phule Krishi Vidyapeeth Rahuri-413 722, Maharashtra	100-105	Medium sized bulbs, pinkish white colour, 22-25 cloves, maturity 140-145 days
Sweta (selection-10)	NA	-do-	100-105	Bigger bulbs, silvery white, 25-26 cloves/ bulb 130-135 days maturity
Ooty-1	Clonal selection from germplasm	Tamil Nadu Agricultural University (TNAU), Coimbatore - 640 003, Tamil Nadu	150-170	Attractive white in colour. Big bulbs, 20-25 cloves. Resistant to thrips and tip drying.
Co-2	NA	-do-	NA	Attractive white bulbs
Co-1	NA	-do-	NA	NA
VL Garlic-1	Clonal selection from EC 158250	Vivekananda Parvatiya Krishi Anusandhan Sansthan, Almora - 263 601	140 to 150 in hill and 90-100 in plains	NA
ARU-52	NA	Vivekanda Parvatiya Krishi Anusandhan Sansthan, Almora - 263 601	NA	NA
T-56-4	Selection	Punjab Agricultural University (PAU), Ludhiana-141 004, Punjab	80-100	Smaller size bulb, white in colour 25-35 cloves, 130-135 days maturity
LGG-1	NA	Punjab Agricultural University (PAU), Ludhiana-141 004, Punjab	NA	NA
Pant Lohit-1 2001	Farmers field primitive cultivar	GB Pant University of Agriculture and Technology, Pant Nagar, Uttaranchal	NA	Bulbs attractive, purple skinned variety, heavy seed potential. Tolerant to purple blotc h disease.
Pant Lohit-2	NA	GB Pant University of Agriculture and Technology, Pant Nagar, Uttaranchal	NA	NA

Source: Johnykallupurackal and Ravindran (2006)

Planting

The crop is sown from September to November in plains of Northern India and in hills from March to April. In South India, crop is planted from August to November. Field is divided into small plots to facilitate proper irrigation. Cloves are dibbled at a spacing of 15 cm between rows and 7.5 cm between plants. Planting could also be done on ridges.

Manuring

Garlic crop responds well to added manures and fertilizers, 40-50 tonnes of organic manure is recommended. Combined application of N, P and K was beneficial for good yield and boldness of cloves. At Ludhiana, 200:50:50 kg NPK/ha gave the highest yield of bulbs. In West Bengal, 60:60:120 kg NPK/ha recorded the highest yield. In Meghalaya region (soils deficient in P and S), bulb yield, nutrient uptake and protein content of garlic were high at 100 kg P_2O_5/ha and 20 kg S/ha (Majumdar *et al.*, 2003).

Irrigation

First irrigation is given just after sowing and then the field is irrigated after every 10-15 days till the season warms up. Pandey and Singh (1990) recommended irrigation at 60 mm CPE for good crop of garlic at Haryana. Frequency of irrigation should be decreased towards maturity of crop as more moisture results in regeneration and emergence of sprouts.

Interculture and weed control

Field should be kept free from weeds by timely weeding and hoeing. Hoeing the crop just before formation of bulbs (about 2½ months from sowing) loosens soil and helps in setting of bigger compact bulbs.

Harvesting

Crop is ready for harvesting when the top turns yellowish or light brownish and shows signs of drying up, 130-180 days after planting depending on cultivars, soil, season etc. Bulbs are lifted, freed from earth and the leaves are tied at top. The bulbs are cured for 3-4 days in shade before storing them. Tops are removed before marketing the produce. Garlic bulbs store well in ordinary well ventilated room. After removing tops, yield/hectare varies from 4 to 10 tonnes, depending on soil, variety and management practices. In hilly areas of Tamil Nadu, garlic bulbs with tops intact are subjected to a smoking process by hanging them from kitchen roof. This procedure improves keeping quality of bulbs.

Diseases, pests and disorders

Blight caused by *Alternaria solani*, is the main disease affecting garlic. Thrips and red spider mites are major pests. Rubbering of garlic bulbs is a physiological disorder in fresh garlic bulbs at harvest. Rubberized bulbs become spongy when dried. The problem is severe in over irrigated and over fertilized fields. Pre-mature sprouting of bulbs is another disorder which leads to splitting of cloves which reduces market quality of bulbs (Peter *et al.*, 1995).

Storage

Pre-harvest maleic hydrazide (MH) sprays were used to prolong storage life of garlic. Spraying of MH 3000 ppm and 2500 ppm, three weeks and two weeks before harvest respectively inhibits sprouting for 300 days and reduce loss of weight of bulbs during storage.

Processed products and uses

Dehydrated garlic, garlic powder, garlic flakes, oil of garlic and garlic oleoresin are the processed products of garlic. Because of the clinging odour of fresh garlic, there is preference for other forms particularly garlic powder among the consumers. Central Food Technological Institute (CFTRI), Mysore, has got a process patent for the production of garlic powder. Recently spray dried garlic preparations are also available.

Garlic is used all over the world as a condiment for flavouring various dishes. Further, raw garlic is used in the manufacture of garlic powder, garlic salt, garlic paste and garlic vinegar. Garlic is also used ground with other spices to act as an antiseptic and to improve flavour and digestion. Garlic oil is a flavouring agent and can be used as an insecticide and also as a biofungicide. Garlic oil is very useful as an adhesive for broken glass. It also possesses antibacterial properties. Medicinal qualities of garlic were known from time immemorial. Modern drug research has proved its efficacy in reducing blood cholesterol, blood sugar and hyper tension. Garlic antibiotic - allicin - was effective on malignant tumors on humans and animals. Therapeutic uses of garlic on digestion and on utilization of food ingredients also have been scientifically proved.

Asafoetida
(*Ferula asafoetida* L., F. Apiaceae)

Ferula is a genus of perennial herbs distributed from Mediterranean region to Central Asia. A few of the species are important as sources of oleo-gum resins used as condiment and in medicine. Asafoetida of commerce is dried latex obtained mainly from living rootstocks or tap roots of several species of *Ferula*. It was introduced in India by Arabian physicians. There are about 60 species of genus *Ferula* which are widely distributed in three main regions viz. Central Asia, Europe and North America. Three species occur in Kashmir. India is the largest importer of asafoetida, mainly from Afganisthan and Iran.

A part of imported resin is re-exported from India to various countries after processing.

Botany and chemical composition

Important species of *Ferula* are the following:

F. asafoetida L., *F. alliaceae* Boiss, *F. rubricaulis* Boiss, *F. foetida* (Bunge) Reget, *F. galbaniflua* Boiss, *F. sumbul* Hook.f., *F. narthex* Boiss, *F. jaeschekeana* and *F. suaveeolens*. Of these, *F. narthex* and *F. jaeschekeana* occur in India.

Types of asafoetida of commerce imported to India are: (a) asafoetida, (b) galbanum and (c) sumbal. Of these asafoetida is the most important one and it is obtained from several species of *Ferula*. 'Galbanum' is the oleo-gum resin obtained from *F. galbaniflua*, a tall herb and a few other species. 'Sumbul' consists of transversely sliced and dried rhizomes and roots of *F. sumbul* and *F. suaveeolens*. *F. narthex* is a perennial odorous herb, 1.5-3.0 m high. Leaves pinnately decompound; secondary and tertiary pinnae decurrent, entire or irregularly crenate-serrate, sheaths large, ovate, upper leaves much reduced, flowers small, yellow, in simple or scarcely compound umbels springing from within the sheaths; tap root thick and branched.

Asafoetida contains resin (40-64%), gum (2.5%) and volatile oil (10-17%). Resin consists chiefly asaresinotannol, free or combined with ferulic acid. Oil of asafoetida is obtained by steam distillation of gum-resin. Chief constituents of oil are disulphides which impart a disagreeable odour to the oil. Noleau *et al.* (1991) reported 13 sulphur containing compounds in asafoetida.

Crop management

Not much information is available on cultivation aspects of asafoetida. There are several grades of asafoetida which come under different classifications. Two major grades are 'Hing' and 'Hingra'. 'Hing' is the exudate of *F. asafoetida* whereas 'Hingra' is the exudate of *F. foetida*. 'Hing' is superior in quality. Among them again, there are several grades. Broadly recognized classes of asafoetida in market are milky white asafoetida and red asafoetida.

Ferula plants when 4-5 years old develop massive tap root or carrot shaped roots 12.5 to 15.0 cm in diameter at the crown. In March-April, just before plants flower, upper part of root is laid bare and stem cut off close to crown. Exposed surface is covered by a dome-shaped structure made of twigs and earth. A milky juice exudes from cut surface. After some days, exudate is scraped off and a fresh slice of root cut when more latex exudes; sometimes resin is removed along with slice. Collection of resin and slicing of root are repeated until exudation ceases (3 months after first cut). Resin is sometimes collected from successive incisions made at junction of stem and tap root with three incisions. A few plants yield about 1 kg or more of gum-resin.

Quality and flavour of asafoetida available in market are affected by a number of factors viz. part of plant from which prepared, season of collection, method of preparation, degree of adulteration, nature of adulterant used etc.

Asafoetida of commerce are available in three forms viz. 'tears', 'mass' and 'paste'. The 'tears' constituting purest form of resin are rounded or flattened, 5-30 mm in diameter and are greyish or dull yellow in colour. Mass asafoetida is the common commercial form which consists of tears agglutinated into a more or less uniform mass, usually mixed with fragments of root, earth etc. Paste form also contains extraneous matter.

Processed forms and uses

Since pure asafoetida is not preferred for use because of its strong flavour/odour it is sold in compounded form. Main raw materials used are gum arabic and starch. Compounded asafoetida types manufactured and marketed in India are in mass form. Adulteration is also common.

Asafoetida is extensively used in India as a food flavourant. It is considered stimulant, carminative, expectorant and antispasmodic. It is one of the most effective remedies for treating flatulence, hysteria and nervous disorders. It is also widely used to treat asthma, chronic bronchitis, cough and cold. Asafoetida is used as an antidote in case of overdosing of opium (Garg, 2003).

Greater Galanga
(*Alpinia galanga* (Linn.) Willd., F : Zingiberaceae)

Greater galanga is a perennial herb, native to China, now cultivated in Indonesian / Malaysian Archipelago, Thailand, Eastern Himalayas and South Western India. Economic part is the dried rhizome or root of plant. *Galanga* is one of the spices which reached European markets relatively early and source of spice remained almost unknown until 1867 when its botanic origin was confirmed in China. Greater galanga has now become naturalized in many parts of India and is extensively cultivated all over India, mostly in shady situations.

Botany and chemical composition

Greater galanga is a perennial herb, 1.8 - 2.4 m high with tuberous aromatic roots. Leaves oblong - lanceolate, acute, glabrous, 30-60 cm long, ligule rounded, flowers greenish white, streaked with red, in dense flowered, 30 cm long panicles, capsules orange or red, globose. Rhizome is 2.5-10.0 cm thick, reddish brown externally and light orange brown inside. It has a tough and fibrous structure and a spicy pungent taste. Rhizomes of other *Alpinia* species are sometimes substituted for genuine drug.

Fresh rhizomes on steam distillation yield an essential oil (0.04%) with a peculiar strong and spicy odour. Oil contains methyl cinnamate (48%), cineol (20-30%) camphor and a-pinene. According to Verghese (2003c) myrcene is one of the dominant constituents in essential oil. Principal odorants of galangal oil are identified as 1,8-cineole, linalool, geranyl acetate, eugenol and chavicol acetate (Verghese, 2003c).

Crop management

Galangal thrives the best in tropical climate with an annual rainfall ranging from 1500-3000 mm in well drained hilly areas. It can also be cultivated as an intercrop in coconut and rubber plantations. Crop is generally propagated through rhizomes. Rhizome bits (5.0 cm) with atleast one shoot are planted with onset of monsoon on raised beds. Spacing adopted is 30 x 20 cm. Sufficient FYM and mulch must be provided. A fertilizer dose of 100:50:50 kg NPK/ha is recommended in 2-3 splits. No serious pests and diseases are reported in galangal.

Harvesting and processing

Optimum stage of harvest for obtaining maximum rhizome and oil yield is 36-42 months after planting though it can be harvested even after 18 months (Saji and Sasikumar, 2003). Shoot portion is cut off and rhizomes are dug out carefully. Roots are removed, rhizomes are cleaned and are cut into 5 cm long pieces and dried under sun light for 3-5 days to attain about 10% moisture. Average yield of rhizome is 23 t/ha.

Uses

Both rhizomes and fruits are used for culinary and medicinal uses. In Indonesia it is used as a condiment. In India, it is well known for its medicinal uses. It is considered as a tonic, stomachic, carminative and stimulant and is used as a fragrant adjunct to complex preparations and also in cough and digestive mixtures. Its chief use is for clearing voice. The drug has an expectorant action and is useful in many respiratory ailments. Galanga oleoresin is used in medicine and in flowering foods.

Saffron

(*Crocus sativus* Linn., F. Iridaceae)

Saffron is one of the oldest and certainly among the world's most expensive spices. True saffron consists of dried trilobed stigmas of *Crocus sativus*, a native of South Europe. It is cultivated in Spain, France, Italy, Greece, Turkey, Persia, India and China. It is one of the oldest commodities of Jammu and Kashmir, spread over 200 villages involving 4000 families (Dhar *et al*., 2000). Recorded accounts of saffron cultivation in Kashmir date back to 550 AD, nearly four centuries earlier than its recorded cultivation in Spain by Arabs. In international market, Spanish saffron has an edge over Indian brands, despite being of the best quality. Price of saffron ranges from Rs.45000-50000/kg.

Botany and chemical composition

Saffron is a small bulbous perennial, 15-25 cm in height with an underground globular corm. Flowers are large, blue or lavender in colour. Plant is hysteranthous, with flowers arising directly from corms. Flowers have trifid or tri-lobed, orange coloured stigmas which, along with style tops, yield saffron of commerce.

Saffron yields an essential oil and fixed oil as high as 1.37 per cent and 13.4 per cent respectively. Saffron contains glycosides, crocin and picrocrocin together with lycopene, b-carotene, g-carotene and zeaxanthin. Crocin is the chief colouring principle in saffron. Chief flavouring component is safranal (Auria *et al.*, 2004).

Crop management

Soil and climate

Saffron thrives well in cold regions with warm or subtropical climate. It requires rich, well drained, sandy or loamy soil. Corms tend to rot in humid and water logged soils. Elevations between 1300-2200 m are ideal for growing saffron. Photoperiod exerts a considerable influence on flowering of saffron; an optimum period of 11 hours illumination is desirable. Dry and moderately humid weather conditions at flowering time are ideal; frost or rain during flowering adversely affect crop yield.

Seed rate and planting

Plant is propagated vegetatively by corms. Corms of about 2.5 cm diameter and above are used for planting. Seed rate is 8 tonnes/ha. Large corms give more flowers and higher stigma yield (Munshi *et al.*, 2004). Mother corms reproduce annually 2-12 cormlets and some of which also produce flowers in following year. Corms once established continue to live for 10 to 15 years, new corms being produced annually and older ones rotting away. Planting is done from July to August in small raised plots surrounded by drains. Corms are planted 15-20 cm apart in rows at a distance of 10 cm. Plots are hoed and weeded, manures and fertilizers are applied at rate of 15-22 tonnes of FYM/ha and 50:10:50 kg of NPK/ha. Immediately after harvest, soil in between rows is dug out and fertilizers applied. Saffron is grown as a rainfed crop in Kashmir and hence no irrigation is given.

Harvesting and processing

Plants flower in October-December. Number of flowers and period of blooming depend upon weather conditions. In first year, yield will be negligible and by third year, good yield is obtained. Flowers are collected daily, early in morning after dew disappears. Flowers are cleaned and style and stigma are separated from perianth. These operations require skilled labour and is usually done by women labourers. About 147,000 to 166,000 flowers give five kg of fresh stigma that go into make a kg of pure dry saffron. Yield of saffron in Kashmir is much lower than those in other countries (1.5 kg/ha). Production of saffron keeps increasing till 8th year and starts declining thereafter and hence replanting is recommended after nine years.

Processed products

Value of saffron depends mainly on its colour, flavour and overall good quality which in turn depends on method by which stigma is dried. Tripartite stigmas plucked from freshly collected flowers and dried in sun constitute saffron of best quality. Stigmas are sundried to 8-12% moisture. This constitutes first grade (Shahi saffron). Whole flowers are dried in sun for 3-5 days, beaten lightly with sticks, passed through coarse sieves. Material which passes through is then thrown into water; petals that float on water are thrown away and parts which sink are collected and dried. This forms second grade saffron (Mongra saffron). Discarded parts of flowers are again subjected to beating process and process of throwing entire pounded mass in water is repeated. Product which sinks is collected, which is very much inferior in value thus constituting third grade (Lacha saffron). Alternatively, during bumper flower harvest, whole flowers are dried in sun and stigmas are picked late by hand from dried flowers.

In Spain, drying process is called 'toasting', stigmas are placed over earthern kilns or pots containing low fire in sieves.

Final product as sold in market is a loosely matted mass of dark, reddish brown flattened stigmas with a characteristic aromatic odour and butter taste. When fresh, they are glossy and unctuous to touch, but after keeping they become dull and brittle. Dry stigmas are stored in cloth bags or butter paper and packed in closed tins or air tight wooden boxes. Present price of saffron is around Rs.50000/kg.

Uses

Saffron is an important food adjunct, i.e. food flavourant and food colourant and is also reputed to possess several medicinal properties. It is an important ingredient of Ayurvedic and Unani systems of medicine in India.

Sweet Flag
(*Acorus calamus* L., F. Araceae)

Sweet flag is a perennial, aquatic herb, native to Asia and North America. It was introduced in Europe by 1557 and has become naturalized there. It is of ancient use in China and is also grown in India, Malaysia and elsewhere in South Eastern Asia. dried Rhizomes constitute the drug "Calamus" of commerce.

Sweet flag is common in Kashmir and Kumaon ranges of Himalayan forests. It is plentyful in marshy tracts of Kashmir, Manipur and Naga Hills. It is cultivated in Karnataka and Andhra Pradesh and traded both for medicinal and condiment purposes.

Botany and chemical composition

Acorus calamus is a semi-aquatic perennial aromatic herb with creeping rhizomes growing wild and also cultivated throughout India. Rhizomes are horizontal jointed, somewhat vertically comprised, spongy within, pale to dark brown or occasionally orange brown in colour, leaves grass - like or sword shaped, long and slender, flowers small, yellow-green in a spadix, berries green, angular, 1-3 seeded, seeds oblong. *Acorus gramineus* Soland ex. Ait., a Japanese species is occasionally met within Sikkim upto an altitude of 1,800 m and in Khasi hills upto 1,500 m. Air-dried plant yields 0.5-0.9 per cent essential oil having asarone as a major constituent. The plant is used for same medicinal purposes as *A. calamus*.

Rhizomes, roots and leaves yield a light brown to brownish yellow volatile oil known as "Calamus oil". Fresh rhizomes yield 1.8 per cent; dry rhizomes 1.5-3.5 per cent; leaves 0.2 per cent and fresh aerial parts 0.12 per cent oil. Important constituents of Indian calamus oil are asarone (upto 82%) and its ß-isomer. Indian calamus oil has a higher content of 'asarone' compared to commercial oils. Other constituents are calamenol, calamine, methyl eugenol, α-pinene and camphene. A bitter aromatic principle, acorin and a flavone diglycoside were separated from the plant.

Crop management

Sweet flag thrives the best in marshy and moist places under variable climates. It prefers an area with well distributed rainfall and plenty of sunshine. Acorus comes up well in clayey loam and light alluvial soils of river banks. It also grows well in mud along banks of pools and sluggish streams. Calamus crop is grown in same way as rice in fields, which are puddled and kept irrigated like a rice field.

Seed and planting

Planting material is the growing ends or tops of previous crop. Two tonnes of rhizomes are required for planting one hectare. Results of study conducted by Philip *et al.* (1989) on effects of planting materials on growth parameters, yield and quality of *A. calamus* are given in Tables 32 and 33.

Table 32. Effect on planting material on growth parameters of *A. calamus*

Treatments	**Height of plant (cm)**	**No. of tillers/ plant**	**No. of leaves/ tiller**	**No. of leaves/ plant**	**Leaf area (cm^2)**
P_1 - Rhizome bits of 3 cm length with growing tops	9.6	6.2	5.5	33.5	73.7
P_2 - Rhizome bits of 6 cm length with growing tops	11.8	7.3	5.6	40.4	75.1
P_3 - Rhizome bits of 3 cm length without growing tops	8.4	4.4	5.5	22.9	50.9
P_4 - Rhizome bits of 6 cm length without growing tops	7.5	5.1	6.0	29.8	56.7
CD (P=0.05)	1.3	1.0	NS	5.8	8.9

Source: Philip *et al.* (1989)

Table 33. Effect of planting material on yield and quality parameters of *A. calamus*

Treatments	**Yield of green rhizomes per plot (g)**	**Yield of dry rhizomes per plot (g)**	**Drying percentage yield per ha (kg)**	**Green rhizome yield per ha (kg)**	**Dry rhizome content (%)**	**Essential oil oil per ha (kg)**	**Yield of essential**
P_1 - Rhizome bits of 3 cm length with growing tops	112.0	62.7	56.4	896	502	1.45	7.3
P_2 - Rhizome bits of 6 cm length with growing tops	172.5	93.9	54.7	1380	751	1.39	10.4
P_3 - Rhizome bits of 3 cm length without growing tops	92.8	51.6	55.8	742	413	1.37	5.7
P_4 - Rhizome bits of 6 cm length without growing tops	128.5	72.0	56.0	1028	576	1.35	7.8
CD (P=0.05)	16.8	9.1	NS			NS	

Source: Philip *et al.* (1989) NS - Not significant CD = Critical difference

Rhizome bits of 6 cm length with growing tops were found ideal under Kerala conditions. Land should be ploughed thoroughly to a fine tilth, watered heavily. Sprouted rhizome pieces are planted at a spacing of 30 x 30 cm and pressed into the mud to a depth of about five cm inside soil. Rhizomes are planted in such a way that plants in second row come in between first row and not opposite to them. Planting season is March-April. One has to ensure that harvest may not coincide with rainy season. Field is kept regularly irrigated and water level maintained as for paddy i.e., at least five cm depth in beginning and increased to 10 cm as plants grow in height.

Manuring and fertilizer application

Twenty five tonnes of FYM and 8:16:20 kg of NPK/ha should be applied as basal dose in March-April. Remaining 17:34:40 kg NPK/ha is to be given in two split doses at four months and eight months after planting. Field is weeded at monthly intervals, care should be taken not to disturb rhizomes while pulling out weeds (Selvi *et al.*, 2003).

Pests and diseases

Mealy bug is the serious pest in sweet flag. No serious disease is reported.

Table 34. Effect of time of harvest on yield and quality of *A. calamus*

Treatments	Yield of green rhizomes per plot (g)	Yield of dry rhizomes per plot (g)	Drying percentage yield per ha (kg)	Green rhizome yield per ha (kg)	Dry rhizome content (%)	Essential oil oil per ha (kg)	Yield of essential
D_1 - Harvesting 8 months after planting	116.8	59.7	51.2	935	478	1.55	7.4
D_2 - Harvesting 9 months after planting	134.0	71.8	53.4	1072	574	1.38	7.9
D_3 - Harvesting 10 months after planting	136.8	79.3	58.0	1094	634	1.33	8.4
D_4 - Harvesting	114.3	69.5	60.2	914	556	1.30	7.2
CD(P=0.05)	16.1	8.4	3.0			0.14	
CD (P=0.05)	16.8	9.1	NS			NS	

Source: Philip *et al.* (1989) NS - Not significant CD = Critical difference

Harvesting

Crop is ready for harvest in 10 to 12 months when tops begin to turn yellow. Irrigation is stopped and field is allowed to dry, retaining just enough moisture to make digging easy. Harvesting should be done carefully as rhizomes are at 60 cm depth and about 30-60 cm spread. Harvested rhizomes are cut into small pieces of 5.0-7.5 cm and all fibrous roots are removed. Rhizome pieces are then washed thoroughly and sun dried. Dried rhizome pieces are then put in gunny bags and rubbed thoroughly to remove leaf scales. Mean yield of fresh rhizomes per hectare is 10 tonnes. The study conducted by Philip *et al.* (1989) revealed that both fresh and dry yield of rhizime were high when the crop was harvested after 10 months. Essential oil content decreased with the delay in harvest (maximum values at 8months after planting).

Processed products and uses

Dry rhizomes are reduced to desired size in a suitable disintegrator and steam distilled to get volatile oil. Oil yield is about 4.6 per cent. Medicinal properties of calamus plant are known since the earliest dawn of human civilization. The oil is reported to cure gastritis and is supposed to promote digestion. Dried rhizome is an aromatic bitter tonic and carminative. It is also useful in stomach disorders, mouth diseases, intestinal worms and for breaking tobacco addiction. The greatest use of oil at present is in perfumery industry as a component of finer perfumes. They are also used in preparation of agarbathies and dhup. Powdered rhizome also possesses insecticidal properties. Rhizome and its oil are used in manufacture of liquor, essence and bitters.

Long Pepper

(*Piper longum* L., F. Piperaceae)

Long pepper is the dried fruit of *Piper longum*, which is a slender aromatic climber with woody roots, occurring in hotter parts of India, from Central Himalayas to Assam, Khasi and Mikir hills, lower hills of Bengal and evergreen forests of Western Ghats from Konkan to Travancore (Kerala). It is indigenous to South Asian countries such as India, Sri Lanka, Bhutan and Nepal. It also occurs in Malaysia, Indonesia and Phillipines. Long pepper was well known in ancient India. It was mainly used as source of pungency, used in flavouring various beverages and dishes. From India, it reached rest of Asia and Mediterranean through "Spice route" and it was used as a spice in all these regions (Ravindran and Balachandran, 2005). Importance of long pepper as a spice diminished with spread of more pungent black pepper. However, long pepper is still used in flavouring dishes in many parts of Central Asia, Afghanistan, Baluchistan, Nepal and Bhutan and in North East India. In rest of India, it is more precious as a medicinal plant and is costlier than black pepper.

Botany and chemical composition

Piper longum which is the true source of Indian long pepper, is a slender creeping shrub, spreading on ground and rooting at nodes. It grows and creeps over small shrubs, rocks etc. but does not climb on trees as in black pepper. Plant produces distinct dimorphic branches, those main branches creeping on ground (orthotrops) have cordate leaves with long petioles (plagiotrops) and those axillary branches that grow erect produce leaves that are sessile or with short petioles. The former shoots are vegetative and grow by activity of terminal bud, while short axillary branches are fruiting branches, produce fruits (spikes) opposite leaves and growth is sympodial. Leaves are simple, alternate and variable in size and shape.

Long pepper is dioecious. Flowers are arranged on a short cylindrical spike, the male spikes are much longer than female spikes. Female spikes are short, stout and flowers fused laterally. Each flower consists of only one ovary that arises from axil of a bract. Fruits are small, closely packed. Fruiting is apomictic and fruits are produced without pollination. Hence male plants are not required for pollination (Ravindran and Balachandran, 2005).

Dried fruits of long pepper, on steam distillation gives 0.7% of an essential oil with spicy odour, resembling that of pepper and ginger oils. Spikes (fruits) contain sylvatin, sesamin, diaeuolemin, piperine, pipalartine, asparinin, pulviatilol, pipericide, guineenside, longamide, piplasterol & dihydropiperonaline.

Crop improvement

Eventhough *P. longum* is recognized as the genuine source of the drug long pepper, spikes of certain other species are also traded as long pepper in the market - such as *P. peepuloides, P. chaba* and *P. mullesca. P. chaba* is commonly known as Java long pepper. Fruits of this species are larger, more conical and turn orange red on ripening and are more pungent with less flavour. *P. peepuloides* looks similar to *P. longum* while *P. mullesca* spikes are globose with minute fruits that are not laterally fused.

Major production areas have their own species adapted clones, such as 'Suali' clone, (of Assam and North East), 'Asli' clone (of Bengal and Assam). The only improved clone is a selection 'Viswam' released from Kerala Agricultural University. It gives an average yield of 317 kg/ha. This variety has prolonged flowering phase and stout leaves, short and thick spikes and is suitable as an intercrop in irrigated coconut and arecanut gardens. Fruits contain 2.83 % alkaloid.

Crop management

Soil and climate

Long pepper thrives well in moist, moderately hot climate, from plains to an elevation of about 80 m. It tolerates temperature up to 32°C; requires annual rainfall of 250-350 cm, distributed evenly. Well-drained, sandy loam is ideal for its cultivation. In its natural habitat, long pepper is an undershrub and when cultivated requires partial shade. It comes up very well as an intercrop in coconut and areca gardens. In North Bengal and Assam, long pepper is grown mostly in areca gardens. When long pepper is grown for its roots, loose soil (like sandy loam) should be selected for ease in harvesting.

Propagation

It is propagated exclusively using three-four node cuttings of orthotropic stem that creeps on ground. Single node-cuttings can also be rooted in poly bags. Direct planting of cuttings in field with onset of rain is the common method adopted by growers.

When long pepper is grown for roots, only male plants are grown, as they are more vigorous growing and produce more quantity of roots.

Land preparation

Land is ploughed a couple of times, levelled and raised beds of 3 meter x 2 m are prepared. Pits are dug on beds at 60 x 60 cm, filled up with topsoil and FYM or compost.

Planting and aftercare

Either four-five node cuttings or poly bag rooted cuttings are used. They are planted at the rate of two cuttings/pit.

When planted as an intercrop (as in Assam and Northern West Bengal where long pepper is an intercrop in areca gardens), pits are dug after a light digging. At the center of four areca plants, a clump of plants four or five are usually planted in a pit. In coconut gardens, usually they are planted in an irregular fashion leaving basins. However, when coconuts are planted at regular intervals, long pepper can be in two or three rows in between two rows of coconuts.

Weeding and watering are essential during initial stages. Cow dung or compost can be applied with onset of rains. Usually chemical fertilizers are not recommended. However, 50 kg N, 20 kg P_2O_5 and 7kg K_2O/hectare can boost yield and returns considerably.

Pests and diseases

Fungal diseases caused by *Phytophthora capsici, Colletrotrichum* and *Fusarium* cause damage to plant. Bordeaux mixture is the only recommended fungicide. Virus disease leads to crinkling, mottling and yellowing. Badna Virus and Cucumber Mosaic Viruses are implicated in this malady.

Mealy bugs and tea mosquito also infect long pepper plants. Insecticides are usually not recommended.

Harvesting and processing

Spikes can be harvested six months after planting. Spikes are harvested when green and they should not be allowed to ripe. Ripe spikes are useless. Spikes reach harvesting stage 60-70 days after their formation. Spikes are picked three-four times a year. Harvested spikes are spread out on mats or cement floor for drying. Sun dried spikes are black in colour. Productivity reaches maximum three or four years after planting and subsequently decreases. Usually plants are uprooted once in every five years, roots are harvested and dried, and field replanted. When grown for roots, male plants are planted and they are harvested 18-24 months after planting.

Uses

Fruits are used as a spice and also in pickles and preserves. 'Thippali' or 'Pipali' is the most extensively used medicinal plant in Ayurvedic system of medicine. It is used in over 320 classical compound medicinal formulations and in many modern herbal formulations. Long pepper is one of the 'Thrikatus' which forms part of many compound formulations. Fruits and roots are credited with numerous medicinal uses & may be used for diseases of respiratory tract, viz. cough, bronchitis and asthma, as counter irritants and as analgesic when applied locally for muscular pains and inflammation, as snuff in coma and drowsiness and internally as a carminative, as a sedative in insomnia and epilepsy as general tonic and haemantinic, as cholagogues in obstruction of bile duct and gall bladder; as an emmenagogue and abortifacient and for miscellaneous purposes as anthelmintic and in dysentery and leprosy. Several studies indicated that constituents of 'Thrikatu' (especially piperine) have bioavailability enhancing activity. Long pepper is also a very important home remedy for many ailments and it is a bioceutical of great importance.

Horse Radish
(*Armoracea rusticana* P.Gaertn., F. Brassicaceae)

Horse radish is a native of South Eastern Europe. It is one of the oldest condiments, long fleshy cylindrical roots or the grated roots are used as condiment particularly with roast beef. It is sometimes grown at higher altitudes in tropics including India. Roots of horse radish tree (*Moringa oleifera*) are some times used as substitute.

Botany and chemical composition

Horse radish is a stout glabrous perennial herb, with long fleshy cylindrical roots, large, oblong secrete leaves and small white flowers. Roots possess an acrid, pungent taste and when scraped or bruised, emit a characteristic pungent odour. Pungency is due to glucoside sinigrin which is broken down in water by enzyme action. Roots also contain allyl isothiocyanate (0.08%). Roots yield an essential oil (0.05-0.2%). It is a rich source of vitamin C, the fresh material contains on an average 302 mg ascorbic acid /00 g.

Crop management

Horse radish is not commercially cultivated in India. In USA, where it is commercially grown, the varieties are classified into two, common and Bohemian. Common type has broad crinkled leaves and produces roots of high quality. Bohemian type has narrow smooth leaves, is disease resistant, but yields poor quality roots. Horse radish is highly sterile and is propagated by root cuttings. Which are planted at 30 cm spacing in trenches. Root are harvested in cold weather when the flavour is at maximum (Ravindran and Pillai, 2004).

Processed products and uses

Horse radish is used as an appetizer condiment. It is credited with digestive and anti-scorbutic properties because of high vitamin C content. Besides, leaves and roots are used as food. It is also used for preparation of 'horse radish cream', 'sauce' or 'relish'. Horse radish is considered stimulant, diaphoretic, diuretic and digestive. Crushed horse radish has an inhibitory effect on growth of micro-organisms.

Hyssop

(*Hyssopus officinalis* L., F. Lamiaceae)

Hyssop is an aromatic perennial shrub, native of Mediterranean region and temperate Asia and naturalized in parts of Europe and America. In India it is found in the Himalayas from Kashmir to Kumaun at altitudes of 2500-3000 m above MSL. Leaves and flower tops constitute the condiment or spice. In Europe, it is mainly grown for its essential oil and to some extent for ornamental purposes.

Botany and chemical composition

Hyssop is an aromatic shrubby perennial 0.3-0.6 m high. Branches erect or diffuse, leaves sessile, linear-oblong or lanceolate obtuse, entire; flowers are bluish purple, in auxillary tufts arranged unilaterally on terminal branches; nutlets are dark brown, narrow, trigonow, smooth.

A volatile oil is obtained from the aerial parts of the plant (0.15-0.30% from and 0.3-08% from fresh dried materials). Hyssop is stated to yield a greyish green dye. The fresh herb contains iodine in a concentration of 14 mg/kg.

Cultivation

The plant thrives in light rich soil in hill stations. It may be propagated by seeds, cuttings and divisions. Maximum yield of oil is obtained right after the blooming. First harvest can be done two years after planting. One hectare with approximately 10,000 plants yield, on an average 3,000 kg of herbage. A field of hyssop has a life span of 6-7 years. The harvested plants are uniformly dried under shade and packed. The oil obtained from all the above ground parts is of uniform quality. Hyssop oil is highly priced and because of this it is occasionally adulterated with spike, lavender or rosemary oils.

Uses

The leaves and flowering tops are used as a condiment and flavouring liqueurs and in perfumes. Hyssop is considered a stimulant, carminative and used against colds, coughs, and lung complaints. Hyssop oil is used in flavouring of beverages, in perfumery and cosmetics and also in medicine.

ŽŽŽ

Garlic

Asafoetida

Greater Galanga

Saffron

Sweet Flag

Long Pepper

Horse Radish

Hyssop

References

Abraham, C. C. and Skaria, B. P. 1995. New record of the root borer *Zeuzera coffeae* Nietner (Cossidae:Lepidoptera) as a pest of allspice (*Pimenta dioica* Linn.) Merril. *Insect Environment* 1(1): 6

Abraham, G. 2001. Increasing productivity of mustard *(Brassica juncea)* through split application of sulphur. *Ind. J. agric. Sci.* 71(10): 674-675

Abraham, Z., Mithal, S. K. and Gautam, P. L. 1999. Genetic resources of *Garcinia* species at NBPGR Regional Station, Thrissur, Kerala, India. (In): *Biodiversity, Conservation and Utilization of Spices, Medicinal and Aromatic Plants.* Indian Institute of Spices Research, Kozhikode, pp. 131-133

Adaniya, S. 1998. Variation in pollen fertility and germinability in ginger (*Zingiber officinale* Rosc.) *J. Japanese Soc. Hort. Sci.* 67: 872-874

Agrawal, S., Sharma, R. K., Bhatt, B. N. and Agrawal, S. 1990. Quality evaluation in coriander. *Indian Cocoa, Arecanut, and Spice J.* 13(4): 137-138

Ahmed, M. A., Hussein, M. S. and El-Sherbeny, S. E. 1989. Yield analysis of fenugreek plants. *African J. Agrl. Sci.* 16(1-2): 163-172

Ahmed-Shah, H., Vedamuthu, P. G. B., Abdul-khader, M. D. and Prakasam, V. 1988. Influence of different levels of potassium on yield and curcumin content of turmeric. *Proceeding of National Seminar on Chillies, Ginger and Turmeric.* APAU, Hyderabad, pp. 109-113

Ajithkumar, K., Jayachandran, B. and Ravi, V. 2002. Influence of shade regimes on photosynthetic rate and stomatal characters of ginger (*Zingiber officinale* Rosc.). *J. Spices and Aromatic Crops* 11: 26-29

Ali, Y., Jain, G. C., Kapdi, S. S., Agarwal, Y. C. and Bhatnagar, S. 1991. Development of brush type ginger peeling machine. *Agricultural Mechanization in Asia, Africa and Latin America* 22(2): 71-73

Ambili, A. 1997. Cyto-morphological investigations in *Piper* spp. M.Sc (Hort.) thesis, Kerala Agricultural University, Vellanikkara, Thrissur, Kerala, India, 77p

Amin, R. S. 1978. Scion wood grafting a new technique for hard wood plants. *Curr. Sci.* 47:468-469

Amma, P. S. 1981. Studies on the effect of containers, potting mixtures and growth regulators on growth and survival of clove seedlings. M.Sc. (Hort.) thesis, Kerala Agricultural University, Thrissur, Kerala, India, 91p.

Anandaraj, M. and Sharma, Y. R. 1992. A new leaf rot in *Pimento dioica. Indian Phytopathology* 45(2): 276-277

Anandaraj, M., Sivaraman, K. and Krishnamoorthy, B. 1989. Effective weed control through weedicides in plantation crops. *Indian Cocoa, Arecanut and Spices J.* 13:63-64

Ancy, J. and Jayachandran, B. K. 1993. Effect of shade and fertilizers on the quality of ginger. *S. Indian Hort.* 41(4): 219-222

Ancy, J. and Jayachandran, B. K. 1998. Enhanced nutrient uptake and rhizome yield in ginger (*Zingiber officinale* R.) under shade. In: *Developments in Plantation Crops Research.* Allied Publishers Ltd., pp. 236-238

Ancy, J. and Jayachandran, B. K. 2000. Assimilate partitioning in ginger as influenced by shade and nutrient levels. *Proceedings of 12th Kerala Science Congress,* Peermedu Development Society Kumily, Kerala, pp. 407-410

Anith, K. N., Das, T. P. M., Jayaraj, M., Vasanthakumar, K. and Aipe, K. C. 2000. Integration of solarization and biological control with a fluorescent *Pseudomonas* spp. for controlling bacterial wilt *Ralstonia solanacearum* (E.F.Smith) Yabuuchi *et al.* of ginger. *J. Biological Control* 14(1) 25-29

Anu, A., Nirmalbabu, K. and Peter, K. V. 2004. Variations among somaclones and its seedling progeny in Capsicum annuum L. *Plaant Cell, Tissue and Organ Culture* 76: 261-267

Anuja, G. 2005. Growth, development and quality of vanilla (*Vanilla planifolia* Andrews) as influenced by organics and growth regulators. M.Sc. (Hort.) thesis, Kerala Agricultural University, Thrissur, Kerala, India, 81p

Anuradha, M. N., Farooqui, A. A., Vasundhara, M., Kathiresan, C. and Srnivasappa, K. N. 2002. Effect of biofertilizers on the growth, yield and essential oil content in rosemary (*Rosemarinus officinalis* L.). Proceeding of National Seminar on the Strategies for Production and Export of Spices, 24 – 26 oct., 2002, IISR, Calicut, India.

Arya, P. S. 2001. *Ginger Production Technology*. Kalyani Publ., New Delhi.

Atakisi, I., Saglam, C., Turhan, H., Arslanoglu, F., Kaba, S. and Onemli, F. 2001. Cultivation of sage (*Salvia officinalis*) in thrac region, Turkey. *Zeitschrift-fur-arznei-and-zewurzpflanzen* 6(1): 15-19

Augustine, A. 2001. Selection of promising lines, production of somaclones and their utilization in paprika (*Capsicum annuum* L.). Ph.D thesis, Universoty of Calicut, Kerala

Auria, M. D., Mauriello, G. and Rana, G. L. 2004. Volatile organic compounds from saffron. *Flavour and Fragrance J.* 19(1): 17-23

Azhakiamanavalan, K and Vadivel, E. 1997. Dryland tamarind PKM-1: pre and post release status. *Proceedings of the National Symposium on Tamarindus indica L.* Forest Dept., Tirupathi, Andhra Pradesh, pp.7-9

Babu, P. 1993. *Effect of shade and mulch on the yield of ginger.* M.Sc (Hort.) thesis, Kerala Agricultural University, Kerala, India. 108p.

Babu, P. and Jayachandran, B. K. 1994. The quality of ginger (*Zingiber officinale* R.) as influenced by shade and mulch. *S. Indian Hort.* 42(3): 215-218

Babu, T. D., Achuthan, C. R., Nazeem, P. A. and Nair, R. V. 2000. Genetic DNA isolation and RAPD analysis in the tree spice - nutmeg (*Myristica fragrans*) *Proceedings of the 87th Indian Science Congress Part IV*, Indian Science Congress Association, Calcutta, pp. 38

Babu, T. D., Nazeem, P. A., Keshavachandran, R., Achuthan, C. R., Girija, D., Sureshkumar, P. and Peter, K. V. 2004. Detection of genetic diversity in piper species using RAPD and AFLP markers. *Procedings of Natioanl Seminar on New Perspectives in Spices, Meicinal and Aromatic Plants.* IISR, Kozhikode, Kerala, India, pp. 1-9

Badoc, A. and Lamarti, A. 1991. A chemotaxonomic evaluation of *Anethum graveolens* L. (dill) of various origins. *J. Essent. Oil Res.* 3(4): 269-278

Bagade, D. S. and Shinde, A. K. 1993. Effect of growth regulators on seed germination and seedling growth of clove. *Indian Cocoa, Arecanut and Spices J.* 16 (3&4) :121-122

Balachandran, S. M., Bhat, S. R. and Chandel, K. P. S. 1990. In vitro clonal multiplication of turmeric (*Curcuma* spp.) and ginger (*Zingiber officinale* Rosc.). *Pl. Cell Rep.* 8: 521-524

Balaji, B. R. 1988. Investigation on nutrition of curry leaf. M.Sc. (Ag) thesis, University of Agricultural Sciences, Bangalore, Karnataka, India, 98p.

Balakrishnamoorthy, G. and Kennedy, R. R 1999. *Improved Techniques in Tree Spices Cultivation.* Horticultural Research Station, Yercaud, Tamil Nadu,

Balakrishnan, K., Anandan, R. and Nanthakumar, S. 1998. Non destructive method of leaf area estimation in nutmeg and clove. *South Indian Hort.* 46 (5 & 6): 364-365

Balashanmugam, P. V., Vijayakumar, R. M. and Subramanyam, K. S. 1990. Effect of zinc and iron on turemeric yield. *South Indain Hort.* 38: 284-285

Bandara, J. M., Chandrasena, K. S. G. and Abeyhoon, A. M. D. 1985. Compatability and pathogenesity of *Phytopthera palmivora* on hosts of high density cropping systems in the upcountry of Srilanka. *J. Agri. Sci.* 22(1): 39-48

Banerjee, D. P., Chatterjee, B. K. and Sen, S. 1982. Air layering of cinnamon. *South Indian Hort.* 30:272-273

Baruah, A. and Nath, S. C. 2001. Best quality tejpat – *Cinnamomum impressinerium* Meissn. Indian Spices 14(5): 5-7

Baruah, A., Nath, S. C. and Boissya, C. L. 1998. Synthetics and diversities of Cinnamon species used as "tejpat" in northeast India. Proceedings of National Seminar on Recent Trends in Spices and Medicinal Plant Research. 2-4 April 1998. Cacutta, India, (Abstr. B-3).

Bauman, D., Hadolin, M., Rizner, H. A. and Knez, Z. 1999. Supercritical fluid extraction of rosemary and sage antioxidants. *Acta Alimentaria Budapest* 28(1): 15-28

Bavappa, K. V. A. and Reuttimann, R. A. 1981. Nutmeg cultivation and processing. *Tech. Bull.* 2.

UNDP/FAO Research Project on Minor Export Crops, Department of Minor Export Crops, Sri Lanka

Beena, S. 1994. Standardisation of top working in nutmeg. M.Sc. (Hort.) thesis, Kerala Agricultural University, Thrissur, Kerala, India, 97p

Belavadi, V. V. and Parvathi, C. 1998. Estimation of honey bee colonies required for effective pollination in cardamom. *Proceedings of National Symposium on Diversity of Social Insects and Other Arthropods and the Functioning Systems.* III Congress of IUSSI, Mudigere, p. 30

Belavadi, V. V., Chandrappa, H. M., Shadakshari, Y. G. and Parvathi, C. 1993. Isolation distance for seed gardens of cardamom (*Elettaria cardamomum* Maton). In: *Pollination in Tropics* (eds. Veeresh, G. K., Umashanker, R. and Ganeshan, K. N.). IUSSI, Bangalore, pp. 241-243

Bennet S. S. R., Durai, A., Nicodemus, A., Gireesan, K., Nagarajan, B., and Varghese, M. 1997. Genetic improvement of *Tamarindus indica* L., for increased productivity. *Proceedings of the National Symposium on Tamarindus indica L.* Forest Dept., Tirupathi, Andhra Pradesh, pp.47-50

Bernath, J., Kattaa, A., Nemeth, E. and Franke, R. 1996. Production-biological investigation of fennel (*Foeniculum vulgare*) populations of different genotype. *Atti del convegno internazionale: Coltivazione e miglioramento di piante officinali* (Italy) 2-3: 287-292

Bharti, V., Prasad, U. K. and Bharti, V. 2001. Response of Indian mustard (*Brassica juncea* L.) to irrigation and Sulphur. *J. Applied Biology* 11(1-2): 27-30

Bhat, S. and Sudharshan, M.R. 2004. Vanilla- Crop improvement at Saklespur. *Spice India* 17(3): 42-45

Bhat, S. K. and Sudharshan, M. R. 1998. Floral biology and fruit growth in vanilla (*V. planifolia* Andr.) *Recent Advances in Plantation Crops Research* (eds. Muraleedharan, M. and Rajkumar, R.) Allied Publishers, Coimbatore, pp. 110-112

Bhat, S. K. and Sudharshan, M. R. 2000. Effect of pollen load on growth and development of vanilla (*Vanilla planifolia* Andr.) fruits. *Proc. Centennial Conference on Spices and Aromatic Plants* (eds. Ramanna, K. V., Santhosh, J. E., Nirmalbabu, K., Krishnamurthy, K. S. and Kumar, A.), September 20-23, Calicut, Kerala, pp. 13-16

Bhonde, S. R. D., Singh, K. and Dwivedi, M. M. 2002. A note on the seed bulbs (Planting material) requirement of different varieties of garlic. NHRDF Newsletter Vol.XXII (4) pp. 10-12

Bhuvaneshwari., Farooqi, A. A. and Sreeramu, B. S. 2001. Anise: production technology. *Spice India* 14(8): 17-19

Bhuyan, 1997. Micropropagation of curryleaf tree by axillary proliferation using intact seedlings. *Plant Genetic Res. Abstr.* 1998 7(2): 951

Bhuyan, S., Prasad, S., Zaren, Y., Yildiz, M. T., Ozcan, E., Guzel, A. and Isik, E. B. 1990. Drying characteristics of ginger and development of small capacity drier. *Proc. International Congress on Mechanization and Energy in Agriculture,* Adana, Turkey, pp. 501-509

Bindu, R. 1997. *In vitro* pollination, embryo rescue and germination studies in ginger (*Zingiber officinale* Rosc.). M.Sc (Hort.) Thesis, Kerala Agricultural University, Kerala, India.

Biswas, A. K., Gupta, U. and Deka, T. N. 2004. Large cardamom research – activities and achievements. *Spice India* 17(3): 35-38

Brar, P. S. and Gill, S. P. S. 2000. Effect of clove size on yield of garlic. *Haryana J. Hort. Sci.* 29: 1-2

Challapilli, A. P., Hulamani, N. C. and Chimmad, V. P. 1992. Studies on evaluation of some promising tamarind (*Tamarindus indica* L.) tree. *My Forest* 28(4): 317-320

Chandra, R. and Govind, S. 1999. Genetic variability and performance of ginger genotypes under mid-hills of Meghalaya. *Indian J. Hort.* 56: 274-278

Chandramani, P., Harris. V., Parasuraman, C., Asokhan, S., Balasubramani, P. and Muthumaralingam, S. 2001. Pests of clove. *Spice India* 14(1): 3

Chandrappa, H. M., Shadakshari, Y. G. Sudharsan, M. R., Dushyandhakumar, B. M., Edison, S. and Shivashankar, K. T. 1998. Breeding studies in cardamom (*Elettaria cardamomum* Maton). In: *Development Studies in Cardamom* (eds. Mathew, N. M. and Jacob, C. K.). Allied Pub., New Delhi, pp. 20-27

Channabasavanna, A. S. 2002. Standardisation and economic analysis of fertilizer levels for coriander (*Coriandrum sativum* L.). *J. Maharashtra Agricultural University* 27(2): 160-162

Chempakam, B. and Peter, K. V. 2000. Processing and chemistry of tamarind. *Spice India* 13(4): 9-13

Chen, Q. Y., Shi, H. and Ho, C. T. 1992. Effects of rosemary extracts and major constituents on lipid oxidation and soybean lipoxygenase activity. *J. American Oil Chemists Society* 69(10): 999-1002

Chengat, T. 1997. *Influence of organic manures on and Azospirillum on growth , yield and quality of ginger* (*Zingiber officinale*). M.Sc. (Hort.) thesis, Kerala Agricultural University, Kerala, India. 109p.

Cheriayan, E. V. 2000. Varaibility in *Capsicum chinense.* M.Sc (Hort) thesis, Kerala Agricultural University, Vellanikkara, Kerala, India

Chezhiyan, N. and Ananthan, M. 1997. Effectr of seed grading on the germination of tree spices clove and nutmeg. *South IndianHort.*45(5&6): 249-251

Choudhary, R. and Chandel, K. P. S. 1995. Studies on germination and cryopreservation of cardamom (*Elettaria cardamomum* Maton) seeds. *Seed Sci. Biotech.* 23:235-240

Chowdhury, A. R. 2000. GC-MS studies on essential oil from Carum caarvi L. *Proc. Centennial Conference on Spices and Aromatic Plants* (eds. Ramanna, K. V., Santhosh,

J. E., Nirmalbabu, K., Krishnamurthy, K. S. and Kumar, A.), September 20-23, 2000. Calicut, Kerala, pp. 305-306

Chundawat, B. S. 1990. *Arid Fruit Culture*. Oxford and IBH Publishing Co. Pvt. Ltd., New Delhi, pp.158-162

Clevely, A. and Richmond, K. 1999. *Growing and Using Herbs*. Sebastian Kelly, Oxford.

Cristopher, T. and Rajan, M. V. 1994. *In vitro* clonal propagation of *Capsicum* species. *Plant Cell, Tissue and Organ Culture* 38(1): 25-29

Damayanthi, K. P. M., Sasikumar, B., and Remashree, A. B. 2003. Reproductive biology and incompatibility studies in ginger (*Zingiber officinale* Rosc.). *Phytomorphology* 53: 123-131

Das, N. 1999. Effect of organic mulching on root-knot nematode population, rhizome rot incidence and yield of ginger. *Ann. Plant Protect. Sci.* 7(1): 112-114

Das, P., Rai, S. and Das, A. B. 1999. Cytomorphology and barriers in seed set of cultivated ginger (*Zingiber officinale* Rosc.). *Iranian J. Bot.* 8: 119-129

Das, T. P. M., Pradeepkumar, T., Mayadevi, P., Aipe, K. C. and Kumaran, K. 2000. Stability analysis of ginger (*Zingiber officinale* Rosc.) genotypes. *J. Spices and Aromatic Crops* 9(2):165-167

Dash, K. and Jana, E. C. 2004. Genetic variability in a collection of turmeric (Curcuma longa L.) genotypes. *Procedings of Natioanl Seminar on New Perspectives in Spices, Meicinal and Aromatic Plants*. IISR, Kozhikode, Kerala, India, pp. 56-62

Dashora, A., Sharma, R. K., Sastry, E. V. D. and Singh, D. 2000. Combining ability analysis in varietal diallel cross of fennel (Foeniculum vulgare). *Proc. Centennial Conference on Spices and Aromatic Plants* (eds. Ramanna, K. V., Santhosh, J. E., Nirmalbabu, K., Krishnamurthy, K. S. and Kumar, A.), September 20-23, 2000. Calicut, Kerala, pp. 78-81

Datta, S., Ghosh, P. and Chatterjee, R. 2001. Production technology of mentha. *Spice India* 14(9): 2-3

Dearnley, T. C. and Duckett, J. G. 1998. Juniper in the lake district national park - a review of condition and regeneration. *Watsonia* 22(3): 261-267

Desikan, K. R. 1980. Jamun. *Dryland Horticulture* :4-5

Devadas, V. S. and Manomohandas, T. P. 1988. Studies on the viability of allspice seeds. *Indian Cocoa, Arecanut and Spices J.* 11(3): 100

Devaraj, P., Chandrasekaran, P., Lakshmidevi, R. and Durairaj, S. 1997. Comparative studies on the growth performances of various tree species in black cotton soil. *Journal of Ecobiology* 9(1): 71-72

Dhar, A. K. and Singh, A. K. 2000. Saffron: biology, utilization, agriculture, production and quality. *J. Medi. Aroma. Pl. Sci.* 22(1b): 355-360

Dwivedi, S., Singh, M., Singh, A. P., Singh, V., Naqvi, A. A., Khanuja, S. P. S., Sushilkumar, Dwievedi, S. and Kumar, S. 2001. Registration of a new variety Pranjal of Mentha x Piperita. *J. medi. arom. Pl. Sci.* 23(2): 99-101

Edison, S., Johny, A. K., Babu, N. K. and Ramadasan, A. 1991. *Spice Varieties.* National Research Centre for Spices, Calicut, Kerala, pp. 23-34

El-Din, A. A. E. 2003. Growth, yield and essential oil of anise in relation to water supply. *Annals of Agricultural Science Cairo* 48(2): 777-785

Farooki, A. A. Sreeramu, B. S. and Srinivasappa, K. N. 2000. Hints on cultivation of thyme - a herb spice. *Spice India* 13(2): 19-21

Farooqui, A. A. and Sreeramulu, B. S. 2001. Cultivation of Medicinal and Aromatic Plants. University Press India Ltd, Hydrabad, India, pp. 433 – 43

Figueiredo, R. D., Marques, M., Nakagawa, J., Figueiredo, R. O. and Ming, L. C. 2004. Composition of coriander essential oil from Brazil. *Acta Horti.*629: 135-137

Gamborg, O.L., Miller, R.A. and Ojima, K. 1968. Nutrient requirements of suspension cultures of soybean root cells. *Exp. Cell Res.* 50: 151-158

Garg, R. B. L. 2003. Asafoetida - an ayurvedic drug. *Spice India* 16(10): 30

Geetha, C. K. and Aravindakshan, M. 1992. A comparison of growth and dry matter production in bush pepper and vine pepper as influenced by applied NPK. *Indian Cocoa, Arecanut and Spices J.* 15: 95-98

Geetha, C. K. and Nair, P. C. S. 1989. Studies on spike shedding in black pepper(Panniyur-1). *South Indian Hort.* 37: 282-286

Geetha, S. P. 2002. *In vitro Technology for Genetic Conservation of some Genera of Zingiberaceae.* Ph.D. thesis, University of Calicut, Kerala, India,

Geetha, S. P., Manjula, C. and Sajina, A. 1995. *In vitro* conservation of genetic resources of spices. *Proc. of 7th Kerala Science Congress*, January 27-29, STEC, Thiruvananthapuram, Kerala, pp. 12-16

Geetha, S. P., Nirmalbabu, K., Rema, J., Ravindran, P. N. and Peter, K. V. 2000. Isolation of protoplasts from Cardamom (*Elettaria cardamom*). *J. Spices and Aromatic Crops* 9(1): 23-30

Geetha, V. 1995. Urigam tamarind-a unique variety. *Spice India* 8:8

Geetha, V. and Prabhakaran, P. V. 1987. Genotypic variability, correlation and path coefficient analysis in turmeric. *Agric. Res. J. Kerala* 25(2): 240-254

George, B. E. 1992. *Evaluation of ginger cultivars for shade tolerance.* M.Sc. (Hort.) thesis, Kerala Agricultural University, Kerala, India.

George, C. K. 1993. Pimento-facts farmers should know. *Indian Cocoa, Arecanut and Spices J.* 16(3 & 4): 113-114

George, C. K. 2004. Star anise. Hand Book pf Herbs and Spices. Vol.II. (ed. Peter, K. V.). Wood Head Publishing Limited, England, pp. 290-293

George, L. and Narayanaswamy, S. 1973. Haploid capsicum through experimental androgenesis. *Protoplasma* 78: 467-470

George, M. V., Mohammed Syed, A. A. and Korikanthimath, V. S. 1984. *Diospyros ebenum* Koenig, an ideal shade tree for cardamom. *J. Plantation Crops* 12(2): 160-163

George, S. T. 1988. Garcinia – the neglected acidic fruit of Kerala. *Indian Cocoa, Arecanut and Spices J.* 11(3):100

George, S. T., Babylatha, A. K., Mathew, L. K. and Geetha, C. K., 1992. Pattern of flowering and flower development in Kodampuli. *Indian Cocoa, Arecanut and Spices J.* 16(2): 68-70

Gnandas, D. 1989. Standardization of index leaf/leaves for assessing the nutritional status of clove in relation to soil fertility. M.Sc. (Hort.) thesis, Kerala Agricultural University, Thrissur, Kerala, India,

Gopalakrishnan, M., Menon, N. and Mathew, A. G. 1982. Changes in the composition of clove oil during maturation. *J. Food Sci. and Tech.* 19(9): 190-192

Gopalakrishnan, R. 2004. Spices in coconut garden – a compatable combination for sustainable income. *Spices India* 17(9): 2-10

Gopalakrishnan, T. R 1985. Inheritance of clusterness, destalkness and deepred colour in chilli. Ph.D thesis, Kerala Agricultural University, Vellanikkara, Kerala

Gopalakrishnan, T. R. and Indira, P. 2006. Chilli – improvement through five decades. *Advances in Spices Research* (eds. Ravindran, P. N., Nirmalbabu, K., Shiva, K. N. and Johnykallupurackal, A.). Agrobios (India), Jodhpur, pp. 515-533

Gopalakrishnan, T. R., Indira, P. and Peter, K. V. 1998. Ujjwala – a chilli variety reisitant to bacterial wilt. Spices India 11(2): 10-13

Gopalam, A. and Ravindran, P. N. 1987. Indexing of quality parameters in black pepper cultivars. *Indian Spices* 22/23: 8-11

Gopalan, A. 1997. Profile of essential chemical constituents in tree spices. *Supplement on Cultivation and Utilization of Aromatic plants*, Regional Research Laboratory, Jammu-Thawi, pp. 405-422

Gopalan, C., Sastri, R. B. and Balasubramanyan, S. C. 1971. *Nutritive Value of Indian Foods.* Health Bull. 23. National Intitute of Nutrition, Hyderabad

Gosavi, P. G. 1988. Studies on drying and utilization of dried kokum (*Garcinia indica* Choisy) and karonda (*Carica carandes* L.). M. Sc. (Ag.) thesis, Konkan Krishi Vidyapeeth, Dapoli, Maharashtra

Goudra, B. K. H and Madalagiri, B. B. 1992. Effect of nitrogen nutrition on yield and yield component of curry leaf (*Murraya koenigii* Spreng). *Indian Cocoa, Arecanut and Spices J.* 16(2): 66-68

Goudra, R. P. H., Shivashankar, K. T. and Gowda, N. 1995. Uses and cultivation of curry leaf. *Spice India* 19:17-18

Govind, S., Chandra, R. and Gupta, P. N. 1995. Response of ginger to P in acid alfisol of Meghalaya. *J. Hill Res.* 8: 274-278

Govinda, S., Gupta, P. N. and Ramachandra. 1990. Response of N and P levels on growth and yield components of turmeric in acid soils of Meghalaya. *Indian J. Hort.* 47: 79-84

Govindarajan, V. S. 1985. Capsicum - production, technology, chemistry and quality. Part 1. History, botany, cultivation and primary processing. CRC. Critical Reviews in Food Science and Nutrition 22(2): 109-174

Gowda, K. K., Melanta, K. R., Prasad, T. R. G., Goada, K. K. and Guru-Prasad, R. R. 1999. Influence of NPK on the yield of ginger (*Zingiber officinale* Rosc.). *J. Plantation Crops* 27(1): 67-69

Gunasekharan, M. and Krishnaswamy. V. 1999. Influence of seed size and orientation of placement on seed quality characteristics of clove. *Indian J. Arecanut, Spices and Medicinal Plants* **1**(3): 94-98

Gunjate, R. T., Karnik A. R. and Limaye, V. P. 1984. Flower and sex forms in kokum. *Indian Cocoa Arecanut and Spices J.* 8(2):32-36

Gupta, C. R. and Aswathi, O. P. 1997. Effect of mulch material on growth and yield of ginger (*Zingiber officinale* Rosc.). *Veg. Sci.* 24: 113-115

Gupta, R. 1982. Studies in cultivation and improvement of dill (*Anethum graveolons*) in India. In: *Cultivation and Utilization of Aromatic Plants.* (eds. Atal, C. K. and Kapur, B. M.). Regional Research Laboratory, Jammu, India, pp. 545-548

Gupta, R. 2004. Dill. In: *Hand Book of Herbs and Spices.* Vol.I. (ed. Peter, K. V.). Woodhead Publishing Limited, England, pp. 173-179

Gupta, R. K. 1993. *Multipurpose Tree for Agroforestry and Waste Land Utilization.* Oxford and IBH Pub. Co., pp.482-484

Haldankar, P. M., Joshi, G. D., Salvi, M. J. and Patil, J. L. 1991. Effect of season and shed provision on softwood grafting in kokam. *Indian Cocoa Arecanut and Spices J.* 14(4):158-159

Haldankar, P. M., Joshi, J. D., Salvi, M. J. and Patil, J. L. 1992. Standardization of softwood grafting in kokam. *J. of Maharashtra Agric. Univ.* 17(2):227-229

Haldankar, P. M., Nagvekar, D. D., Desai, A. G. and Gunjate, R. T. 1996. Studies on air layering on cinnamon. *Indian Cocoa Arecanut and Spices J.* 20(3): 89-90

Haldankar, P. M., Nagvekar, D. D., Desai, A. G., Patil, J. L. and Gurijate, R. T. 1995. Air layering in allspice. *Indian Cocoa, Arecanut and Spices J.* 19(2): 56-57

Haldankar, P. M., Nagvekar, D. D., Desai, A. G., Patil, J. L. and Gunjate, R. T. 1999. The effect of season and the root stocks on approach grafting in nutmeg . *Indian Arecanut Spices and Medicinal Plants J.***1**(2): 52-55

Haldankar, P. M., Nagvekar, D. P., Patil, J. L. and Gunjate, R. T. 1994. Varietal screening for yield and quality in cinnamon. *Indian Cocoa Arecanut and Spices J.* 28(3):79-81

Haldankar, P. M., Salvi, M. J., Joshi, G. D. and Patil, J. L. 1993. Factors influencing softwood grafting in kokam (*Garcinia indica* Choisy). *Indian Cocoa, Arecanut and Spices J.* 17(1&2):15-18

Halesh, D. P., Gowda, M. C., Farooqui, A. A., Vasundhara, M. and Srinivasappa, K. N. 2000. Influence of nitrogen and phosphorus on growth, yield and nutrient content of fenugreek (*Trigonell foenum – graecum* L.). *Proc. Centennial Conference on Spices and*

Aromatic Plants (eds. Ramanna, K. V., Santhosh, J. E., Nirmalbabu, K., Krishnamurthy, K. S. and Kumar, A.), September 20-23, 2000. Calicut, Kerala, pp. 191-194

Hatmode, C. N., Rathod, D. L., Ingole, A. S., Deotale, R. D. and Chore, C. N. 2001. Effect of different levels of fertilizers on biochemical, yield and contributing parameters in mustard. *J. Soils and Crops* 11(2): 229-234

Hegde, K. R., Sulikeri, G. S. and Hulamani, N. C. 1990. Effect of growth regulator and pre-sowing treatment on rooting of cinnamon air layers. *South Indian Hort.* 37: 329-333

Heine, H. 1998. Variety trials with biennial caraway (*Carum carvi* L.). *Zeitschrift fur Arznei and Gewurzpflanzen* (Turkey) 3(2): 67-69

Hena, K.K. 2005. Morphological and molecular characterization of variability in *in vitro* derived seedlings of vanilla (*Vanilla planifolia* Andrews). M.Sc.(Ag.) thesis, Kerala Agricultural University, Thrissur. 64p

Houle, G. and Babeux, P. 1994. Variation in rooting ability of cuttings and seed characteristics of five populations of *Juniperus communis* var. Depressa from subarctic Quebec. *Canadian J. Botany* 72(4): 493-498

Ibrahim, K. K., Pillai, V. S. and Sasikumaran, S. 1988. Expression and association of spike character in black pepper. *Agric Res. J. Kerala* 26: 127-129

Ibrahim, M. E. 1999. Evaluation of lovage (*Levisticum officinale.* Koch) as a new aromatic plant under Egyptian cultivation conditions. *Egyptian J. Hort.* 26(2): 177-186

ICRI (Indian Cardamom Research Institute). 2002. *Annual Research Report.* Indian Cardamom Research Institute, Myladumpara. pp.15-16

IISR, 1997. *Ann. Rep. 1996-1997*, Indian Institute of Spice Research, Calicut, Kerala, India,

IISR. 2002. Annual Report for 2001. Indian Institute of Spices Research, Calicut, India.

Indira, P. 1997. Diversity interrelationships among *Capsicum* spp. and froms and development of paprika. Ph.D thesis, Kerala Agricultural University, Vellanikkara, Kerala

Inocencio, C., Rivera, D., Alcaraz, F. and Barberan, F. A. 2000. Flavonoid content of commercial capers (*Capparis spinosa*, *C. sicula* and *C.orientalis*) produced in mediterranean countries. *European Food Research and Technology* 212(1): 70-74

Irulappan, I. and Anbu. A. 1994. Under exploited tropical fruit trees with promising economic value. In: *Trees and Tree Farming.* Peekay Tree Crops Development Foundation, pp.230-235

Jasural, S. C., Mishra, V. K. and Verma, K. S. 1993. Intercropping ginger and turmeric with poplar (*Populas deltoides* "G-3" Marh.). *Agroforestry Systems* 22(2): 111-117

Jayachandran, B. K. 1989. *Induced Mutations in Ginger.* Ph.D. thesis, Kerala Agricultural University, Kerala, India.

Jayachandran, B. K., Ancy, J.,Babu, P., Nizam, S. A. and Mridula, K. R. 1998. Under the coconut tree in India, ginger has made in the shade. *Agroforestry Today* 10(3): 16-17

Jayachandran, B. K., Bai, M. M., Salam, M. A. and Mathew, K. P. 1991. Performance of ginger under shade and open conditions. *Indian Cocoa, Arecanut and Spices J.* 15(2): 40-41

Jayachandran, B. K., Bai, M. M., Salam, M. A. and Mathew, K. P. 1992. Storage of seed ginger. *Spice India* 5(12): 4-8

Jayaraj, P. 1990. Effect of potassium on mitigating the effect of shade in intercrops. M.Sc. (Ag.) thesis, Kerala Agricultural University, Kerala, India, 103p.

Jayasree, S., Wahid, P. A. and Kamalam, N. V. 1988. Absorption of soil applied phosphorus by black pepper vine and support tree in relation to the root activities. *J. Plantation Crops* 16: 73-87

Jimenez, N. G., Alaonso, M. J. P. and Negueruela, A. V. 2000. Chemical composition of fennel oil, *Foeniculum vulgere* Miller. from Spain. *J. Essential Oil Research* 12(2): 159-162

Johny, A. K. and Ravindran, P. N. 2005a. Over 225 hi-yielding spice varieties in India – Part II. *Spice India* 18(5): 40-45

Johny, A. K. and Ravindran, P. N. 2005b. Over 225 hi-yielding spice varieties in India – Part III. *Spice India* 18(6): 39-52

Johnykallupueackal, A. and Ravindran, P. N. 2006. High yielding spices varieties developed in India. *Advances in Spices Research* (eds. Ravindran, P. N., Nirmalbabu, K., Shiva, K. N. and Johnykallupurackal, A.). Agrobios (India), Jodhpur, pp. 93-139

Jose, L. 2001. Genetic variability in chilli (*Capsicum annuum* L.) with wmphasis to reaction to leafcurl virus. M.Sc thesis, Kerala Agricultural University, Vellanikkara, Kerala

Joseph, A. and Jayachandran, B. K. 1996. Nutrient requirement of ginger (*Zingiber officinale* Rosc.). under shade. *Indian Cocoa, Arecanut and Spices J.* 20(4): 115-116

Joseph, B., Joseph, D. and Philip, V. J. 1996. Plant regeneration from somatic embryos in black pepper. *Plant Cell, Tissue Culture and Organ Culture* 47: 87-90

Joseph, J. 1980. The nutmeg - its botany, agronomy, production, composition and uses. *J. of Plantation Crops* 8(2):61-72

Joseph, J. 1981. Floral biology and variation in cinnamon. *PLACROSYM IV*, pp. 431- 434

Josephrajkumar, A., Chakrabarty, R. and Thomas, J. 2005. Occurrence of trypsin like protease in cardamom (*Elettaria cardamom* Maton). Indian J. Biochem. Biophysics 42: 243-245

Joshi, G. D., Nadkarni, H. R. and Patil, R. S. 1999. Processing of kokum (*Garcinia indica*) and citrus. (In): *Post Harvest Handling and Processing of Horticultural Crops.* Directorate of Extn. Edn., Dr. Balasaheb Sawant Konkan Krishi Vidyapeeth, Dapoli, Dist. Ratnagiri, pp.20-21

Joshi, S. V. and Patil, S. R. 2002. Isozyme banding pattern in parents and hybrids of mustard. *J. Soils and Crops* 12(2): 225-230

Joshy, A. C. 1946. A note on the development of the pollen in *Myristica fragrans* and the affinities of the family Myristicaceae. *J. Indian Bot. Soc.* 25:139-143

Joshy, H. B. 1983. *The Sylviculture of Indian Trees* vol. IV. Controller of Publications, GOI, Delhi, pp.232-236

Joy, P. P., Thomas, J., Mathew, S. and Jose, J. 1996. Elite cinnamon types for leaf oil and eugenol. *Proceedings of the 8th Kerala Science Congress.* Cochin University, Kochi, Kerala, pp.143-144

Kalra, A., Gupta, A. K., Katiyar, N., Srivastava, R. K. and Kumar, S. 2003. Screening of *Coriandrum sativum* accessions for seed and essential oil yield and early maturity. *Plant Genetic Resources Newsletter* 133: 19-21

Kandiannan, K., Koya, K. M. A. and Peter, K. V. 1999. Mixed cropping of ginger (*Zingiber officinale* Rosc.) and maize (*Zea mays* L.). *J. Spices and Aromatic Crops* 8(2): 219-220

Kandiannan, K., Sivaraman, K., Thankamani, C. K. and Peter, K. V. 1996. Agronomy of ginger (*Zingiber officinale* Rosc.). *J. Spices and Aromatic Crops* 5: 1-27

Kannan, K. 1971. Some problems of field planting of nutmeg. *Arecanut and Spices Bull.* 3 (1): 6-8

Kannan, K. 1973. Top working as a means of converting male into female plants in nutmeg. *Arecanut and Spices Bull.* 4(4): 14-16

Kannan, K. and Balakrishnan, S. 1967. A note on viability of cinnamon seeds. *Madras agric. J.* 54: 78-79

Karale, A. R., Kavlgud, S. N. and More, T. A. 1997. Studies on tamarind in Maharashtra. *Proceedings of the National Symposium on Tamarindus indica L.* Forest Dept., Tirupathi, Andhra Pradesh, pp.10-12

Karnik, A. R. 1978. Studies on flowering and fruiting in kokam (*Garcinia indica* Choisy). M.Sc. (Ag) thesis, Konkan Krishi Vidyapeeth, Dapoli, Maharashtra

Karunakaran, P., Nair, M. C. and Das, L. 1993. Grey blight disease of cinnamon (*Cinnamomum verum*) leaves. *J. Spices and Aromatic Crops* 2 (1&2): 66-67

Karuppaiah, P, Prakash, M. and Manivannan, K. 1997. Studies on the biochemical constituents of *Tamarindus indica* tender leaves. *Proceedings of the National Symposium on Tamarindus indica L.* Forest Dept., Tirupathi, Andhra Pradesh, pp.235-237

KAU. 2003. *Package of Practices Recommendations " Crops" –* 2003. Kerala Agricultural University, Kerala, India

Kaushal, P. S. 1994. Studies in the western Himalayan juniper: I. Distribution pattern and taxonomic considerations. Research Bulletin of the Punjab University, *Science* 44(1-4): 53-62

Kennedy, R. R, Thangaraj, T., Subramanian, C. and Balakrishnamoorthy, G. 1998. Variability studies in tamarind. *Proceedings of the National Conference on Recent Trends in Spices.*

Kennedy, R. R. and Balakrishnamoorthy, G. 2000a. *Integrated Nutrient Management in Clove.* Directorate of Research, TNAU, Coimbatore, p. 3

Kennedy, R. R. and Balakrishnamoorthy, G. 2000b. Improved cultivation technique for cinnamon (Tamil). *Spice India* 13(7): 17-20

Kennedy, R. R. and Nageswari, K. 2000. Measuring tree girth in tree spices. *Udhyaniki Jeevan* 7(2): 5

Kennedy, R. R. and Thangaraj, T. 1996. Survey, collection and quality evaluation of tamarind trees in Coimbatore. Topical Research Report submitted to Department of Spices and HC & RI, TNAU, Coimbatore

Kennedy, R. R., Thangaraj, T. and Balakrishnamoorthy, G. 1997. Variabbility studies in tamarind. *Udhyaniki Jeevan* 4 (23): 8

Keshavachandran, R. and Khader, M. D. 1989. Tissue culture propagation of turmeric. *S. Indian Hort.* 37(2): 101

Keskar, B. G., Karale, A. R., Dhawale, B. C. and Choudhary, K. G. 1989. Improvement of tamarind by selection. *Maharashtra J. Hort.* 4: 821-824

Khan, I. A. and Ahmad, Z. 2000. Studies on cultivation, seed production and nutrition quality of parsley (*Petroselinum hortense*) in mid hills of Uttarakhand. *Bioved* 11(1-2): 33-35

Khan, K. F. and Krishnaswamy, A. R. 1953. A new essential oil of nutmeg (*Myristica fragrans* Houtt.) and its possible economic uses. *Sci. Cult.* 19:186-188

Khan, K. F. and Krishnaswamy, A. R. 1953. A new essential oil of nutmeg (*Myristica fragrans* Houtt.) and its possible economic uses. *Sci. Cult.* 19:186-188

Khan, M. M. A., Samiullah., Afaq, S. H. and Afidi, M. M. R. K. 1992. Yield and quality of fennel (*Foeniculum vulgare* Mill.) in relation to basal and foliar application of nitrogen and phosphorus. *J. Plant Nutrition* 15(11): 2502-2515

Khan, N. A. and Ahsan , N. 2000. Evaluation of yield potential of defoliated mustard cultivars. *Tests of Agrochemicals and Cultivars* 21: 33-34

Khan, N. A., Ansari, H. R., Khan, M., Mir, R. and Samiullah. 2002. Effect of phytohormones on growth and yield of Indian mustard. *Ind. J. Pt. Physiol.* 7: 75-78

Khandakar, U. R. and Nigam, K. B. 1996. Effect of farmyard manure and fertility level on growth and yield of ginger (*Zingiber officinale* Rosc.). *Indian J. Agric. Sci.* 66(9): 549-550

Kokkini, S., Karousou, R., Dardioti, A., Krigas, N. and Lanaras, T. 1997. Autumn essential oils of Greek oregano. *Phytochemistry* 44(5): 883-886

Kolak, I., Satovic, Z. and Rozic, I. 2002. Anise (*Pimpinella anisum* L.). *Sjemenarstvo* (Croatian) 19(1-2): 93-99

Kole, P. C. and Mishra, A. K. 2002. Divergence analysis in fenugreek (*Trigonella foenum-graecum* L.). *Research on Crops* 3(1): 197-199

Konde, B. K., Patil, R. B. and Ruikar, S. K. 1988. Mineral uptake by ginger as influenced by N levels and inoculation with diazeotrophic cultures. *Proceedings of International Congress of Plant Physiology*, New Delhi, India, pp. 15-20

Korikanthmath, V. S., Mathew, M. J., Mulge, R., Hegde, R. and Hosmani, M. 1999a. Studies on selection of elite lines of CL-37 seedling progeny of cardamom. *J. Spices and Aromatic Crops* 8:57-61

Korla, B. N. 1999. Evaluation of ginger clones for quality attributes under rainfed and irrigated conditions. *Hort. J.* 12: 39-44

Korla, B. N. and Rastogi, K. B. 1977. Path coefficient analysis in chilli. Punjab Hort. J. 17: 155-156

Korla, B. N. and Tiwari, S. K. 1999. Performance of ginger genotypes under rainfed and irrigated conditions. *Hort. J.* 12(2): 67-70

Korla, B., Rattan, R. S. and Dohroo, N. P. 1990. Effect of mulches on rhizome growth and yield of ginger (*Zingiber officinale* Rosc.). *South Indian Hort.* 38(3): 163-164

Kostrezewa, E. and Karwowska, K. 1994. Studies on the taste and aroma components of tarragon cultivated in Poland. *Prace Instytutow I Laboratoriow Badawczych Przemyslu Spozywczego* (Polish) 48:43-57

Kotgirwar, V.V., Chaudhari, M. H., Damke, M. M. and Jadhao, B. J. 1998. Effect of planting methods and clove size on growth, yield and quality of garlic cv. Godawari. PKV Res. J. 21(2): 182-184

Krishnamoorthy, B. Gopalan, A. and Alesham, J. 1988. Quality parameters of cinnamon (*Cinnamomum verum*) in relation to flush colour. *Indian Cocoa, Arecanut and Spices J.* 12(2):38

Krishnamoorthy, B. 1987. Nutmeg. *Planters Chronicle* 82(3):83-84

Krishnamoorthy, B. 1998. Clove. *Planters Chronicle* 83(6): 198-200

Krishnamoorthy, B. 2000. Sex conversion in nutmeg. *Spice India* 13(5): 11-12

Krishnamoorthy, B. and Rema, J. 1988. Nursery technique in tree spices. *Indian Cocoa, Arecanut and Spices J.* 11(3): 83 - 84

Krishnamoorthy, B. and Rema, J. 1991. Allspice. *Spice India* 4(10): 9-10

Krishnamoorthy, B. and Rema, J. 1994. Air layering in cassia (*Cinnamomum aromaticum*). *J. Spices and Aromatic Crops* 3(1): 48-49

Krishnamoorthy, B. and Rema, J. 2004. Allspice. In: *Hand Book of Herbs and Spices . Vol.II.* (ed. Peter, K. V.). Woodhead Publishing Linmited, London, pp. 117-138

Krishnamoorthy, B., 1986. Pollen germination and tube growth in nutmeg. *Indian Cocoa, Arecanut and Spices J.* 10(1): 10-11

Krishnamoorthy, B., Rema. J., Zachariah, J. T., Abraham, J. and Gopalan. A. 1996. Navashree & Nithyashree-two new high yielding and high quality cinnamom (*Cinnamonum verum*) selections. *J. Spices and Aromatic Crops* 18(3): 79-81

Krishnamoorthy, B., Sasikumar, B., Rema, J., Sayed, A. A. M. and Abraham, J. 1991. Variability and association in nutmeg. *Indian Cocoa, Arecanut and Spices J.* 14(3): 121-122

Krishnamurthy, K., Khan, M. M., Avadhani, K. K., Venkatesh, J., Siddaramaiah, A. L., Chakravarthy, A. K. and Gurumurthy, B. R. 1989. *Three Decades of Cardamom Research at R.R.S., Mudigere. Tech. Bull.*, University of Agricultural Sciences, Bangalore, India.

Krishnarajan, I. 1991. Curry leaf cultivation. *Spice India* 3:5-6

Kulwal, V., Taydeg, S., Chaudari, M. H. and Verma, R. M. 1997. Soft wood grafting – an easy method of propagation of tamarind. *Proceedings of the National Symposium on Tamarindus indica L. Forest Dept., Tirupathi, Andhra Pradesh, pp.* 94-97

Kumar, A., Anandaraj, M. and Sarma, Y. R. 2003. Rhizome solarization and microwave treatment: Ecofriendly methods for disinfecting ginger seed rhizomes. In: *Bacterial Wilt and Ralstonia solanacearum species Complex* (eds. Allen, C., Hayward, A. C. and Prior, P.). American Phytopathological Society Press, USA.

Kumar, K. B., Kumar, P. P., Balachandran, S. M. and Iyer, R. D. 1985. Developmant of clonal plantlets in cardamom (*Elettaria cardamomum*). *J. Plantation Crops* 13:31-34

Kumar, N., Pugalendhi, L. and Thamburaj, S. 1992. An observation on germination of clove seeds in Shevroys. *Spice India* 5(2): 7-8

Kumar, P. S. S., Kalyanasundaram, D., Kavith, S. and Kumar, G. S. 2002. Intercropping in coconut to improve the farm productivity. *Indain Coconut J.* 32(5): 13-15

Kumar, S. 1999. A note on economically important Zingiberaceae of Sikkim Himalaya. In: *Biodiversity, Conservation and Utilization of Spices, Medicinal and Aromatic Plants,* ISS (IISR), Calicut, Kerala, India, pp. 201-207

Kumar, S. S. N. 1983. Colletotrichum blight of cinnamon. *Indian Cocoa, Arecanut and Spices J.* 6(4): 106

Kumar, S., Singh, J. and Sharma, A. 2004. Bay leaves. In: *Hand Book of Herbs and Spices.* Vol.I. (ed. Peter, K. V.). Woodhead Publishing Limited, England, pp. 52-59

Kurian, A. and Valsala, P. A. 1995. Evaluation of turmeric types for yield and quality. *J. Trop. Agric.* 33: 75-75

Kurian, A., Valsala, P. A. and Nair, G. S. 1997. *In situ* green manure production as mulch material in ginger. *J. Trop. Agri.* 35: 50-58

Kurien, S. and Nair, P. C. S. 1988. Effect of pruning on yield of pepper (*Piper nigrum* L.). *Agric. Res. J. Kerala* 26: 137-139

Kuruvilla, K. M., Radhakrishnan, V. V. and Sudharshan, M. R. 2003. Planting materials: Vanilla – the Prince of Spices. Spices Board, Cochin, pp. 10-19

Kuruvilla, K. M., Radhakrishnan, V. V., Madhusoodanan, K. J. and Potty, S. N. 1996. Floral biology of vanilla. *Spice India* 9(8): 20-22

Kuruvilla, K.M., Vadivel, V.,Radhakrishnan, V.V. and Madhusoodanan, K.J. 2004. Crop Improement programmeson vanilla. *Spice India.* 17(3): 39-41

Kuruvinashetty, M. S., Haridasan, P. and Iyer, R. D. 1982. Tissue culture studies in turmeric. *Proc. Nat. Sem. Ginger and Turmeric,* Kasaragod, India, pp. 39-41

Lad, S. K. 1993. A case of seed setting in cultivated turmeric types (*Curcuma longa* Linn.). *J. Soils Crops* 3(1): 78-79

Larik, A. S. and Rajput, L. S. 2000. Estimation of selection indices in *Brassica juncea* L. and *B. napus* L. *Pakistan J. Bot.* 32(2): 323-330

Lawate, H. S., Joshi, M. S. and Sardeshpande, J. S. 1986. Leaf spot disease of nutmeg incited by *Cladosporium oxysporium. Indian Cocoa, Arecanut and Spices J.* 10(1): 28

Lewis, Y. S. 1973. The importance of selecting the proper variety of a spice for oil and oleoresin extraction. *Proc. Conference on Spices.* T. P. I. London, pp. 183-185

Lewis, Y. S. and Neelakantan, S. 1964. The chemistry, biochemistry and technology of tamarind. *J. Sci. Ind. Res.* 23:204-206

Lissamma, J., Nazeem, P. A., Mini, S. T., Shaji, P. and Mini, B. 1996. *In vitro* techniques for mass multiplication of black pepper (*Piper nigrum* L.) and ex vitro performance of the plantlets. J. *Plantation Crops* 24(Supplement): 511-516

Lokesh, A. R., Sarala, B. S. and Chavan, R. 1997 Evolutionary significance of style orientation in *Tamarindus indica* L. and possible evolutionary constraints for low reproductive success. *Proceedings of the National Symposium on Tamarindus indica L.* Forest Dept., Tirupathi, Andhra Pradesh, pp.212-217

Madhusoodanan, K. J., Radhakrishnan, V.V. and Sudharshan, M. R. 2003. Botany. In: *Vanilla the Prince of Spices* (eds.Thomas, J. and Rao, Y. S.). Spices Board, Cochin, India, pp 5-10

Madhusoodanan, K. J., Sudharsan, M. R. and Kuruvilla, K. M. 1998 Improved crop variety for higher yield and better capsule quality in cardamom. *Spice India* 1(6): 10-13

Madhusudanan, K. N. and Babu, G. 1994. Recalcitrancy, viability and germination of nutmeg seeds. *J. Plantation Crops* 22:25-29

Mahadeva, P. 1993. Studies on dehydration for oil extraction in curry leaf varieties. *Karnataka J. Agrl. Sci*

Majumdar, B., Venkatesh, M. S., Kumar, K. and Patiram, J. 2003. Response of garlic (Allium sativum L.) to phosphorus and sulphur application in acid alfisol of Meghalaya. *J. Spices and Aromatic Crops* 12(2): 183-186

Malhotra, S. K. 2004. Minor seed spices 1 – Ajowan, dill, celery and anise. In : *Fifty years of Spices Research in India* (ed. Ravindran, P. N.). IISR, Calicut, Keerala, India,

Malhotra, S. K. and Vijay, O. P. 2003. Plant genetic resources of seed spices in India. *Seed Spices Newsletter* 3(1): 1-4

Malhotra, S. K. and Vijay, O. P. 2004. Ajowan. . In: *Hand Book of Herbs and Spices.* Vol.II. (ed. Peter, K. V.). Woodhead Publishing Limited, England, pp. 107-116

Malik, M. Vyas, P., Rangaswamy, N.S. and Shivanna, K. R. 1999. Development of two new cytoplasmic male-sterile lines in Brassica juncea through wide hybridisation. *Plant Breeding* 118(1): 75-78

Manju, P. R. 2001. Genetic cataloguing of hot chilli. M.Sc thesis, Kerala Agricultural University, Vellanikkara, Kerala

Manjunath, B. L., Balbatti, M. S., Dhandar, D. G. and Mathew, N. M. 1998. Intercropping in coconut for higher monetary returns. In: *Developments in Plantation Crops Research. Proceedings of 12th Symposium on PlantationCrops* (eds. Kuruvilla, J. C., Licy, J., Joseph, T., Meenattoor, J. R. and Thomas, K. K.), PLACROSYM XII, Rubber Board, Kottayam, India, pp. 199-202

Manjunath, M. N., Sattigeri. V. D., Rao, R. S. N., Usharani, M. and Nagaraja, K. V. 1991. Physiochemical composition of tamarind powder. *Ind. Fd. Packer*. 45:39-42

Mankar, S. M., Deutale, R. D., Wandile, R. M. and Dighe, R. S. 2003. Influence of zinc sulphate on growth and yield of mustard (*Brassica juncea*). *J. Soils and Crops*13 (2): 378-382

Manoj, A. M. 1990. Qauality characteristics of clove and nutmeg at different stages of maturity. M.Sc. (Hort.) thesis, Kerala Agricultural University, Vellanikkara, Thrissur, Kerala, India, 104p.

Manure, G. R., Shivaraj, B., Farooqui, A. A. and Surendra, H. S. 2000. Yield attributes, seed, essential oil yield and oil content of coriander (*Coriandrum sativum* L.). *Proc. Centennial Conference on Spices and Aromatic Plants* (eds. Ramanna, K. V., Santhosh, J. E., Nirmalbabu, K., Krishnamurthy, K. S. and Kumar, A.), September 20-23, 2000. Calicut, Kerala, pp. 139-140

Markose, B. L. 1996. Genetic and biochemical basis of resistance to bacterial wilt in chilli. Ph.D thesis, Kerala Agricultural University, Vellanikkara, Kerala

Marymathew, K., Rao, Y. S., Pradipkumar, K., Madhusoodanan, K. J. and Potty, S. N. 1999. *In vitro* culture systems in vanilla In: *Plant Tissue Culture and Biotechnology, Emerging Trends* (ed. Kishore, K. P. B.). Universities Press (I) Ltd, Hyderabad, pp. 171-179

Mathai, C. K. and Nair, B. P. 1990. Biomass production levels in relation to economic yield in black pepper varieties. *J. Plantation Crops* 18: 125-128

Mathew, D. 2004. Genetic divergence, morphological and molecular characterization in *Capsicum* spp. Ph.D thesis, University of Agricultural Sciences, Bangalore, Karnataka

Mathew, L. 1992. Viability and germination studies in nutmeg seeds (*Myristica fragrans*). *Indian Cocoa, Arecanut and Spices J.* 16(1):21-23

Mathew, M. K., Rao, Y. S., Pradipkumar, K., Joseph, S., Lakshmanan, R. and Madhusoodanan, K. J. 1999.

Micropropagation of curry leaf (*Murrya Koenigii*) *J. Spices and Aromatic Crops* 8(1): 77-79

Mathew, P. A. 1985. A case of reversible plagiotrophy in nutmeg. *J. Plantation Crops* 13:147-151

Mathew, P. A. and Joseph, J. 1982. Epicotyl grafting. *J. Plantation Crops* 10:61-63

Mathur, R., Dangi, R., Dass, S. C. and Malhotra, R. C. 2000. The hottest chilli variety in India. *Curr. Sci.* 79(3): 287-288

Mehta, K. G. and Patel, R. H. 1983. Stability papameters for rhizome yield in *Curcuma longa*. *Indian Cocoa, Arecanut and Spices J.* 6: 98-99

Menon, K. P. G. 1975. Indian turmeric wins world market. *Indian Spices* 12(2): 6-13

Menon, R. and Nair, P. C. S. 1987. Studies on spike shedding in black pepper var. Panniyur-1. *South Indian Horticulture* 35: 438-441

Menon, R., Valsala, P. A. and Nair, G. S. 1992. Evaluation of open pollinated progenies of turmeric. *S. Indian Hort.* 40(2): 90-92

Mheen, H. J. V. and Mheen, H.J. V. 1996. Crop research on root-herbs angelica, levisticum and valerian. *Verslag-Proefstation-voor-de-Akkerbouw-en-de-Groenteteelt-in-de-Vollegrond* (Dutch) 219: p.79

Miniraj, N. 2000. Research priorities in cardamom areas. *Spice India* 13(7): 1-5

Miniraj, N. and Murugan, M. 2000. Ecological decline of cardamom hills - an analysis. In: *Spices and Aromatic Plants*, ISS, IISR, Calicut, pp. 172-176

Miniraj, N., Murugan, M. and Joseph, C. R. 2000. Evaluation of cardamom (*Elettaria cardamom* Maton) germplasm. *J. Spices Arom. Crops* 9(1): 55-56

Mockute, D., Bernotiene, G. and Judzentiene, A. 2003. The essential oil of the aerial parts of cultivated *Origanum vulgare* L. in Lithuania. *J. Essential Oil Bearing Plants* 6(2): 109-115

Mohan, E., Guruprasad, T. R., Herle, P. S., Seetharamu, G. K. and Krishnamurthy, M. K. 2004. Studies on biomass production, harvest index, concentration and uptake of NPK in different turmeric cultivars. Indian J. Arecanut, Spices and Medicinal Plant 6(4): 119-122

Mohanalakshmi, M., Thangaraj, T. and Shanmugasundaram, K. 2000. Effect of maturity and season for rooting of curry leaf cuttings. *Proceedings of the Centennial Conference on Spices and Aromatic Plants.* IISR, Calicut

Mohanbabu, N and Muthuswamy, S. 1984. Influence of Potassium on the quality of turmeric. *South Indian Hort.* 32: 343-346

Mohankumar, G. N., Mokashi, A. N., Narayanaswamy, P., Nalini, P., Devar, K. V., Reddy, D. N. R. and Nalawadi, U. G. 1984. Studies on the floral biology of cinnamon. *Indian Cocoa, Arecanut and Spices J.* 8(4): 100-102

Mohanty, D. C. 1979. Genetic variability and inter-relationship among rhizome yield and yield components in turmeric. *Andhra Agric. J.* 26: 77-80

Mohanty, D. C. and Panda, B. S. 1991. High yielding mutant V1K1-3 ginger. *Indian Cocoa, Arecanut and Spices Journal* 15: 5-7

Mohanty, D. C., Naik, B. S. and Panda, B. S. 1990. Ginger research in Orissa with special reference to its varietal and cultural improvement. *Indian Cocoa, Arecanut and Spices J.* 14(2): 61-63

Mohanty, D. C., Sarma, Y. N., Panda, B. S. and Edison, S. 1993. Studies on fertilizer management and seed rates in ginger variety – Surachi. *Indian Cocoa, Arecanut and Spices J.* 16(3-4): 101-104

Murashige, T and Skoog, F. 1962. A revised medium for rapid growth and bioassey with tobacco tissue culture. *Physiol. Plant* 15: 473-497

Murthy, K. N. 1997. Status paper on tamarind tree improvement works in Karnataka. *Proceedings of the National Symposium on Tamarindus indica L.* Forest Dept., Tirupathi, Andhra Pradesh, pp.20-25

Murugan. 2002. Molecular charecterization of *Piper* species using RAPD technique. M.Sc.(Hort) theses, KAU, Vellanikkara, Thrissur, Kerala, India, 78p.

Muruleedharan, A. and Balakrishnan, S. 1972. Studies on the performance of some varieties of turmeric (*Curcuma* spp.) and its fertilizer requirements. *Agric. Res. J. Kerala* 10: 112-115

Muruleedharan, V. K., Velayudhan, K. C. and Thomas, T. C. 1980. Priliminary studies on the germplasm collection of *Curcuma*. Proceedings of the National Seminar on Ginger and Turmeric., CPCRI, Calicut, pp. 55

Muthulakshmi, P. 1998. Variability analysis in *Garcinia cambogia* Desr. (Malabar Tamarind). M.Sc. (Hort.) thesis, Kerala Agricultural University, Thrissur, Kerala, India, 124p.

Muthulakshmi, P. and George, S. T. 1998. Kodampuli an excellent botanical therapy against fat. *Agro. India* 11(6): 40

Muthulakshmi, P. and George, S. T. 1999. Kudampuli – a multipurpose fruit tree. *Indian Hort.* 44(2): 7-8

Muthulakshmi, P., George, S. T. and Mathew, L. K. 2001. Quality parameters of Kodampuli rind under different drying conditions. *Spice India* 14(1): 14-16

Muthuswamy, S. and Ahmed Shah, H. 1982. Comparative quality evaluation of Salem and Erode turmeric types. *Indian Cocoa, Arecanut and Spices J.* 5: 77

Nachegouda, V. 1997. Propagation of tamarind by layering and greenwood cleft grafting. *Proceedings of the National Symposium on Tamarindus indica L.* Forest Dept., Tirupathi, Andhra Pradesh, pp.85-88

Nadgauda, R. S., Khuspe, S. S. and Mascarenhas, A. F. 1982. Isolation of high curcumin varieties of turmeric from plant tissue culture. *Proceedings of V Annual Syposium on Plantation Crops* (ed. Iyer, R. D.), CPCRI, Kasargod, pp. 143-144

Nadgauda, R. S., Mascarenhas, A. F., Hindre, R. R. and Jaganathan, V. 1978. Rapid multiplication of turmeric (*Curcuma longa* L.) plants by tissue culture. *Indian J. Exp. Biol.* 16:120-122

Nadgauda, R., Mascarenhas, A. F. and Madhusoodanan, K. J.1983. Clonal multiplication of cardamom (*E. cardamomum* Maton) by tissue culture. *J. Plantation Crops* 11: 60-64

Nagaraja, K. V., Manjunath, M. N. and Nalini, M. L. 1975. Chemical composition of commercial tamarind juice concentrate. *Ind. Fd. Packer* 29:17-20

Nagawekar, D. D., Desai, A. G., Josh, G. D., Magdam, M. B. and Khan, H. H. 2002. Performance of spice crop as intercrops in coconut plantation under Konkan condition. Plantation Crops Research and Development in the New Millennium (ed. Rethinam, P., Khan, H. H., Reddy, V. M., Mendal, P. K. and Suresh, K.) Coconut Development Board, Kochi, India, pp. 333-336

Nageswari, K., Pugalendhi, L. and Azhakiamanavalan, R.S. 2000. Propagation studies in cinnamon (*Cinnamonum Zeylanicum*). *Spice India* 13(2): 11-12

Naidu, C. V., Sivaramaprasad, N. V. and Reddy, R. T. V. 1997. Studies on market survey and selection of high yielding plus trees of *Tamarindus indica* in Andra Pradesh. *Proceedings of the National Symposium on Tamarindus indica* L. Forest Dept., Tirupathi, Andhra Pradesh, pp.198-211

Nair, M. K. 1978. Clove and nutmeg. *Indian Fmg*. 28(4): 10-13, 35

Nair, M. K., Premkumar, T., Sarma, Y. R. and Ratnambal, M. J. 1977. Prospects and problems of tree spices cultivation in India. *Indian Spices* 14(2&3): 2-9

Nair, M. N. C., Hariharan, R. and Koshi, M. P. K. 1979. Effect of presowing treatment on the sprouting of clove seeds. *Indian Cocoa, Arecanut and Spices J.* 3(2): 33

Nair, N. C. and Bahl, P. N. 1956. Vascular anatomy of the flower of *Myristica malabarica* Lank. *Phytomorphology* 6:127-134

Nair, N. C. and Pillai, N. V. 1959. Morphology of the flower and aril in *Myristica fragrans* Van Houtten. *Sci. and Cult.* 25:211-213

Nair, R. R. and Gupta, S. D. 2003. Somatic embryogenesis and plant regeneration in black pepper (*Piper nigrum* L.). *Direct Somatic Embryogenesis from Tissues of Germinating Seeds and Ontogeny of Somatic Embyoes. J. Hort. Sci. Biotech.* 78(3): 416-421

Nair, S. 1970. Two spice crops – clove and cinnamon. *Arecanut and Spices Bull.* 1(3):8-11

Najda, A., Wolski, T., Dyduch, J. and Baj, T. 2003. Determinaiton of quantitative composition of poliphenolic compounds occuring in anatomically different parts of Levisticum officinale Koch. Electronic J. Polish Agri. Universities Series Hort. 6(1): http: 11[www.ejpan.media.pl/series/volume 6/issue 1/horticulture/art –02.html.]

Nambiar, M. C. 1979. Morphological and cytological investigations in the genus *Curcuma* Linn. Ph.D. Thesis, University of Bombay, Bombay, 95p.

Nath, P. D. 1993. Effect of shade and treatment for rhizome rot of ginger. *Ann. of Agric. Res.* 14(3): 327-328

Nath, S. C. amd Baruah, A. K. S. 1994. Eugenol as the major component of the leaf oils of *Cinnamomum impressinervium* Meissn. *J. Essent. Oil Res.* 6(2): 211-212

Navaneetha, N., Palaniswamy, K. P., Khader, A. and Kumar, N. 1990. Studies on rooting of cutting of *Tamarindus indica. South Indian Hort.* 38(4): 220-224

Nawale, R. N. and Karmarkar, M. S. 1997. Softwood grafting technique for rapid multiplication of high yielding clones in kokum (*Garcinia indica Choisy*). *Indian Cocoa, Arecanut and Spices J.* 21(3): 76-77

Nawale, R. N., Nadkarni, H. R. and Sabale, S. B. 1995. Effect of foliar application of plant growth regulators on clove seedlings. *Spice India* 8(2): 11-12

Nawale, R. N., Parulekar, Y. R. and Magdum, M. B. 1997. Kokum (*Garcinia indica* Choisy) cultivation in konkan region of Maharashtra. *Indian Cocoa, Arecanut and Spices J.* 21(2): 42-43

Nazeem, P. A. 1979. Studies on growth, flowering, fruitset and fruit development in nutmeg. M.Sc. (Hort.) thesis, Kerala Agricultural University, Thrissur, Kerala, India, 103p.

Nazeem, P. A. 1989. Deficiency symptoms of mineral nutrients in clove (*Syzygium aromaticum*). Ph. D. thesis, Kerala Agricultural University, Thrissur, Kerala, India, 135p.

Nazeem, P. A. and Menon, R. 1994. Blossom biological and hybridisation studies in turmeric (Curcuma spp.). *S. Indian Hort.* 43(3): 161-167

Nazeem, P. A., Joseph, L., Thampi, M. S., Sujatha, R. and Nair, G. S. 1993. *In vitro* culture system for indirect organogenesis for black pepper (Piper nigrum L) *Golden Jubilee*

Symposium on Horticultural Research - Changing Scenario, Bangalore, 24-28 May 1993 (Abst) p. 250

Nazeem, P. A., Sudhadevi, P. K. and Nair, P. C. S. 1992. *Tree Spices* (In Malayalam), Directorate of Extension, Kerala Agricultural University, Mannuthy, Kerala,

Nazeema, K. K. 1992. Standardisation of softwood and epicotyl grafting in *Garcinia cambogia.* M.Sc. (Hort.) thesis, Kerala Agricultural University, Thrissur, Kerala, 129p.

Negi, K. S., Srivastava, K. S., Muneem, K. C. and Pant, V. K. 2004.Prospects of the garden sage as a new essential oil yielding crop for Uttaranchal. *J. Economic and Taxonomic Botany* 28(1): 13-17

Nirmalbabu, K. 2003. Molecular Characterisation and preparation of molecular in black pepper. Final report submitted to National Agricultural Technology Project, ICAR, New Delhi, pp. 50

Nirmalbabu, K., Geetha, S. P., Manjula, C., Sajina, A., Minoo, D., Samsudeen, K., Ravindran, P. N. and Peter, K. V. 1996. Biotechnology-its role in conservation of genetic resources of spices. *Biotechnology for Development* (eds. Das, M. R. and Satish Mundayoor.). State Committee on Science, Technology and Environment, Thiruvananthapuram, Kerala, pp. 198-212

Nirmalbabu, K., Geetha, S. P., Minoo, D., Ravindran, P. N. and Peter, K. V. 1999. *In vitro* conservation of cardamom (*Elettaria cardamomum* Maton) germplasm. *Plant Gen. Resources News letter* 119:41-45

Nirmalbabu, K., Minoo, D., Pillai, G. S. and Jayakumar, V. N. 2006. Progress of spices technologies in India. *Advances in Spices Research* (eds. Ravindran, P. N., Nirmalbabu, K., Shiva, K. N. and Johnykallupurackal, A.). Agrobios (India), Jodhpur, pp. 169-194

Nirmalbabu, K., Ravindran, P. N. and Peter, K. V. 1997. *Protocols for Micropropagation of Spices and Aromatic Crops.* Indian Institute of Spice Research, Calicut, India.

Nirmalbabu, K., Rema, J., Ravindran, P. N. and Peter, K. V. 1993. Micropropagation of black pepper and related species - its potential in crop improvement. *Golden Jubilee Symposium on Horticultural Research - Changing Scenario,* Bangalore, 24-28 May 1993 (Abst) p. 250

Nirmalbabu, K., Ravindran, P. N. and Sasikumar, B. 2003. Field evaluaton of tissue cultured plants of spices and assessment of their genetic stability using molecular markers. Final Report Submitted to Department of Biotechnology, Governamant of India, 94p.

Nizam, S. A. and Jayachandran, B. K. 1977. Effect of seed rhizome size and varieties on the quality of ginger (*Zingiber officinale* Rosc.). *Proceedings of 13th Kerala Science Congrress.* STEC, Trivandrum, Kerala, pp. 457-458

Nizam, S. A. and Jayachandran, B. K. 2001. Performance of mini-seed rhizome as planting material in ginger (*Zingiber officinale* R.). *Proceedings of 13th Kerala Science Congress,* Kerala Forest Research Institute, Thrissur, Kerala, pp. 457-458

Noleau, I., Richard, H. and Peyroux, A. S. 1991. Volatile compounds in leek and asafoetida. *J. Essential Oil Research* 3(4): 241-256

Nybe, E. V. 2001. *Three Decades of Spices Research at KAU.* Kerala Agricultural University, Thrissur, India, 131p.

Nybe, E. V. 2004. Technology Mission on Black Pepper. Final Report, Department of Plantaion Crops and Spices, College of Horticulture, Kerala Agriculture University: 183

Nybe, E. V. and Miniraj, N. 2005. Harness the potential of spices as intercrops. *Spice India* 18(9): 2-9

Nybe, E. V. and Nair, P. C. S. 1986. Nutrient deficiencies in black pepper (*Piper nigrum* L.) 1. Nitrogen, Phosphorus and Potassium. *Agric. Res. J. Kerala* 24: 132-150

Nybe, E. V. and Nair, P. C. S. 1987a. Nutrient deficiencies in black pepper (*Piper nigrum* L.) 2. Calcium, Magnesium and Sulpher. *Agric. Res. J. Kerala* 25: 52-65

Nybe, E. V. and Nair, P. C. S. 1987b. Nutrient deficiencies in black pepper (*Piper nigrum* L.) 3. Iron, Manganese and Copper. *Agric. Res. J. Kerala* 25: 147-158

Nybe, E. V. and Nair, P. C. S. 1987c. Nutrient deficiencies in black pepper (*Piper nigrum* L.) 4. Zinc and Boron. *Agric. Res. J. Kerala* 25: 191-200

Nybe, E. V., Nair, P. C. S. and Wahid, P. A. 1989. Relationship of foliar nutrient levels with yield of black pepper (*Piper nigrum* L.). *Tropical Agric.* 66: 345-349

Nybe, E. V., Peter, K. V. and Miniraj, N. 2004. Integrated cropping in coconut involving spice crops. *Indian Coconut J.* 34(10): 3-9

Olszewska, I. K. and Jatoszewska, I. 1999. Chemical differences between tarragon cultivated in Poland and French Tarragon. *Herba Polonica* 45(3): 173-178

Ozcan, M. and Chalchat, J. C. 2002. Composition of the essential oil of *Oreganum majorana* and *O. vulgare* subsp hirtum growing wild in Turkey. *J. Spices and Aromatic Crops* 11(1): 45-49

Padmini, K., Venugopal, M. N. and Korikanthimath, V. S. 1999. Biodiversity and conservation of cardamom (*Elettaria cardamomum* Maton). In: *Biodiversity, Conservation and Utilization of Spices,*

Medicinal and Aromatic Plants. (eds. Sasikumar, B., Krishnamoorthy, B., Rema, J., Ravindran, P. N. and Peter, K. V.). IISR, Calicut, pp. 73-78

Padmini, K., Venugopal, M. N., Korikanthimath, V. S. and Anke Gowda, S. J. 2000. Studies on compound panicle type in cardamom (*Elettaria cardamomum* Maton). In: *Recent Advances in Plantation Crops Research* (eds. Murulidharan, N. and Rajkumar, R.). Allied Pub., New Delhi, pp. 97-99

Padnis, N. A. and Choudhri, K. G. 1971. Sex determination in the seedling stage of nutmeg (*Myristica fragrans* Houtt.). *Trop. Sci.* 8: 265-274

Pandey, A. K. 1999. Response of mango ginger to nitrogen and plant spacing under terrace system of cultivation in mid-altitude areas of Mizoram. *J. Hill Res.* 12: 77-78

Pandey, U. B. 2004. Garlic. In: *Hand Book of Herbs and Spices.* Vol.I. (ed. Peter, K. V.). Woodhead Publishing Limited, England, pp. 180-194

Pandey, U. C. and Singh, J. 1990. Effect of nitrogen and irrigation levels on garlic cv HG-I. National Symposium on Onion and Garlic. 2-3 June 1990. Dr. Y. S. Parmar Univ. Hort. Solan, H. P.,

Pandey, V. S., Pandey, V. P. and Singh, P. K. 2003. Path analysis in turmeric (*Curcuma longa* L.). *Proceedings of National Seminar on New Perspectives in Spices, Medicinal and Aromatic Plants.* 22-29 November 2003, Goa, India, pp. 32-33

Pank, F., Bluthner, W. D., Kruger, H., Junghanns, W. and Overkamp, J. 2001. Successful breeding leverages the annual caraway (*Carum carvi* L. var. annum Hort). *Zeitschrift fur Arznei and Gewurzpflanzen* (Turkey) 6(3): 109-114

Parab, G. V. 2000. Evaluation of pathogenesis related proteins in releation to *Phytophthora* foot rot in black pepper. M.Sc. (Hort.) thesis, Kerala Agricultural University, Vellanikkara, Thrissur, Kerala

Pareek, O. P. 1988. Present status and future needs for genetic resources in arid zone fruits. (In): *Plant Genetic Resources – an Indian Prospective*, pp.320-334

Pareek, O. P. and Sharma, S. 1993. Genetic resources of under exploited fruits. (In): *Advances in Horticulture: Fruit Crops*- Volume-I. Chadha, K. L. and Pareek, O. P. (Eds). Malhothra Publishing House, New Delhi, India, pp.189-225

Pascual, B., Bautisla, A. S., Ferreros, W., Lopez, G. S. and Moranto, J. V. 2003. Analysis of germination of caper seeds as influenced by position of fruit on the mother plant, fruit formation stage and fruit weight. *J. Hirt. Sci. Biotech.* 78(1): 73-78

Patel, D. V., Kuruvilla, K. M. and Madhusoodanan, K. J. 1998. Correlation studies in small cardamom (*Elettaria cardamomum* Maton). In: *Development Studies in Cardamom* (eds. Mathew, N. M. and Jacob, C. K.). Allied Pub., New Delhi, pp. 16-19

Patil, J. L., Haldankar, P. M., Jamadagni, B. M. and Salvi, M. J. 1991. Influence of intercropping with tree spices on yield of coconut (*Cocos nucifera* L.). *Indian Coconut J.* 22(1): 15-18

Patil, S. G., Hebbara, M. and Devarnavadagi, S. B. 1996. Screening of multiple trees for saline vertisols and their bioameliorative effects. *Annals of Arid zone* 35(1): 57-60

Patiram, Upadhaya, R. C. and Singh, L. N. 1995. An appraisal of ginger (*Zingiber officinale* Rosc.) production in Sikkim, India. *J. Spices and Aromatic Crops* 4: 111-118

Paul, R. 2000. Parasexual hybridizaton of *Piper nigrum* through protoplast fusion. M.Sc.(Hort) theses. KAU., Vellanikkara, Thrissur, Kerala, India, 61p.

Paul, S. 1992. Evaluation of turmeric cultivars for shade tolerence. M.Sc (Ag.) theses, Keerala Agricultural University, Vellanikkara, Thrissur, Kerala, India

Paul, S. C. and Sahoo, S. 1993. Selection of elite cinnamon plants for quality bark production. *J. Econ. Tax. Bot.* 17(2):358-355

Paul, S. C., Sahoo, S. and Patra, P. 1998. Essential oil yield and quality of improved cinnamon genotypes. *Proceedings of the National Seminar on Potential Aromatic Plants of India, Exploration and Utilization.* R. R. L, Bhubaneswar, p.49

Paules, D. 1975. Tamarind. *Agricultural Agro Industry J.* 8:35-37

Paulette, E. and Rocca, G. D. 2003. Laural: Laurl nobilis L. – Laorraceae. *Sherwood For. Alb. Oggi*. 9(11): 39-42

Pawar, H. K. 1990. Use of plastic as a mulch in scheduling of irrigation to ginger in semi-arid climate. *Proceedings of 11th International Congress on the Use of Plastic in Agriculture*, pp. 99-109

Pawar, H. K. and Patil, B. K. 1990. Effect of manuring and period of harvesting on production of ginger and its quality. *J. Maharashtra Agric. Univ*. 15(1): 12-24

Phadnis, N. A. and Chowdhari, K. G. 1971. Sex determination in the seedling stage of nutmeg (*Myristica fragrans* Houtt.). *Trop. Sci*. 13:265-274

Philip, J. 1978. Morphological studies and quality evaluation of turmeric (*Curcuma longa* L.) types. M.Sc. (Hort.) thesis, Kerala Agricultural University, Thrissur, Kerala, India, 110p.

Philip, J. 1983. Studies on growth, yield and quality of turmeric. *Indian Cocoa, Arecanut and Spices J*. 7(1): 8-11

Philip, J. 1986. Mineral nutrition of nutmeg (*Myristica fragrans*) in relation to deficiency symptoms and flowering. Ph.D thesis, Kerala Agricultural University, Thrissur, Kerala, India, 231p.

Philip, J. and Nair, P. C. S. 1983. Morphological and yield characters of turmeric types. *ibid* 6(3): 61-63

Philip, J. and Nair, P. C. S. 1986. studies on variability, heritability and genetic advances in turmeric. *Indian Cocoa, Arecanut and Spices J*. 10(2): 29-30

Philip, J. and Nair, P. C. S. 1986. Studies on variability, heritability and genetic advances in turmeric. *Indian Cocoa, Arecanut and Spices J*. 10(2): 29-30

Philip, J., Nair, P. C. S., Nybe, E. V. and Mohankumar, N. 1982. Variation of yield and quality of turmeric. Proceedings of National Seminar on Ginger and Turmeric. CPCRI, Kasargod, pp. 42-46

Philip, J., Nair, P. C. S., Nybe, E. V. and Mohankumar, N. 1989. Variation of yield and quality of turmeric cultivars. *Proc. Nat. Sem. Ginger and Turmeric*. (eds. Nair, M. K., Premkumar, T., Ranindran, P. N. and Sarma, Y. R.). Central Plantation Crops Research Institute, Kasaragod, pp. 43-46

Philippose, E. 1986. Evaluation of productivity of chilli hybrids. M.Sc thesis, Kerala Agricultural University, Vellanikkara, Thrissur, Kerala, India

Pillai, K. S. 1972. The clove. *Arecanut and Spices Bull*. 4(2): 1-5

Pillai, K. S. 1973a. The clove. *Arecanut and Spices Bull*. 4(3): 9-14

Pillai, K. S. 1973b. The clove. *Arecanut and Spice Bull*. 4(4): 1-4

Ping, Y. V., Lei, S., Jun, X. W., Ping, L. X., Xing, Z. and Jing, J. M. 1999. Regeneration of transgenic pepper plants resistant to TMV and CMV and its resistance identification. *Acta Agric.Boreali Sinica* 14(93): 103-108

Pinzaru, G. and Toma, M. 1994. Determining the longevity of plantations of sage (*Salvia officinalis*) grown for leaves in the roman region. *Cercetari Agronomice in Moldova* (Romanian)27(3-4): 197-201

Ponnuswamy, V., Irulappan, I., Annadurai, S. and Vadivel, E. 1982. Variability studies in cinnamon (*Cinnamomum zeylanicum* Breyu.) *South Indian Hort.* 30:159-160

Potty, N. N., Radhakrishanan, T. C. and Ashokan, P. K. 1979. A note on the early growth and performance of six varieties of pepper in the multistoreyed cropping programme in coconut garden. *Agric. Res. J. Kerala* 17(1): 151-152

Potty, S. N. 2003. Introduction. In: *Vanilla the Prince of Spices* (eds.Thomas, J. and Rao, Y. S.). Spices Board, Cochin, India, pp. 1-5

Potty, S. N. and Krishnakumar, V. 1999. Approaches for organic farming in spices. *Spice India* 12: 7-9

Potty, S. N. and Krishnakumar, V. 2003. Agrotechniques. In: *Vanilla the Prince of Spices* (eds.Thomas, J. and Rao, Y. S.). Spices Board, Cochin, India, pp 19-29

Pradeepkumar, T. 1990. Interspecific hybridization in *Capsicum* spp. M. Sc thesis, Kerala Agricultural Univsersity, Vellanikkara, Thrissur, Kerala, India

Pradeepkumar, T., Korichaloo, J. L. and Archak, S. 2001. Molecular characterization of *Piper nigrum* cultivars using RAPD markers. *Curr. Sci.* 8(13): 246-248

Pradhan, K. 1997. Vegetable propagation of tamarind (*Tamarindus indica*) by air-layering for its orcharding in rainfed areas. *Proceedings of the National Symposium on Tamarindus indica L.* Forest Dept., Tirupathi, Andhra Pradesh, pp.75-80

Prakasan, V. 1991. Red leaf spot of cinnamon in lower pulney hills of Tamil Nadu. *Indian Cocoa Arecanut and Spices J.* 14(3): 123

Prasad, B. L. S., Gowda, C. M., Vasundhara, M., Farooqui, A. A. and Srinivasappa, K. N. 2000. Influence of growth regulators and methods of application on rooting of thyme (*Thymus vulgaris* L.) cuttings. *Proc. Centennial Conference on Spices and Aromatic Plants* (eds. Ramanna, K. V., Santhosh, J. E., Nirmalbabu, K., Krishnamurthy, K. S. and Kumar, A.), September 20-23, 2000. Calicut, Kerala, pp. 202-206

Pruthy, J. S. 1976. *Spices and Condiments*. National Book Trust of India, New Delhi, 260p.

Pruthy, J. S. 1992. Post harvest technology of spices-pre-treatments, curing, cleaning, grading and packing. *J. Spices and Aromatic Crops* (1): 1-29

Pruthy, J. S. 1993. *Major Spices of India: Crop Management, Post Harvest Technology*. Indian Council of Agricultural Research, New Delhi, 514p.

Pruthy, J. S. 2001. Minor Spices and Condiments – Crop Management and Post Harvest Technology. Indian Council of Agricultural Research, New Delhi, 782p.

Pruthy, J. S. and Krishnamurthy, S. 1984. A study on factors governing the recovery and quality of pectin from nutmeg waste (rind). *Indian Cocoa Arecanut and Spices J.* 8(3): 75-78

Pruthy, J. S., George, N. A. and Bhat A. V. 1978. Post harvest technology of cinnamon and utilization of cinnamon waste (leaves and twigs). *Indian Cocoa, Arecanut and Spices J.* 1(4): 87-92

Pugalendhi, L., Nageswari, K. and Ahakiamanovalan, R. S. 1997. Black pepper – suitable as intercrop in coffee plantations of Shevroys. *Spice India* 10(7): 3-4

Pujari, P. D., Patil, R. B. and Sakpal, R. T. 1987. Studies on growth, yield and quality components in different turmeric varieties. *Indian Cocoa, Arecanut and Spices J.* 11: 15-17

Purseglove, J. W., Brown, E. G., Green, C. L. and Robbin. 1981. *Spices Vol. II.* Longman, New York, pp. 15-17

Purushotham, K. and Rao, S. B. S. N. 1990. Propagation of tamarind by veneer and soft wood grafting. *South Indian Hort.* 38(4): 225

Pushpangadan, P. and Bradu, B. L. 1995. Basil: *Advance in Horticulture.* Vol. II. *Medicinal and Aromatic Plants.* (eds. K. L. Chadha and R. Gupta). Malhotra Publishing House, New Delhi, pp. 628-655

Putievsky, E., Dudai, N., Ravid, U. and Pudulosi, S. 1997. Cultivation, selection and conservation of *Oregano* species in Israel. *Proceedings of the IPGRI International Workshop on Oregano*, 8-12 May 1996, CIHEAM, Valenzano, Bari, Italy, pp. 103-110

Radha, C., Agarwal, Y. C., Bhatnagar, S. and Mehta, A. K. 1993. Applicztion of abrasive and Iie peeling of ginger at individual farmer's level. *Agricultural Mechanization in Asia, Africa and Latin America* 24(2): 61-64

Radhakrishnan, V. V. 1992. Spice bark. *Spice India* 5(4): 11-13

Radhakrishnan, V. V., Madhusoodhan, K. J., Priya, P. M., Kuruvilla, K. M. and Thomas, J. 2004. Crop improvement studies on cardamom. *Spice India* 17(3): 19-21

Radusiene, J., Karpaviciene, B. and Radaitiene, D. 2002. Primary evaluation of *Oreganum vulgare* germplasm in Lithuania. *Sodininkyste ir Darzininkyste* 21(1): 21-32

Rai, S. and Anita, G. 1997. Mother rhizome extraction of ginger (*Zingiber officinale* Rosc.) – an age old practice in Sikkim and Darjeeling hills. *Environ. and Ecol.* 15(4): 910-912

Rai, S. and Hossain, M. 1998. Comparative studies of three traditional methods of seed rhizome storage of ginger (*Zingiber officinale* Rosc.) practiced in Sikkim and Darjeeling hills. *Environ. and Ecol.* 16(1): 34-36

Rai, S., Das, A. B. and Das, P. 1999. Variation in chlorophyll, carotenoids, proteins and secondary metabolites amongst ginger (*Zingiber officinale* Rosc.) cultivars and their association with rhizome yield. *Newzealand J. Crop and Hort. Sci.* 27: 79-82

Rajan, V. R. 1997. Biotechnology of spices, medicinal and aromatic plants. *Proc. Nat. Sem. Biotechnology of Spice and Aromatic Plants* (eds. Edison, S., Ramanna, K. V., Sasikumar, B., Babu, K. D. and Eapen, S. J.), Indian Institute of Spice Research, April 24-25, 1996, Calicut, India, pp. 25-28

Rajappan, K., Ramaraj, B. and Natarajan, S. 2000. Diseases of clove. *Spice India* 13(5): 20-22

Rajgopal, C. K. and Sadanandan, A. K. 1985. Assessment of soil zinc and copper as reflected by the leaf composition of pepper (*Piper nigrum* L.) PLACROSYS – IV, *Proc. Indian Society of Plantation Crops* (ed. Sethuraj, M. R.), Kasargod, Kerala, India, pp. 325-330

Raju, K. V. 1990. An economic analysis of cultivation and export of coriander. *Agriculture Situation in India* 45(8): 523-526

Raju, K. V. and Reni, M. 2004. Kokam and cambodge. . In: *Hand Book of Herbs and Spices.* Vol.I. (ed. Peter, K. V.). Woodhead Publishing Limited, England, pp. 207-215

Raju, N. K. 1995. Curry leaf. *Kissan World*, p.10

Ramachandran, K. and Nair, P. N. C. 1992. Induced tetraploidy of ginger (*Zingiber officinale*.Rosc). *J. S[ices and Aromatic Crops* 1:39-42

Ramnivas., Singh, D. K. and Goutham, R. A. 1998. A note on the effect of clove, planting distance and GA on the growth and yield of garlic. Orissa J. Hort. 26(2): 72-74

Ranaware, V. S., Nawale, R. N., Khandekar, R. G. and Magdum, M. B., 1995. Effect of season on air layering of cinnamon. *Indian Cocoa, Arecanut and Spices J.* 19(3):81-84

Rani, P. U. 1997. Application of principal component analysis in chilli (*Capsicum annuum* L.). *Mysore J. Agric. Sci.* 31(2): 137-142

Rani, T. G. 1994. Induction of orthotrops in vegetatively propagated nutmeg (*Myristica fragrans*). M.Sc. (Hort.) thesis, Kerala Agricultural University, Thrissur, Kerala, India, 99p.

Rao, M. R. and Rao, D. V. R. 1988. *Proc. Nat. Sem. Chillies, Ginger and Turmeric.* Spices Board and Andhra Pradesh Agricultural University, Hyderabad, pp. 84-96

Rao, M. R. and Rao, D. V. R. 1994. Genetic resources of turmeric. *Advances in Horticulture Vol. IX*, (eds. Chaha, K. L. and Rethinam, P.). Malhotra publishing House, New Delhi, pp. 131-149

Rao, M. R. N. 1991. Prospects of nutmeg, clove and cinnamon cultivation in Andaman and Nicobar Islands. *Indian Cocoa, Arecanut and Spices J.* 14:118-120

Rao, M. R., Reddy, K. R. C. and Subbarayadu, M. 1975. Promising turmeric types of Andhra Pradesh. *Indian Spices* 12(2): 2-13

Rao, N. S. K., Narayanaswamy, S., Chacko, E. K. and Doreswamy, M. E. 1998. Regeneration of plantlets from callus of *Elettaria cardamomum* Maton. *Proc. Indian Acad. Sci.* 91(B): 37-41

Rao, Y. S., Mary-Mathew, K. and Pradeep-Kumar, K. 2005. Vanilla – the tissue culture way. *Spice India* 18(5): 29-34

Rao, Y. S., Mary-Mathew, K., Pradipkumar, K., Lakshamanan, R., Madhusoodanan, K. J. and Thomas, J. 2004. ICRI inroads into tissue culture in spices. *Spice India* 17(3): 22-26

Rao, Y. S., Mathew, M. K., Kumar, P. K., Joseph, S., Laxmanan, R., Madhusoodhanan, K. J. and Potty, S. N. 1997. Tissue culture studies on tree spices. *Proceedings of the National Symposium on Biotechnology of Spices, Medicinal and Aromatic Crops (BIOSAP)*. IISR, Calicut, pp.36-44

Rapisarda, A., Iauk, L. and Ragusa, S. 1998. Biometric analysis of the fruits of subspecies and varieties of *Foeniculum vulgere* Mill. (Apiaceae). *Rivista Italiana Messina* (Italy) 24: 19-22

Rasalam, S. J. 1978. *Practical Research in the field of Agriculture* (Budding, Grafting, Layering) (In Malayalam). Atma Nilayam, Parassala, Trivandrum, 36p.

Rathiah, Y. 1986. Turmeric cultivation in Assam. *Indian Cocoa, Arecanut and Spices J.* 9: 91-93

Rathore, A. L., Pal, A. R. and Sahu, K. K. 1998. Tillage and mulching effect on water use, root growth and yield of rainfed mustard and chickpea grown after lowland rice. *J. Sci. Food and Agric.* 78(2): 149-161

Ratnambal, M. J., Ravindran, P. N., Nair, M. K. and Nirmalbabu, K. 1990. Two high yielding selections of Karimunda. *Spice India* 3(11): 9-11

Ravikumar, P. 2002. Rosemary. *Spice India* 15(10): 16-19

Ravindran, D. S. 1989. Vegetative propagation of fruit yielding tree spp. *My Forest* 25(4): 57-60

Ravindran, P. N. 1991. Studies on black pepper and some of its wild relatives. Ph.D. thesis, University of Calicut, Kerala,

Ravindran, P. N. 2000. Introduction: *Black pepper – Piper nigrum* (ed. Ravindran, P. N.). Harwood Academic Publishers, Australia, pp. 1-22

Ravindran, P. N. and Balachandran, I. 2005. Thippali - the Indian long pepper. *Spice India* 18(11): 32-42

Ravindran, P. N. and Madhusoodhan, K. J. 2002. *Cardamom – The Genus Elettaria*. Taylor & Francis, London, 374p.

Ravindran, P. N. and Nirmalbabu, K. 1994. Chemotaxonomy of South Indian Piper. *J. Spices and Aromatic Crops* 3: 6-13

Ravindran, P. N. and Nirmalbabu, K. 2005. *Ginger – The Genus Zingiber*. CRC Press, London, 552p.

Ravindran, P. N. and Peter, K. V. 1995. Biodiversity of major spices and their conservation in India. *Proceedings of South Asia Coordinators Meeting on Plant Genetic Resources* (eds. Arora, R. K. and Rao, V. R.). IPGRI, New Delhi, pp. 75-86

Ravindran, P. N. and Pillai, G. S. 2004. Under utilized herbs and spices. . In: *Hand Book of Herbs and Spices*. Vol.I. (ed. Peter, K. V.). Woodhead Publishing Limited, England, pp. 53-103

Ravindran, P. N., Balakrishnan, R and Nirmalbabu, K. 1992. Numerical taxonomy of South Indian *Piper* L. (Piperaceae) 1. Cluster analysis. *Rheedea* 2: 55-61

Ravindran, P. N., Sasikumar, B., George, J. K., Ratnambal, M. J., Nirmalbabu, K. and Zachariah, T. J. 1994. Genetic resources of ginger (*Zingiber officinale* Rosc.) and its conservation in India. *Plant Genetic Resources Newsletter* 98: 1-4

Ravindran, S., Manilal, K. S. and Ravindran, P. N. 1993. Dermal morphology of *Cinnamomum* sp. and its taxonomic significance. *J. Pl. Anat. Morph.* 6(1): 42-50

Reddy, D. V. R., Gopalasundaran, P. and Biddappa, C. C. 2000. Cultivating clove in coconut garden. *Indian Hort.* 45(2): 29-31

Reena, J. and Bagyaraj, D. J. 1990. Growth stimulation of *Tamarindus indica* by selected VAM fungi. *World J. of Microbiol. and Biotech.* 6:59-63

Regeena, S. and Ravi, S. 1995. Curry leaf cultivation in the homesteads of south Kerala. *Spice India* 9:11-13

Reghunath, B. R. 1989. *In vitro* studies on the propagation of cardamom (*Elettaria cardamomum* Maton). Ph.D. thesis, Kerala Agric. Univ. Trichur, India.

Rema, J. and Krishnamoorthy, B. 1989. Economic uses of tree spices. *Indian Cocoa, Arecanut and Spices J.* 12(4): 120

Rema, J. and Krishnamoorthy, B. 2000. Garcinia species of economic importance – distribution and uses. *Indian Spices* 37(1): 20-23

Renjith, 1999. Response of turmeric, *Curcuma domestica* Val. to *in vivo* and *in vitro* pollination. M.Sc. (Hort.) thesis, Kerala Agricultural University, Thrissur, Kerala, India, 185p.

Rethinam, P. and Edison, S. 1991. Trees with spicy twang. *Indian Fmg.* 41(8): 17-24

Rivera, D., Inocencio, C., Obon, C., Carreno, E., Reales, A. and Alcaraz, F. 2002. Archaeobotany of capers (Capparis) (Capparaceae). *Vegetation History and Archaeobotany* 11(4):295-314

Rivera, D., Inocencio, C., Obon, C.and Alcaraz, F. 2003. Review of food and medicinal uses of Capparis L. subgenous Capparis (Capparaceae). *Economic Botany* 57(4): 515-534

Roy, A., Chatterjee, R., Hassan, A. and Mitra, S. K. 1999. Effect of Zn, Fe and B on growth, yield and nutrient content in leaf of ginger. *Indian Cocoa, Arecanut and Spices J.* 15(4): 99-101

Rudramurthy, L. K. 1993. Studies on clonal multiplication in curry leaf. *Karnataka J. of Agri. Sci.* 3

Sabale, S. B. 1991. Studies on effect of seed storage, seed treatment, germination media and foliar application of growth regulators on germination and seedling growth in clove (*Eugenia caryophyllata* Thunle). M.Sc. (Agri.) thesis, Konkan Krishi Vidya Peeth, Dapoli, Maharashtra

Sabale, S. B., Nadkarni, H. R. and Nawle, R.N. 1992. Effect of seed storage on germination and seedling establishment in clove. *Indian Cocoa, Arecanut and Spices J.* 16 (1): 26-27

Sabale, S. B., Nawle R. N., Jamadagni, B. M. and Nadkarni, H. B. 1995. Effect of different media on germination and seedling growth of clove. *Indian Cocoa, Arecanut and Spices J.* 19(2): 43-45

Sadanandan, A. K. 1994. Nutrition of black pepper. *Advances in Horticulture. Vol. 9 – Plantation and Spice Crops* (PART I) (eds. Chadha, K. L. and Rethinam, P.), ICAR, New Delhi, pp. 423-456

Sadanandan, A. K., Abraham, J. and Anandaraj, M. 1992. Impact of high production technology on productivity and disease of black pepper (*Piper nigrum* L.). *J. Plantation Crops* 20(Suppl): 384-387

Sadanandan, A. K., Hamza, S., Krishnamurthy, K. S., Kandiannan, K. and Korikanthimath, V. S. 1988. Effect of organic farming on nutrient uptake, yield and quality of ginger. *Proceedings of the National Seminar*, ISS (IISR), Calicut, pp. 89-94

Sadanandan, A.R. and Hamza, S. 1998. Studies on nutritional requirement of bush pepper (*Piper nigrum* L.) for yield and quality. *Development in Plantation Crops Research* (eds. Mathew, N.N., Kuruvilla, J.C., Licy, J., Joseph. T. and Meenattoor, J.R.). Allied Publishers Ltd, New Delhi, pp.223-227

Saha, A. and Kole, P. C. 2001. Genetic variability in fenugreek grown in sub-humid lateritic belt of West Bengal. *Madras Agrl J.* 88(4-6): 345-348

Sahoo, A., Paul, S. C. and Patra, S. 2000. Quality cinnamon production in India. *J. Med. Aromatic Plant Sci.*

Sahu, S. K. and Mitra, G. N. 1992. Influence of physiochemical properties of soil on yield of ginger and turmeric. *Fertl.News* 37: 59-63

Saifudin, N. 1981. Foliar diagnosis yield and quality of turmeric (*Curcuma longa*) in relation to nitrogen, phosphorus and potassium. M.Sc (Ag), Thesis, KAU, Vellanikkara, Karala, India, 98p.

Saikia, L. and Shadeque, A. 1992. Yield and quality of ginger (*Zingiber officinale* Rosc.) varieties grown in Assam. *J. Spices and Aromatic Crops* 1: 131-135

Saini, G. S. 2004. Grow celery - this way. *Spice India* 17(11): 41-43

Saini, G. S. 2005. Scientific method to grow fenugreek. *Spice India* 18(1): 39-42

Saji, K. V. and Sasikumar, B. 2003. Ubiquitous greater galangal. *Spice India*. 16(8): 20

Sajina, A., Mini, P. M., John, C. Z., Nirmalbabu, K., Ravindran, P. N. and Peter, K. V. 1997. Micropropagation of large cardamom (*Amomum subulatum* Roxb.). *J. Spices and Aromatic Crops* 6:145-148

Sajina, A., Minoo, D., Geetha, S. P., Samsudeen, K., Rema, J., Nirmalbabu, K., Ravindran, P. N. and Peter, K. V. 1996. Production of synthetic seeds in spices. In: *Biotechnology of Spices* (eds. Ramana, K. V.

Sasikumar, B., Nirmalbabu, K. and Eapen, S. J.). Indian Society for Spices, Calicut, pp. 65-69

Sajina, A., Minoo, D., Geetha, S. P., Samsudheen, K., Rema, J., Nirmalbabu, K. and Ravindran, P. N. 1997. Production of synthetic seeds in a few spices crops. In: *Biotechnology of Spices, Medicinal and Aromatic Plants* (eds. Edison, S., Remana, K. V., Sasikumar, B., Nirmalbabu, K. and Eapen, S. J.). Indian Society of Spices, Calicut, pp. 65-69

Sambandamoorthy, S. and Vadivel, E. 1996. Growing tamarind in dry land. *The Hindu*, 11, July, 1996

Sampathu, S. R. and Krishnamoorthy, N. 1982. Processing and utilization of kokam (*Garcinia indica*). *Indian Cocoa, Arecanut and Spices J.* 6(1): 12-13

Sankar, C. R. and Swamy, S. M. 1988. Influence of light and temperature on leaf area index, chlorophyll content and yield of ginger. *J. Maharashtra Agric. Univ.* 13(2): 216-217

Sankar, V., Lawande, K. E. and Tripathi, P. C. 2005. Mother clove selection in garlic – a review. *Spice India*. 18(9): 30-33

Sankaraiah, V. and Reddy, I. P. 1987. Studies on the effect of different manorial and soacing levels with local and high yielding turmeric varieties. *Indian Cocoa Arecanut and* Spices *J*. 11: 94-95

Sankaranarayanan, R., Vijayakumar, M. and Rangaswamy, P. 1994. Cow urine for ideal seed germination in tamarind. *Indian Hort*. 38(4): 15

Santapau, H. and Henry, A. N. 1973. *A Dictionary of Flowering Plants in India*. Council of Scientific and Industrial Research, New Delhi, 198p.

Sarma, Y. R., Ramachandran, N. and Anandaraj, M. 1991. Management of Phytopthora infections in black pepper. *Diseases of Black Pepper* (eds. Sarma, Y. R. and Premkumar, T.). National Research Centre for Spices, Calicut, pp. 55-101

Sasikumar, B. and Veluthambi, K. 1996a. Strain and vector specificity in *Agrobacterium* – black pepper interaction. *J. Plantation Crops* 24(suppl): 597-602

Sasikumar, B. and Veluthambi, K. 1996b. Tranformation of black pepper (*Piper nigrum*) USING *Agrobacterium* Ti plasmid based vectors. *Indian Perfumer* 40(1): 13-16

Sasikumar, B., Chempakom, B., George, J. K., Remashree, A. B., Devasahayam, S., Damayanthi, K. P. M.,

Ravindran, P. N. and Peter, K. V. 1999. Characterization of two interspecific hybrids in *Piper*. *J. Hort. Sci. Biotech*. 74: 125-131

Sasikumar, B., Ravindran, P. N. and George, J. K. 1992. Breeding ginger and turmeric. *Indian Cocoa, Arecanut and Spices J*. 19(1): 10-12

Sasikumar, B., Saji, K. V., George, K. J. and Haridas, P. 2004. Three new black pepper clones from IISR, Kozhikode. *Spice India* 17(7): 2

Sastri, R. L. N. 1954. On the vascular anatomy of the female flowers of *Myristica fragrans* (Agst.). *Proceedings of Indian Sci. Congr.* Part III, Indian Science Congress Association, Calcutta, pp.132-133

Sastri, R. L. N. 1955. Structure and development of the nutmeg seed. *Curr. Sci.* 24:172-173

Sastry, D. E. V., Kumar, B. L. and Singh, D. 2000. Association studied for yield and its attributes in M_4 generation of fenugreek. *Proc. Centennial Conference on Spices and Aromatic Plants* (eds. Ramanna, K. V., Santhosh, J. E., Nirmalbabu, K., Krishnamurthy, K. S. and Kumar, A.), September 20-23, 2000. Calicut, Kerala, pp. 75-77

Satheesan, K. N. 2000. Physiological and biochemical aspects of flowering, berry set and development in black pepper (*Piper nigrum* L.). Ph.D. thesis, Kerala Agricultural University, Thrissur, Kerala, India, 166p

Satheesan, K. N., Rajagopalan, A., Sukumarapillai, V., Jagdeesh Kumar, T. N., Mammootty, K. P. and Unnikrishnan Nair, P. K. 1997. Influence of irrigation levels on yield, quality and incidence of pests and diseases in black pepper (*Piper nigrum* L.) (Abstract). *National Seminar on Water and Nutrient Management for Sustainable Production and Quality of Spices*, Madikeri, p. 21

Satisha, J., Melanta, K. R. and Venkatesha, J. 1997. Effect of age of rootstocks on success of softwood grafting in tamarind. *Current Research* 26(6/7): 93-94

Sawant, D. S., Haldankar, P. M., Nagvekar, D. D. and Rajput, J. C. 1999. Screening of kokam (*Garcinia indica* Choisy) genotypes. *Indian J. of Arecanut Spices and Medicinal Plants* 1(2): 55-59

Saxene, P., Gill, R., Rashid, A. and Maheswari, C. S. 1981. Isolation and cultures of protoplasts of *Capsicum annuum* L. and their regeneration into plants flowering *in vitro*. *Protoplasma* 108: 357-360

Sebastian, A., Sujatha, V. S., Nybe, E. V., Nair, G. S. and Augustine, A. 1996. Isoenzyme variation and species relationship in the genus *Piper*. *J. Tropical Agric.* 34: 85-92

Seena, P. G. 1996. Incorpuration of bacterial wilt resistance in green chilli (*Capsicum annuum* L.). M.Sc. thesis, Kerala Agricultural University, Vellanikkara, Thrissur, Kerala, India

Selvarajan, M., Chezhiyan, N., Muthulakshmi, P. and Ramar, A. 2002. Evaluation of coriander genotypes for growth and yield. *South Indian Horticulture* 50(4-6): 458-462

Selvi, B. S., Selvaraj, N., Nageswari, R. and Raghu, R. 2003. Sweet flag. *Spice India* 16(8): 4

Sengupta, K. and Datta, A. K. 2001. Three productive mutants in coriander. *J. Hill Res.* 14(1): 6-10

Shah, S. C., Gupta, L. K., Bhujwan, H. S. and Bhujwan, S. H. 1996. Effect of auxins on rooting of stem cuttings of rosemary (*Rosemarinus officinalis* L.) Recent Hort. 3(1):126-8

Shaji, P., Anandaraj, M. and Sarma, Y. R. 1996. Comparative study of protoplast isolation and development in *Piper nigrum* (black pepper) and *P. colubrinum*. In *PLACROSYM XII*, Rubber Research Institute of India, Kottayam, Kerala, India, pp. 51-53

Shankar, G. 2003. Exploration of induced variability for crop improvement in ginger (*Zingiber officinale* Rosc.) M.Sc (Hort.) thesis, Kerala Agricultural University, Kerala, India. 175p.

Shanmugam, P., Vijayakumar, M. and Shanmugasundaram, K. A. 2000. Standardisation of planting material with reference to the stages of sprouting in turmeric (Curcuma longa L.). *Proc. Centennial Conference on Spices and Aromatic Plants* (eds. Ramanna, K. V., Santhosh, J. E., Nirmalbabu, K., Krishnamurthy, K. S. and Kumar, A.), September 20-23, 2000. Calicut, Kerala, pp. 113-117

Shanmughavelu, K. G. and Rao, M.V. N. 1977. *Spices and Plantation Crops*. Popular Book Depot, Madras-34, 298p.

Sharma, G. C. and Bajaj, B. K. 1998. Effect of intercropping bell pepper with ginger on plant parasitic nematode populations and crop yields. *Ann. of Appl. Bio.* 133(2): 199-205

Sharma, H. K. and Singh, O. C. 2000. Clove – a spice of significance. *Spice India* 13(1): 2-3

Sharma, M. M. and Sharma, R. K. 2004. Coriander. . In: *Hand Book of Herbs and Spices*. Vol.II. (ed. Peter, K. V.). Woodhead Publishing Limited, England, pp. 145-161

Sharma, P. K., Kumar, S. and Yadav, G. L. 2000. Influence of sowing time on the performance of ajowan (Trachyspermum ammi) under rainfed conditions. *Proc. Centennial Conference on Spices and Aromatic Plants* (eds. Ramanna, K. V., Santhosh, J. E., Nirmalbabu, K., Krishnamurthy, K. S. and Kumar, A.), September 20-23, 2000. Calicut, Kerala, pp. 210-211

Sharma, S. and Bharadwaj, R. 1997. Tamarind - a suitable fruit crop for dry arid regions. *Proceedings of the National Symposium on Tamarindus indica L.* Forest Dept., Tirupathi, Andhra Pradesh, pp.4-6

Sharma, S., Dohoo, N. P. and korla, B. N. 1997. Effect of VAM inoculation and other field practices on growth parameters of ginger. *J. Hill Res.* 10(2): 74-76

Sheeba, P. T. 1996. Induction of autotetraploidy in ginger. M.Sc (Hort.) thesis, Kerala Agricultural University, Kerala, India. 73p.

Sheela, K. B. 1998. Genetic improvement of bird pepper (*Capsicum frutescens* L.) by selection. Ph.D thesis, Kerala Agricultural Unversity, Vellanikkara, Thrissur, Kerala, India

Sherly, R. 1994. Growth, flowering, fruit set and fruit development in Kodampuli (*Garcinia cambogia*). M.Sc. (Hort.) thesis, Kerala Agricultural University, Thrissur, Kerala, India, 102p.

Shinde, N. N., Ingle, G. N., Shinde, B. N. and Chavan, S. D. 1997. Tamarind - past, present and future in Marathwada region. *Proceedings of the National Symposium on Tamarindus indica L.* Forest Dept., Tirupathi, Andhra Pradesh, pp.16-19

Shivegowda, S. T., Mythili, J. B., Saiprasad, G. V. S., Gowda, T. K. S., Anand, L. and Gowda, R. 2002. *In vitro* regeneration and transformation in chilli pepper (*Capsicum annuum* L.). *J. Hort. Sci. Biotech.* 77(5): 629-634

Shylaja, M. 1984. Studies on Indian cinnamomums. A study on major species occurring in Kerala. Ph.D. thesis, Calicut University, Kerala, India,

Shylaja, M. R., Sreekandan Nair, G., Nazeem, P. A., Mallika, V. K. and Mathew, M. K. 1996. *In vitro* screening of black pepper for tolerence to *Phytopthora capsici*. *J. Plantation Crops* 24(Suppl.): 171-178

Singh, A. K. 2001. Correlation and path analysis for certain metric traits in ginger. *Ann. Agri. Res.* 22: 285-286

Singh, B. G., Vijayachandran, S. N., Mahadevan, N. P. and Santhi, K. 1997. Analysis of variation in fruit and seed characteristics in *Tamaridus indica* Linn.. *Proceedings of the National Symposium on Tamarindus indica L.* Forest Dept., Tirupathi, Andhra Pradesh, pp.69-74

Singh, D. and Gangawar, B. 1991. Management practices for higher productivity of coriander in India - a review. *Agricultural Reviews Karnal* 12(1): 15-21

Singh, D. K., Sinha, S. K. and Singh, V. N. 2000. Control of bacterial wilt of ginger (*Zingiber officinale*) with antibiotics. *J. Res.* 12: 41-43

Singh, D., Rajput, S. S. and Sastry, E. V. D. 2000. A new method of emasculation of fennel (Foeniculum vulgare). *Proc. Centennial Conference on Spices and Aromatic Plants* (eds.

Ramanna, K. V., Santhosh, J. E., Nirmalbabu, K., Krishnamurthy, K. S. and Kumar, A.), September 20-23, 2000. Calicut, Kerala, pp. 82-83

Singh, J. K., Mishra, P. J. and Singh, R. A. 1998. Nutrient uptake, moisture use and economics of Indian mustard *Brassica juncea* L. Czerny. and Cosson. Varieties at rainfed conditions. *Envi.Ecol.* 16(1): 54-57

Singh, K. 1982. Cultivation of ginger and turmeric in Meghalaya. In: *Ginger and Turmeric, Proc. National Seminar*, (eds. Nair, M. K., Premkumar, T., Ravindran, P. N. and Sarma, Y. R.)., Central Plantation Crops Research Institute (CPCRI), Kasargod, India, pp. 224-226

Singh, L., Mishra, V. K., Chougule, A. B. and Singh, L. 1996. Effect of size of planting material and spacing on yield and quality of garlic variety Agrifound White. NHRDF News letter 16(3): 1-3

Singh, M., Chauhan, J. S., Kumar, P. R. and Singh, M. 2002. Growth and yield performance of different released varieties of mustard under irrigated and rainfed conditions. *New Agriculturist* 13(1-2): 19-13

Singh, N. B. 2003. Accomplishment and challenges in rapeseed-mustard research. *Brassica* 5(3-4): 1-11

Singh, P. P., Singh, V. B., Singh, A and Singh, H. B. 1999. Evaluation of different ginger cultivars for growth, yield and quality character under Nagaland condition. *J. Medi. Aroma. Pl. Sci.* 21(3): 716-718

Singh, S. P. 1992. *Fruit Crops for Waste Land.* Scientific Publishers, Jodhpur, pp.215-221

Singh, V., Dubey, O. P., Nair, R. C. P. and Pillai, G. B. 1978. Biology and bionomics of insect pests of cinnamon. *J. Plantation Crops* 6(1): 24-27

Sitepu, D. and Kasim, R. 1991. Black pepper diseases in Indonesia and their control strategy. *Diseases of Black Pepper* (eds. Sarma, Y. R. and PremKumar, T.). National Research Centre for Spices, Calicut, India, pp. 13-28

Sivaraman, K. 1992. Studies on the productivity of turmeric, maize and onion intercropping systems under varied population and nitrogen levels. Ph.D thesis, TNAU, Coimbatore, Tamil Nadu, India, 220p.

Spices Board. 1992. Report of the National Commission on Vanilla. Spices Board, Cochin, India

Spices Board. 1995. *Dried Ginger for Exports. Guidelines on Quality Improvement.* Spice Board, Cochin, India,

Spices Board. 2000. Pungency level of Indian chillies. WWW. Indianspices. Com.

Sreeja, V. S. 1999. Characterisation of dioecy and standardisation of propagation through cuttings in allspice. M. Sc. (Hort.) thesis, Kerala Agricultural University, Thrissur, Kerala, India, 68p.

Sreekala, G. S. 1999. *Biomass production and partitioning of photosynthates in ginger under different shade levels*. Unpublished M.Sc. (Hort.) Thesis, Kerala Agricultural University, Kerala, India.

Sreekala, G. S. and Jayachandran, B. K. 2001. Rhizome character of ginger as influenced by shade. *Proceedings of 13th Kerala Science Congress*, Kerala Forest Research Institute, Thrissur, Kerala, pp. 386-387

Sreekala, G. S. and Jayachandran, B. K. and Viji, M. M. 2001. Effect of shade on chlorophyll pigments of ginger cv. Rio-de-Janeiro. *Proceedings of 13th Kerala Science Congress*, Kerala Forest Research Institute, Thrissur, Kerala, pp. 460-461

Srimathi, P. and Malarkodi, K. 1998. Improved production technology for cumin. *Spice India* 11(12): 4

Srinivasan, V. and Rema, J. 2002. Cultivation practices for sustainable production of black pepper, ginger and turmeric. *Spices Production Technology* (eds. Korikamath, V. S., Faleiro, J. R., Manjunath, B. L. and Ashok Kumar, J.). ICAR Research Complex for Goa, Ela, Old Goa, pp. 21-29

Srinivasan, V., Hamza, S., Krishnamurthy, K. S. and Thankamani, C. K. 2003. Threshold level of soil zinc for optimum production of ginger (*Zingiber officinale* Rosc.). In: *Seminar on New Perspectives in Spices, Medicinal and Aromatic Plants*, Nov. 27-29, 2003, pp. 69-70 (Abst).

Sriram, T. A. 1977. Tree spices, retrospect and research needs. *Arecanut and Spices Bull.* 11(3): 83-84

Srivastava, J. P., Kamaluddin, Srivastava, S. B. L. and Tripathi, S. M. 2000. Path analysis in coriander (*Coriandrum sativum* L.,). *Proc. Centennial Conference on Spices and Aromatic Plants* (eds. Ramanna, K. V., Santhosh, J. E., Nirmalbabu, K., Krishnamurthy, K. S. and Kumar, A.), September 20-23, 2000. Calicut, Kerala, pp. 71-74

Srivastava, S. B. L., Kamaluddin, Tripathi, S. M. and Srivastava, J. P. 2000. Genetic divergence in coriander (*Coriandrum sativum* L.). *Proc. Centennial Conference on Spices and Aromatic Plants* (eds. Ramanna, K. V., Santhosh, J. E., Nirmalbabu, K., Krishnamurthy, K. S. and Kumar, A.), September 20-23, 2000. Calicut, Kerala, pp. 68-70

Stephen, F. 2002. Organics and biofertilzers in improving yield and quality of black pepper (*Piper nigrum* L.). M.Sc. (Hort.) thesis, Kerala Agricultural University, Vellanikkara, Thrissur, 91p.

Stephen, F., Nybe, E.V. and Mathew, J. 2004. Biofertilizers and organics on soil micronutrients status and yield of black pepper. *Proceedings of Natioal Seminar on Advances in Coastal Agriculture and Value Addition from National Perspectives.* CPCRI, Kasaragod, p. 40.

Subramaniyam. 1986. Studies on growth and development of turmeric (*Curcuma longa*). M.Sc (Ag) theses, TNAU, Combatore, Tamil Nadu, India, 110p.

Subramanyam, E., Rageswari, V., Chezhiyan, N. and Shiva, K. N. 2004. Effect of Azospirillum and graded levels of nitrogenous fertilizer on growth and yield of turmeric (*Curcuma longa* L.). *Procedings of Natioanl Seminar on New Perspectives in Spices, Meicinal and Aromatic Plants.* IISR, Kozhikode, Kerala, India, pp. 158-161

Sujatha, K., Muralidharan, P., Latha, P. and Nayar, N. K. 1994. Evaluation of ginger varieties as an intercrop in coconut gardens. *South Indian Hort.* 42(5): 339

Sundararaju, J. S. and Varadarajan, E. N. 1956. Propagation of nutmeg on different rootstocks. *South Indian Hort.* 4:85-86

Suresh, M. 2002. Rosemary - the spice, scented herb. *Spice India* 15(3): 18-19

Suryanarayana, B. C. 2004. Vanilla - Orchid of Commerce. Surabhi Prakashana, Bangalore, India, 71p.

Sushama, P. K. and Jose, A. I. 1994. Nutrition of ginger. In: *Advances in Horticulture Vol. 9.* (eds. Chadha, K. L. and Rathinam, P.)., Malhotra Publishing House, New Delhi, pp. 545-560

Sushama, P. K., Jose, A. I. and Pillai, V. S. 1989. Standardisation of period of sampling for foliar diagnosis in pepper in relation of nitrogen, phosphorus and potassium. *Agric. Res. J. Kerala* 22: 31-36

Tai, T. and Staskawiez, B. J. 2000. Construction of yeast artificial chromosome librarary of pepper (*Capsicum annuum* L.) and identification of clones from the Bs 2 resistance locus. *Theoretical and Applied Genetics* 100(1): 112-117

Tayde, G. S. and Deshmukh, V. D. 1986. Yield of turmeric as influenced by planting material and nitrogen levels. *P K V Res. J.* 10: 63-65

Thakur, S. K. and Sharma, S. K. 1997. Response of ginger to N and P in subtropical zone of H. P. *Indian J. Agric. Sci.* 31: 195-198

Thankamani, C. K. 2000. Influence of soil moisture regimes on growth and yield in bush pepper (Piper nigrum L.). Ph.D theses, Kerala Agricultural University, Vellanikkara, 186p.

Thanuja, 2003. Physiomorphological and biochemical responses of black pepper (*Piper nigrum* L.) to irrigation, pruning and hormone application for flushing, flowering and berry set. Ph.D. thesis, Kerala Agricultural University, Kerala, India, 233p.

Thimmaraju, K. R., Reddy, N. M. A., Rao, V. V. and Sulladmath, U. 1978. Studies on the floral biology of tamarind. *Mysore J. Agric. Sci.* 11: 293-298

Thomas, J. 2003. Mint (Japanese mint). *Spice India* 16(3): 20

Thomas, J. and George, C. K. 1992. Black pepper diseases in India and their management strategies. *Proceedings of International Workshop on Black Pepper Diseases* (eds. Wahid, P., Sitepu, D., Deciyanto, S. and Supeerman, U.). Research Institute for Spice and Medicinal Crops, Borgor, Indonesia, pp. 79-89

Thomas, J. and Vijayan, A. K. 2004. Biological control of fungal diseases. *Spice India* 17(3): 11-16

Thomas, J., Kuruvilla, K. M. and Hrideek, T. K. 2004. 'Mali mulaku' - Idukky's exclusive chilli for export. *Spice India* 17(4): 47-49

Thomas, P. 1997. Characterization of dioecy in nutmeg (*Myristica fragrans*). M.Sc. (Hort.) thesis, Kerala Agricultural University, Thrissur, Kerala, India, 70p.

TNAU. 1996. *Quinquennial Report on Spices (1992-1996)*. Horticultural Research Station, Tamil Nadu Agricultural University, Yercaud, Tamil Nadu,

Tripathi, S. M., Kamaluddin, Srivasttava, S. B. L. and Srivastava, J. P. 2000. Variability, heritability and correlation studies in coriander (*Coriandrum sativum* L.,). *Proc. Centennial Conference on Spices and Aromatic Plants* (eds. Ramanna, K. V., Santhosh, J. E., Nirmalbabu, K., Krishnamurthy, K. S. and Kumar, A.), September 20-23, Calicut, Kerala, pp. 13-16

Usha, K. and Singh, B. 1996. Selection of tamarind tree types for higher yield. *Crop Improvement* 23 (2): 199-202

Vadiraj, 2004. New concepts in cardamom drying. *Spice India* 17(3): 27-28

Vadivel, E. V., Ponnuswamy, V., Irulappan, I. and Dharmaraj, G. 1981. Vegetative propagation in cinnamon, *South Indian Hort*. 29(4): 231-232

Vadivel, V. and Kuruvilla, K. M. 2003. Marjoram. *Spice India* 16(3): 21-22

Valsala, P. A. 1994. Standardization of *in vitro* pollination and fertilization for genetic variability in *Zingiber officinale* Rosc. Ph.D. thesis, Kerala Agricultural University, Kerala, India, 133p.

Valsala, P. A., Abraham, K. and Nair, G. S. 1988. Nutritional and agronomic aspects of ginger and turmeric with reference to Kerala spp. – a review. *Proceedings of National Seminar on Chillies, Ginger and Turmeric*. APAU, Hyderabad, Andhra Pradesh, India, pp. 114-120

Valsala, P. A., Amma, S. P. and Devi, P. K. S. 1990. Feasibility of growing Daincha in the interspace of ginger beds. *Indian Cocoa, Arecanut and Spices J*. 14(2): 65-66

Varadarasan, S. 2003. Medicinal use of rosemary and sage. *Spice India* 16(3): 38

Varadarasan, S., Gopakumar, B. and Chandrasekar, S. S. 2003. Pests and their management In: *Vanilla the Prince of Spices* (eds.Thomas, J. and Rao, Y. S.). Spices Board, Cochin, India, pp. 42-55

Varghese, C. A., Ipe, V. C. and Giridharan, M. P. 1990. Grow nutmeg for higher profiles. *Indian Cocoa, Arecanut and Spices J*. 16 (1): 31-32

Velayudhan, K. C., Muralidharan, V. K., Amalraj, V. A., Goutham, P. L., Mandal, S. and Dineshkumar. 1999. *Curcuma* genetic resources. *Scientific Monograph No. IV*. National Bureau of Plant Genetic Resources, New Delhi, India, p. 149

Velayudhan, K. C., Muralidharan, V. K., Amalraj, V. A., Rana, R. S., Bhag Singh and Thomas, T. A. 1994. *Curcuma Genetic Resources*. National Bureau of Plant Genetic Resources, New Delhi, p. 73

Venkitaswamy, R. 2002. Mixed cropping of black pepper in coconut garden in Pollachi tract of Tamil Nadu. *Proceedings of National Seminar on Strategies for Icreasing Production and Export of Spices*, IISR, Kozhikode, p. 19 (Abstract)

Verghese, J. 2000a. Nutmeg and mace- I, general back ground. *Spice India* 13(12): 2-4

Verghese, J. 2000b. Sweet marjoram. *Spice India* 13(5&6): 8-10, 6-9

Verghese, J. 2001. Nutmeg and mace-IV, essential oils of *Myristica fragrans* Houtt. *Spice India* 14(5): 2-4

Verghese, J. 2003a. Savory. *Spice India* 16(3): 12-15

Verghese, J. 2003b. Poppy seed. *Spice India* 16(6): 8-9

Verghese, J. 2003c. Greater galangal. *Spice India* 16(8): 10-16

Verma, L. R., Chauhan, K. P. S. and Bhonde, S. R. 2004. Studies on comparative performance of seed material on yield and quality of garlic. NHRDF News letter, 25 (1): 11-13

Vijayasree, P. S. 2001. Refinement of in vitro and in vitro pollination techniques in turmeric. M.Sc. (Hort.) thesis, Kerala Agricultural University, Thrissur, 105p.

Vilasini, T. N. 1996. Effectiveness of soil solarisation for the control of soft rot disease in ginger. M.Sc (Ag.). thesis, Kerala Agricultural University, Kerala, India, 98p.

Wang, X. Y., Cheng, B. S., Zhang, G. S., Wang, X. Y. and Cheng, B. S. 1993. Effect of zinc on nitrozen metabolism in ginger seedlings. *J. Shandog Agric. Univ.* 24(2): 206-210

Wealth of India. 1985. *Raw materials.* Vol II. CSIR, New Delhi

Werker, E., Putievsky, E., Ravid, U., Dadai, N. and Katzir, I. 1994. Glandular hairs, secretary cavities and the essential oil in leaves of tarragon (*Artemisia dracunculus* L.). *J. Herbs, Spices and Medicinal Pts* 2(3): 19-32

Wilson, H. and Ovid, A. 1993. Growth and yield responses of ginger (*Zingiber officinale* Rosc.) as affected by shade and fertilizer applications. *J. Plant Nutr.* 16(8): 1539-1545

Yadav, B. D. and Khurana, S. C. 2000. Effect of sowing dates and planting methods on plant growth and seed yield of fennel (*Foeniculum vulgare* Mill.). *Proc. Centennial Conference on Spices and Aromatic Plants* (eds. Ramanna, K. V., Santhosh, J. E., Nirmalbabu, K., Krishnamurthy, K. S. and Kumar, A.), September 20-23, 2000. Calicut, Kerala, pp. 195-198

Yadav, R. K. 1999. Genetic variability in ginger (*Zingiber officinale* Rosc.). *J. Spices and Aromatic Crops* 8: 81-83

Yadav, S. K., Chander, K. and Singh, D. P. 1994. Response of late-sown mustard (*Brassica juncea*) to irrigation and nitrogen. *J. Agricultural Sci.* 123(2): 219-224

Index

D

E

F

W